NUCLEAR SYSTEMS
RELIABILITY
ENGINEERING
AND
RISK ASSESSMENT

NUCLEAR SYSTEMS RELIABILITY ENGINEERING AND RISK ASSESSMENT

edited by

J. B. FUSSELL
University of Tennessee

and

G. R. BURDICK
EG&G Idaho, Incorporated

The papers in this volume were presented at the International Conference on Nuclear Systems Reliability Engineering and Risk Assessment held in Gatlinburg, Tennessee, June 20-24, 1977. The Conference was sponsored jointly by the Electric Power Research Institute, the U. S. Energy Research and Development Administration, and the Department of Nuclear Engineering, University of Tennessee. Reproduction in whole or part is permitted for any purpose of the United States Government.

1977

SOCIETY FOR INDUSTRIAL AND APPLIED MATHEMATICS
33 South 17 Street, Philadelphia, Pennsylvania 19103

Society for Industrial and Applied Mathematics
33 South 17th Street, Philadelphia, Pennsylvania 19103

Library of Congress Catalogue Card Number: 77-91478

TABLE OF CONTENTS

PREFACE

Interest in nuclear systems reliability engineering and risk assessment has increased substantially during the last decade. Much of this widespread interest was stimulated by the Reactor Safety Study (WASH-1400) that was published in 1975. The International Conference on Nuclear Systems Reliability Engineering and Risk Assessment, held during June 20-24, 1977, brought together recent developments in the areas of reliability engineering and risk assessment that are pertinent to nuclear and other systems. These developments include detailed technical advancements as well as aspects that are of a political, social, and programmatic nature.

An objective of the Conference was to exchange among those concerned with reliability and risk of nuclear systems the latest information concerning theory as well as applications. This objective was accomplished through an integrated program of 34 presentations, informal conversations, and a panel discussion. This volume consists of the papers presented at the Conference.

The Conference was highlighted by a panel discussion concerning "Open Problems in Nuclear Systems Reliability Engineering and Risk Assessment." Members of the panel were:

PREFACE

- F. R. Farmer, United Kingdom Atomic Energy Authority
- F. X. Gavigan, U. S. Energy Research and Development Administration
- G. S. Lellouche, Electric Power Research Institute
- S. Levine, U. S. Nuclear Regulatory Commission
- H. J. Otway, Joint IAEA/IIASA Research Project
- N. C. Rasmussen, Massachusetts Institute of Technology
- M. J. Smith, Mechanics Research, Inc.
- C. Volta, Commission of European Communities

In keeping with the successful policy to encourage an open discussion by the panel, no written record was kept of the exchanges.

No simultaneous sessions were held at the Conference; each attendee was able to be present at all presentations. In general, no effort was made to organize sessions that had a theme. This volume is, however, arranged in topic areas with rather broadly-encompassing titles.

The Conference was sponsored by:

- The Electric Power Research Institute
- The Energy Research and Development Administration
- The University of Tennessee.

Indirect support came from many sources, especially the Idaho National Engineering Laboratory through its prime contractor, EG&G Idaho, Inc.

We wish to thank many individuals who, in addition to the speakers and panelists, contributed to the success of the Conference.

PREFACE

(a) Session Chairmen:

A. R. Buhl, Clinch River Breeder Reactor Plant Project

J. H. Carlson, U. S. Energy Research and Development
Administration

R. A. Evans, IEEE Transactions on Reliability

B. J. Garrick, Pickard, Lowe and Garrick, Inc.

J. D. Griffith, U. S. Energy Research and Development
Administration

P. F. Pasqua, The University of Tennessee

F. N. Peebles, The University of Tennessee

P. Rubel, Oak Ridge National Laboratory

(b) F. X. Gavigan and J. D. Griffith, U. S. Energy Research and
Development Administration, for their direct efforts in sponsoring the
Conference.

(c) G. S. Lellouche and M. Levenson, Electric Power Research
Institute, for their direct efforts in sponsoring the Conference.

(d) P. F. Pasqua and F. N. Peebles, The University of Tennessee,
for their support.

(e) Manuscript Review Committee Members:

J. H. Carlson, U. S. Energy Research and Development
Administration

L. L. Conradi, Clinch River Breeder Reactor Plant
Project

J. E. B. Ecker, Hanford Engineering Development
Laboratory

K. A. El-Sheikh, General Electric Company

J. F. Marchaterre, Argonne National Laboratory

D. M. Rasmuson, EG&G Idaho, Inc.

P. Rubel, Oak Ridge National Laboratory

J. I. Sackett, Argonne National Laboratory

W. E. Vesely, U. S. Nuclear Regulatory Commission

D. P. Wagner, The University of Tennessee

J. Weisman, University of Cincinnati

(f) University of Tennessee Conference Program Committee
Members:

J. S. Arendt

D. J. Campbell

C. L. Cate

J. J. Rooney

D. P. Wagner

(g) Support activities for the Conference by University of
Tennessee personnel:

S. L. Baker

V. Guthrie

S. Simpkins

(h) EG&G Idaho, Inc. personnel who provided support to

G. R. Burdick, Coeditor:

F. J. Balkovetz

H. W. Campen

S. Cohen

E. P. Eales

Special thanks go to Norm Rasmussen, who provided a social high-
light of the Conference with his banquet presentation concerning humor-
ous experiences he had during the preparation of the Reactor Safety
Study.

J. B. Fussell, G. R. Burdick, August 1977

Topic 1
POLICY AND SOCIAL ASPECTS

DISCUSSION BY THE EDITORS

Topic 1
Policy and Social Aspects

The papers in this section are involved with the attitudes of individuals concerning nuclear industry policy toward reliability and risk assessment and the social aspects of risk assessment. The first paper in this section, by S. Levine and W. E. Vesely reflects a maturity gained from direct experience with the Reactor Safety Study [1]. The paper expresses a cautious attitude, that is perhaps shared by many individuals within the U. S. Nuclear Regulatory Commission, concerning the results of nuclear system probabilistic and risk assessments that are carried out at this time. High quality performance associated with application of existing and new methodologies is stressed as being important in the effort to gain acceptance of probabilistic methods in the nuclear industry. The need for standardized methodologies and data bases is also emphasized.

The second paper, by F. X. Gavigan and J. D. Griffith, is concerned with using probabilistic methods in high level decision making concerning design and safety research and development. While the paper emphasizes application concerning the liquid metal fast breeder reactor safety program, the approach presented is applicable to other programs. The Line of Assurance approach to programmatic allocation of safety efforts that is presented has gained favor in several areas.

3

In the next paper, F. R. Farmer gives perspective to risk assessment. He notes the problem associated with comparing risks of high and low consequences as well as proposing a quantitative "target" risk of accidental death from the nuclear industry. Many "case in point" examples are given in the paper. The paper is oriented toward allocating reliability requirements in an integrated fashion based on risk.

H. J. Otway, J. Linnerooth and F. Niehaus discuss social aspects of risk assessment in the final paper in this section. The authors point out that the process of carrying out a risk assessment is perhaps more valuable than calculated results because the effort is interdisciplinary to the point of involving social systems. The paper points out that the public attitude toward technology and its risk is more than a matter of education.

REFERENCE

[1] U. S. Nuclear Regulatory Commission, <u>Reactor Safety Study</u>,
        WASH-1400, October 1975.

PROSPECTS AND PROBLEMS IN RISK ANALYSES: SOME VIEWPOINTS

S. LEVINE AND W. E. VESELY*

Abstract. Present problem areas in risk analysis are outlined and promising utilizations are described. Some of the specific problems involve lack of standardization of models, data, and quantitative approaches. The promising areas include performances of generic analyses and sensitivity evaluations. Specific applications being performed within the U. S. Nuclear Regulatory Commission are described.

1. Introduction. The publication of the Reactor Safety Study (WASH-1400) represented somewhat of a milestone in risk assessment efforts, at least in its size and scope [1] . While it is clear that the techniques and concepts used in the study are becoming understood by an increasing number of people in the scientific community, it is difficult to estimate with precision the degree to which the Reactor Safety Study has been more generally accepted. There are those with vested interests who attest to its excellence and validity and others who declaim its deficiencies and lack of validity. It is unfortunate that this attention results in some misuse of the Study by either overstating its applicability and rigor or by overstating criticisms of its validity.

---

* U.S. Nuclear Regulatory Commission, Washington, DC 20555

On an overall basis, however, there appears to be a strong consensus that, whatever deficiencies may exist in WASH-1400, its general
methodology should be used to perform risk assessments of all elements
of the nuclear fuel cycle. The Transactions of the 1976 ANS Winter
Meeting [2] and the IEEE Special Issue on Nuclear System Reliability
and Safety [3], for example, contain papers which address possible
further extensions which might be carried out. While some advocate
use of such studies immediately, there are questions about the proper
time at which such studies should be done. As part of its fuel cycle
analyses, the U.S. Nuclear Regulatory Commission (NRC) is examining the
applicability of WASH-1400 methodology to the risk assessment of transportation, storage, and disposal. The lack of existing facilities and
data, however, can be a significant impediment to the performance of
meaningful risk assessments in these areas. The same comment also
applies to LMFBR and HTGR risk assessments.

WASH-1400 concentrated upon the nuclear power plant and the
systems involved. While WASH-1400 expressed the need for caution in
applying its techniques and stated that additional methodological
development would be needed if quantitative techniques were to be used
routinely, it recognized the fact that the techniques can be used as
another effective tool to help decision-making processes. It is clear
that designers can, with proper care, use these techniques in evaluating
the reliability of various design configurations of safety systems. It

is also clear that the techniques can be used to supplement safety
evaluations. However, we see no need for, or safety benefit to be
derived from, complete risk and reliability analyses of entire applications for nuclear power plant licenses. It is our view that safety
decision-making is a complex matter and should not lightly be changed,
especially where a good safety record has been achieved as is so far
the case for U.S. commercial nuclear power plants.

There is some existing opinion that it is necessary to define an
acceptable level of safety for nuclear power plants and then to determine whether each plant meets that standard. We are not aware of any
large scale technology in the U.S. where such determinations have been
made, nor are we aware of successful schemes to allocate the risks
among the various elements in a particular endeavor. Indeed the
determination of acceptable levels of risk on a societal basis for any
endeavor seems a formidable task. Although we feel that WASH-1400
made a first step in quantitative risk assessment, the quantification
of benefits and the comparison of risks and benefits in commensurate
terms appear to be extraordinarily difficult and require many years of
research. Some of the analytical problems are addressed in the papers
presented at the Asilomar Conference on Risk-Benefit Methodology and
Application [4]. Furthermore, there would be problems of comparable or
greater difficulty in achieving broad scale public acceptance of such
analyses.

It is our view that risk analysis techniques, when applied on a
limited scale, can be extremely useful for determining whether or not
some system failure, some procedure, or some issue is an important
safety matter. Some criterion must be used to judge the importance,
and the results of WASH-1400 can be of help in selecting this criterion. For example, safety issues can be examined with regard to
their impact on the core melt probability predicted in WASH-1400.
WASH-1400 can thus serve as a relative criterion for interpreting the
impacts of calculated risks. The examples and discussions in the
subsequent sections will further expound on this use of risk analyses.

To proceed with a wide application of these techniques, better
systematization of models and analyses, more precise data, and in some
areas, the development of additional modeling are needed. The impre-
cision in existing data can be an unsurpassable obstacle if precise
reliability evaluations are to be performed and if they are to apply to
systems at specific plants. In order to correct failures which are
assessed to be too frequent, reactor designers need causal data
detailing the specific cause of failure (e.g., whether the failure was
due to human failure, manufacturing error, etc.) However, this causal
information is not necessarily needed in licensing evaluations since it
is the impact of failures on safety that is of principal interest and
not necessarily the detailed means of correction. The Nuclear Plant
Reliability Data System (NPRDS [5]), has been established by the

industry as a means to collect data with greater precision and applicability to nuclear power plant safety. If the broader use of reliability techniques is to occur, it is vital that all utilities having nuclear power plants participate fully and effectively as soon as possible. Less than effective voluntary participation in NPRDS will prevent acceptance of the utilization of the techniques and it may become necessary to make participation a government requirement.

In the following discussion, present problems in risk analysis will be covered in more detail and promising uses will be described. Examples of what can be done with risk analysis, and examples of what needs to be done, will be covered. Actual utilizations of risk analysis techniques within the Nuclear Regulatory Commission will be highlighted as some of the examples of what can be done.

2. Systematization of Models and Analyses. At the present time many of the models, data, and quantitative approaches used in risk analyses are in a state of development and cannot be routinely used. Substantial additional efforts are needed to standardize these areas if risk analyses techniques are to be used with confidence on a broad scale. System models, such as fault trees, need to be codified as to kinds of failure included and the detail of failures covered. Quantification approaches need to be categorized as to the treatment of failure, repair, and maintenance. Some of the problems, and some

recent developments, are described in the Proceedings of the Meeting of the CSNI Task Force on Problems of Rare Events [6] and in the SIAM Proceedings on Reliability and Fault Tree Analysis [7].

With regard to systematization of models, when only component failure contributions are included in a fault tree, then what might be termed the "design-based" unavailability is evaluated. The "operation-based" unavailability, which is actually achieved in plant operation, may be significantly higher because of human errors and common cause failures which are not included in the fault tree. Even though these "hardware fault trees" are incomplete they are of utility in identifying hardware problems and in comparing different designs. However, to the extent that different designs can be influenced by different contributions from human errors, testing, maintenance, and common cause failures, care must be exercised in the conclusions drawn from such efforts.

In WASH-1400, the dominant component contributors to calculated system unavailability were active components, such as pumps, circuit breakers and valves. The active components could therefore be isolated in a system schematic and a system model be constructed considering only these components. This "active component model" would be particularly useful if the impact of testing schemes on system unavailability is being evaluated.

When considering the actual unavailability achieved in operation, WASH-1400 found that the dominant contributors (the most important cut sets) generally consisted of single active components, double active components and single passive components (pipes, wiring, etc.). The human error contributions were principally associated with test and maintenance activities, and common cause failures were often associated with calibrations and operation procedures. The number of simultaneous failures considered, the types of human errors considered, and the kinds of common cause failures included could all be codified into specified models and treatments.

In the same way that different failure models can be constructed, different quantitative approaches can be used. The quantification of human errors, the calculation of common cause failure probabilities, the handling of time trends (e.g., wear-out) and the estimations of uncertainties all need to be systematized into acceptable and applicable treatments. The models and quantitative analyses do not necessarily have to be all inclusive as long as any incompleteness or impreciseness is defined and understood. The lack of systematization of models and analyses is one of the reasons risk analysis is still an art instead of a science.

3. Data and Methodology. There seems to be universal agreement that more data should be collected and analyzed; however, the outflow of useable data is still slow, arduous, and often of low quality as

mentioned earlier with regard to the NPRDS system. With regard to
data utilization, it is necessary to develop operations evaluations
programs to ensure that evaluated data are readily available to users.
It is sometimes forgotten that the meaningful output of a data collection program is reliability and safety information and is not a
statistical estimate of a failure rate. The failure rate is only an
intermediate result.

Risk analyses can be classified as being either generic or specific. A generic analysis considers problems which pertain to a class
or population as a whole, such as the class of light water reactors.
A specific risk analysis on the other hand considers a specific problem pertaining to a specific system. The data requirements for the
two types of risk analyses can be quite different. For the generic
risk analysis, population or average data, having large uncertainties,
can be used successfully. For the specific analysis, data directly
applicable to the specific system are required to explicitly account
for unique contributions to system unavailability.

Even though specific system designs were analyzed in WASH-1400,
it would be classed as a generic analysis since generic data were used
and the results applied to a population of 100 reactors. In this
sense, most risk analyses today are generic analyses like WASH-1400.
To be able to account for uniqueness and to be able to better identify
failure interrelationships which exist, more data should be collected

and evaluated, including not only component failure data, but also,
and very importantly, human error data and common cause failure data.

With regard to methodology, human error modeling and common cause
failure modeling need further development. External events, such as
seismic effects, need to be quantified in a more complete and comprehensive manner. The effect of design errors, manufacturing defects,
and quality control errors need to be more explicitly modeled in a
risk analysis framework. The general area of mechanical failures needs
further development so as to obtain applicable probabilistic models.
The dominating variables in accident and consequence phenomena need to
be identified to obtain simple, useable models.

Reliability analysis is generally concerned with system operability or unavailability. The question of functionability, i.e.,
whether the system performs its required function when it operates,
is generally not treated probabilistically in such analyses. It is
possible that, in some cases, functional analyses could show the likelihood of functionability failure to be higher than operability
failure, thus invalidating a conventionally done reliability analysis.
Fortunately, most functionability analyses are done very conservatively
so that this is not likely to happen. However, there are cases where
conservatism in functionability analyses (i.e., the selection of
extremely conservative input parameters to the analysis) imposes

severe penalties on systems; it appears that more effort should be
given to this area to avoid unnecessary penalties.

4.   Insights Gained from Generic Risk Analyses. Even though
present risk assessments tend to be generic, they can be very useful
and can give important information to supplement decision-making
processes. As an example of practical utilizations, an analysis was
performed by the Nuclear Regulatory Commission to investigate the risk
from seismically-induced fires [8]. The question was whether fire
protection systems should be designed to seismic Class 1 requirements
and thus be able to withstand a severe earthquake. This upgrade in
fire protection systems would reduce the risks from those fires which
might follow an earthquake.

The analysis that was performed indicated that the probability
of a seismically-induced fire was small compared to the probability
of a randomly-induced fire occurring from causes not associated with
an earthquake. The randomly-induced fire severe enough to damage the
engineered safeguards functions (ESF), but not necessarily causing a
core melt, was estimated as having a probability of 10^{-3} per reactor
year, with a factor of 10 error spread based on available data. The
probability that an earthquake would cause a fire which was not
extinguished and which could potentially cause similar ESF damage was
estimated to be approximately 7×10^{-6} per reactor year. While
designing the fire protection system to Class I would reduce this

failure probability, even with the existing seismic uncertainties, this probability was already small compared to 10^{-3}.

The analysis performed was straightforward and relatively simple. Since events of similar consequence were being examined, the relative probabilities of the events were the only critical factors. Even though the analysis was simple, the perspective from the analysis served as one of the bases which has led the Nuclear Regulatory Staff to decide that fire protection systems can be designed to seismic Class II rather than Class I.

As another example of the utilization of risk analysis insights, probabilistic and consequence arguments were used as part of the review of the fifteen issues raised by members of the Regulatory Staff [9]. The issues involved questions on current practices and approaches used in licensing. A number of the issues involved purely procedural matters, however, risk analyses arguments were used to help show that nine of the issues involved accident sequences which did not significantly affect the accident risks as computed by WASH-1400.

One of these nine issues, for example, involved certain valves which were assumed to operate in licensing evaluations of postulated steam line breaks. One specific accident scenario of concern was a main steam line rupture followed by a main steam isolation valve failure followed by failure of those valves which were standardly assumed to operate. The approximate probability of this scenario was assessed to be less than 10^{-7} per reactor year which was much smaller

than the probability of accident sequences of similar consequences

calculated in WASH-1400. In other specific analyses related to this

general issue, consequences, e.g., iodine releases, were also deter-

mined to be so small as to not impact the probability versus conse-

quence distributions calculated in WASH-1400.

These are but a few examples of the way in which the Reactor

Safety Study techniques can be applied in the licensing process. We

believe that this type of application will increase with time.

5. Some Extensions of WASH-1400 Methodology. An important

example of a potential application of WASH-1400 techniques which the

Nuclear Regulatory Commission is presently investigating is the pre-

diction of detailed pressure vessel failure probabilities from opera-

tional pressure transients. The interest in this matter arose as a

result of pressure transients being experienced by PWRs during startup

and shutdown. A computer code called OCTAVIA [10], has been developed

and is being examined for possible utilization purposes. To use

OCTAVIA, a given PWR pressure vessel is first described by inputing

its detailed physical and fracture characteristics, such as the vessel

dimensions, ultimate strength and copper content. The operating

environment is described by inputing a neutron fluence value and an

actual temperature.

From the vessel and operational input data, OCTAVIA computes the vessel failure pressures for different sized flaws that can exist in the vessel beltline. OCTAVIA next uses historical data to estimate the probability that a pressure transient will occur and will have a given maximum pressure. The probability of pressure vessel failure is then finally computed to be the probability that a given flaw will exist times the probability that a pressure transient will occur and have a maximum pressure exceeding the failure pressure.

The above approach is attractive since it is straightforward and allows system and operational variables to be directly evaluated as to their effect on the vessel failure probability. Physical theories, experimental research results, operational data, and probabilistic calculations are all brought together in the code. Linear elastic theory serves as the principal basis for the failure pressure evalua-tions, data collected in NRC's research program are used for the vessel toughness calculations, and operational plant data serve as the bases for the transient rates and achieved pressures.

Any of the vessel characteristics, operational variables, or transient characteristics can be changed to investigate possible licensing actions. Using present operational characteristics, the Surry reactor vessel was analyzed as a check case to compare with the predicted probability range of 10^{-8} to 10^{-6} per reactor year given in WASH-1400. The results from OCTAVIA were within this range. Because of aging, the failure probability will increase with time and staff

actions are underway which should reduce this failure probability.

In regard to the prediction of system unavailabilities, when

generic assessments are used to gain insights on a specific system or

a specific reactor, then care must be taken since the specific reactor

or system may not be like the averaged, generic quantification but

instead may be unique in certain respects. Instead of quantifying

the unique aspects of a system, the most one can often do is to

quantify using generic information and then assess error spreads to

cover possible deviations of the individual system. Tradeoff and

sensitivity evaluations can be performed and are often the most

meaningful evaluations because only relative effects are being investi-

gated. The potential improvements from changes in testing and main-

tenance schedules are prime examples of tradeoff and sensitivity

studies as are comparisons of different designs and investigations of

design modifications.

As a specific example of systems evaluations, the FRANTIC com-

puter code [11], is being developed within the Commission to calculate

the detailed unavailability of a safety system. FRANTIC calculates

not only the average unavailability but also the time-dependent

instantaneous unavailability of a system. Even though a system has

a low average unavailability, say averaged over a year, at particular

times during the year the system may have a high instantaneous unavailability. The time-dependent instantaneous unavailability thus gives a detailed picture of the readiness of the system should an accident occur at a given time.

The FRANTIC code includes detailed effects of periodic system testing. The testing characteristics considered by FRANTIC include the test interval, the test duration time, the repair time or allowed downtime, the test override capability, the test efficiency, and the human-caused failure probabilities associated with the test. As an example, the auxiliary feedwater system analyzed in WASH-1400 has been evaluated using the FRANTIC code. Using the data given in WASH-1400, the results from FRANTIC agreed with the results in WASH-1400.

A sensitivity study was also performed using different testing schemes to evaluate their effect on the feedwater unavailability. By optimizing the testing schemes, it was found that the system could potentially be improved; by using optimum testing the average unavailability could be decreased by as much as a factor of 20 and the peak, instantaneous unavailability could be decreased by as much as a factor of 40. The optimum scheme involved optimizing the scheduling of tests and did not involve any more frequent testing. In fact, from this simple example, the general insight was obtained that large benefits may be obtained from staggering tests not only within the same system, but across systems in the same accident sequence. The FRANTIC code thus offers promising prospects and is being reviewed for

potential, future utilizations. In this regard, a Research Review
Group has recently been formed within the NRC to coordinate probability and risk analyses related to this area. Necessary methodological
developments are being reviewed, data requirements are being investigated, and implementation into actual operation is being considered.
Probabilistic bases for testing intervals, allowed downtimes, and
test staggering are among the specific areas being considered.

6. Summary and Conclusions. We have discussed some of the
developments and improvements needed in the present state of the art
of application of risk analysis techniques if the techniques are to
be applied more broadly in a meaningful and acceptable way. As indicated, much useful work still remains to be done in the risk analysis
field.

We have shown ways in which limited but meaningful risk analyses
can be performed at the present time and can be utilized to assist the
licensing process. Because of the insights it gives, risk analysis
techniques will surely continue to expand and grow in the future. It
will take the sustained effect of many individuals and organizations
to aid in its further development.

REFERENCES

[1] Reactor Safety Study - An Assessment of Accident Risks in U.S.
        Commercial Nuclear Power Plants, WASH-1400, (NUREG-75/014),
        October 1975.

[2] Transactions of the American Nuclear Society, 1976 International Meeting, November 1976.

[3] IEEE Special Issue on Nuclear System Reliability and Safety, T-R 76, August 1976.

[4] D. OKRENT, Editor, Risk-Benefit Methodology and Application, Papers Presented at the Asilomar Conference, UCLA-ENG-7598, December 1975.

[5] Reporting Procedures Manual for the Nuclear Plant Reliability Data System, Southwest Research, San Antonio, Texas 78284.

[6] Proceedings of the Meeting on Problems of Rare Events in the Reliability Analysis of Nuclear Power Plants, CSNI Report No. 10, June 1976.

[7] R. E. BARLOW and J. B. FUSSELL, Editors, Reliability and Fault Tree Analysis, SIAM, Philadelphia, 1975.

[8] "Examination of the Seismic Design Basis for Fire Protection Systems," NRC Memo from S. Levine to R. E. Heineman, March 26, 1976.

[9] Staff Discussion of Fifteen Technical Issues Listed in Attachment to November 3, 1976 Memorandum from Director, NRR to NRR Staff, NUREG-0138, November 1976.

[10] W. E. VESELY, E. K. LYNN, and F. F. GOLDBERG, OCTAVIA, A Computer Code to Calculate PWR Pressure Vessel Failure Probabilities From Operationally Caused Pressure Transients, NUREG-0258 (in preparation).

[11] W. E. VESELY and F. F. GOLDBERG, FRANTIC, A Computer Code for Time Dependent Unavailability Analysis, NUREG-0193 (in preparation).

THE APPLICATION OF PROBABILISTIC METHODS TO SAFETY R&D AND DESIGN CHOICES

F. X. GAVIGAN AND J. D. GRIFFITH*

Abstract. The Liquid Metal Fast Breeder Reactor (LMFBR) safety program is committed to identifying and exploiting areas in which probabilistic methods can be developed and used in making reactor safety R&D choices and optimizing designs of safety systems. Emphasis will be placed on a positive approach of solidifying and expanding our knowledge. This will provide the groundwork for a consensus on FBR risk. The management structure which will be used is based on a mechanistic approach to an LMFBR Core Disruptive Accident (CDA) with risk partitioned into "Lines of Assurance," i.e., independent, phenomenologically-based barriers which will impede or mitigate the progression and consequences of accident sequences. Quantitative determination of the probability of breach of these barriers through the completion of work identified for each Line of Assurance will allow the quantification of the contribution to risk reduction associated with the success of each barrier. This process can lead to better use of resources by channeling R&D in directions which promise the greatest potential for reducing risk and by identifying an orderly approach to the development and demonstration of design features which will keep LMFBR risks at an acceptable level.

1. LMFBR Safety Background. The Liquid Metal Fast Breeder Reactor (LMFBR) safety program has been in existence for more than 20 years. Over the years the depth and scope of knowledge has steadily increased with the application of increasing resources. As the energy problem in the U.S. becomes more urgent, the need for the LMFBR in the long term is emphasized, as is the concomitant need for resolution of the key safety issues.

---

*U.S. Energy Research and Development Administration, Washington, DC

A major safety difference between light water reactors (LWRs) and LMFBRs lies in the potential for energetic core-disruptive accidents (CDAs). Unlike LWRs, LMFBRs can be very sensitive to dimensional changes or relocation of core materials, since the intact LMFBR core is not in its most reactive configuration. Therefore, it is theoretically possible that rearrangement of geometry can lead to prompt-critical reactivity excursions and to hydrodynamic disassembly of the reactor, as first discussed by Bethe and Tait in 1956.[1] Ever since the Bethe-Tait study, the assessment of CDAs including re-criticality events has been a major consideration in LMFBR safety analysis and development.

Early R&D efforts concentrated on the energy available in superheated sodium systems and the causes for and sites of super-heat. These efforts were successfully completed when it was shown that significant amounts of energy were not available from this source in a practical reactor system. Attention then turned to the magnitude and sign of the Doppler Effect contributions to reactivity. In the SEFOR program, it was shown that the Doppler Effect is inherently negative. Subsequently, work has been focused on the study of other means of obtaining high energy release such as sodium void and recriticality in disrupted cores.

However, this places one in an extremely difficult research
arena in which the main goal is to do research on rare events - in
effect, to prove the existence of a negative. This opens the door
to answering "what if" questions rather than building a positive
case on what is known. In dealing with low probability CDA phenomena,
it is apparent that establishing statistically the frequency of
occurrence of the prototypical conditions which must occur to attain
high energy releases from phenomena such as recriticality is not
feasible. It appears that arguments must center around an uncertainty
approach, i.e., decisions must be made based on the current state of
knowledge, an approach which one must use in the face of a lack of
full scale, prototypical accident data.[2]

The difficulty was highlighted in the Recriticality Conference
held in April 1976, called to confront the dilemma of proving the
existence, or lack thereof, of a rare event, namely, recriticality.[3]

The purpose of the conference was to review the available evidence
and to discuss whether or not it was sufficient to conclude that an
accident involving an energetic recriticality could be considered
"incredible," i.e., of so low a probability as not to warrant
consideration in design. Should the consensus be negative, the group
was requested to consider what experiments and/or theoretical studies
were needed to change the consensus. The meeting was the first known

to be called to decide whether or not sufficient evidence existed to
prove that a particular postulated or hypothetical reactor event
could no longer be considered credible.

An important consensus that developed was that it was not
possible to agree unanimously, from consideration of first principles,
given the existing state of knowledge, that an energetic recriticality
was incredible. However, commitment to resolve with R&D the stated
uncertainties appeared to be a commitment to an open-ended effort,
illustrated by the ongoing fruitless search to find an energetic Fuel-
Coolant Interaction (FCI) with an oxide fuel under conceivable reactor
conditions. On the other hand it did not appear that it would be a
fruitful R&D approach to try to investigate the upper limit of
energetics possible, given that no participants could postulate an
investigable sequence of events which would not violate the laws of
nature while at the same time giving a high energy release.

In short, the dilemma is that energetic recriticality is not
accepted as impossible and therefore R&D could continue as long as
one could hypothesize interesting and inventive ways of varying
accident conditions and energy releases. Yet there are no specific
mechanistically definable scenarios that one can study to place an
upper bound on the energetics. Many of the phenomena and/or scenarios
are conjecture and are not subject to any upper limit other than the
imagination of the proposer.

A need exists then for an approach to placing these accidents in proper perspective relative to their real hazards and for development of design measures to reduce their probability of occurrence or their probable consequences.

2. R&D Program Based on Risk. Two of the problems of establishing incredibility are the cost of demonstrating the frequency of complex probabilistic events and the philosophical difficulty of fully comprehending the meaning of probabilities in the range of 10^{-7} to 10^{-9}. These problems can be solved only by a form of decomposition of events which breaks the CDA sequence into independent, smaller, more understandable packages which can be defined probabilistically.

In this approach, to be developed more fully later, the four basic events which form a sequence leading to the release of large amounts of radioactivity from the containment building are defined as (1) initiation of fuel melting, (2) whole core melting, (3) release from primary containment, and (4) release to the environment. For each event there is a corresponding Line of Assurance (LOA), the components of which are all those design features and natural phenomena which taken together prevent the breaching of the LOA.

The LOAs are defined as:

(1) Prevent CDAs

(2) Limit Core Damage

(3) Control CDA Progression

(4) Attenuate Radiological Consequences

It is necessary for risk-based R&D planning to assign failure probability goals to each LOA. High probability failures which breach the LOA are precluded by appropriate design considerations. Low probability events which breach the LOA are identified. When such events are identified, and accepted, no further work is done to reduce their probability of occurrence. Rather, their failure responses are identified and accepted as an input for the next LOA. With this approach, each LOA will have a limited but demonstrable probability of breach. All events which exceed the capability of the LOA are handled by further LOAs.

The result of the probabilistic LOA approach based on decomposition of events as described above allows arguments to be centered not around the uncertain probabilities of rare events but around the demonstrability of a positive understanding of natural occurrences and engineered systems which have failure probabilities in the range of 10^{-2} and 10^{-3}. We believe that this will permit partition of the problem in such a way as to allow rational discussion, more orderly R&D planning, and identification of packages of work with associated probability goals, the achievement of which is demonstrable.

Let us examine, for example, Line of Assurance 3, "Control CDA
Progression." A positive understanding can be developed at a given
uncertainty level. One of the arguments that must be developed to
construct this Line of Assurance is that the energy releases are not
large enough to damage the barrier being defended.

We can define the phenomena associated with grossly disrupted
core geometries that must be investigated to develop an understanding
of potential energy releases (e.g., heat transfer, FCI gravity slump-
ing of fuel, etc.). Figures 1 and 2 indicate a method of defining
damaging energy releases for one of these phenomena, the FCI.

Fig. 1

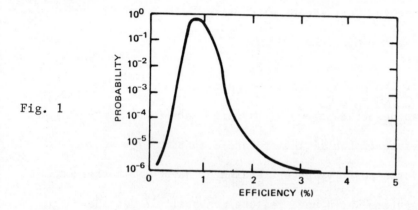

Figure 1 represents the type of graph one could construct of the
probable efficiency of energy conversion during an FCI under postulated
accident conditions (i.e., sodium temperature of 315 to 538°C, oxide
fuel temperatures of ∿538°C, and postulated mixing typical of
transient overpower or transient undercooling events). This curve,
which shows the uncertainty of the efficiency of the reaction, can

be generated from the in- and out-of-reactor experimental R&D programs and theoretical studies.

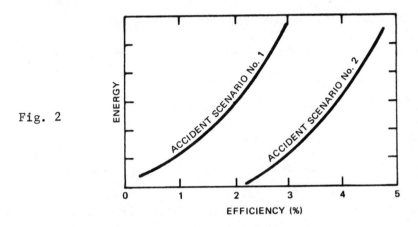

Fig. 2

Figure 2 shows the type of graph one could construct to show the potential for energy releases as a function of FCI energy-conversion efficiencies and mechanistically conceivable reactor accident scenarios. These curves, with uncertainties, can be generated from an understanding of potential accident sequences. Figure 3 combines Figures 1 and 2 to develop a curve of probability vs. energy for FCI-induced events.

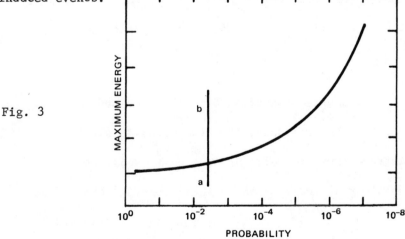

Fig. 3

We believe that probability vs. consequences curves for other relevant phenomena can be developed by a similar process. These curves would permit one to "...define the complete and consistent set of essential physical effects in sufficient detail so that they can be addressed by knowledgeable experts and so that the possibility and probability of their occurrence, separately and in required combinations, can be evaluated experimentally," as suggested in W. H. Hannum's introductory remarks at the recriticality conference.[3]

The positive risk-based approach described above requires that one draw a line such as ab on Fig. 3 and then study and understand the phenomena with probabilities higher than the line to demonstrate lack of sufficient energy to damage the barrier. We would accept the failure of the barrier for postulated phenomena with probabilities lower than the line and would not study the phenomena any further. Instead, we would move to Line of Assurance 4 and develop arguments as to the attenuation of radiological products for these arbitrarily postulated or low-probability events. A similar approach would be used for LOAs 1 and 2.

A matter of considerable significance is the placement of the line. The problem can be approached from two directions. First, what is a realistic technical goal? Second, what is necessary to produce a reasonable risk goal? Both of these considerations will be discussed.

3. What is Realistic? With respect to what is a realistic technical goal, one is largely influenced by the results of similar situations in the LWR industry in its attempts to prove a negative hypothesis. No test or series of tests has been convincing in such cases when the hypothesis involves complex phenomena, many of which have statistical characteristics. Consider the problem of the FCI that is postulated to occur at the end of saturated blowdown after a postulated pipe rupture in an LWR. The postulated events that lead to this FCI are:

1. Primary system pipe ruptures.

2. Pressurized primary system blows down.

3. Emergency core cooling system fails.

4. Subsequently core melts when core is uncovered.

5. Melting core drops into water remaining in bottom of reactor vessel.

6. Fuel-coolant interaction.

Research and development efforts directed toward determining the likelihood of the postulated FCI resulted in the findings that (1) small samples (1 to 10 g) of molten UO_2 dropped into water did not fragment and (2) explosions could not be produced in saturated water. Typical of the supporting evidence was the work by Flory, Paoli, and Mesler at Kansas University.[4] In these tests, heating was accomplished by an induction coil, and the hot molten metal was dropped into a plexiglass tank containing water. Pressure disturbances were monitored, and the interactions were photographed with a high-speed camera. Tests included a wide range of metal and bath temperatures. Metals studied included Pb (m.p. = 337°C), Sn (m.p. = 232°C), Bi (m.p. = 271°C), Zn (m.p. = 416°C), Cu (m.p. = 1083°C), Al (m.p. = 660°C), Hg (m.p. = 2.8°C), Wood's metal, and Cerrobend. Numerous experimental observations were made concerning fragmentation mechanisms. The temperature of the quench water was varied from ~8 to 100°C. The conclusions of these tests were: "Near the freezing point of water the violence of fragmentation is not a function of bath temperature, while above 25°C (77°F) the violence and extent of fragmentation decreased rapidly to zero at 90 to 100°C (194 to 212°F). Tests with several metals dropped into liquid nitrogen at its boiling point gave no fragmentation whatever, giving supporting evidence to the results of the water experiments, namely, that metals will not fragment in a saturated liquid."[4]

Yet in spite of the fact that the saturated water conditions prevailing at the end of a postulated LWR blowdown had been observed to preclude the occurrence of vapor explosions, the authors of the Reactor Safety Study conservatively assigned a probability of 10^{-2} to this event.

When one examines the applicability of the supporting data and relative complexity of the phenomena involved in the above example, compared to the problem under discussion, one is persuaded that a probability of 10^{-2} to 10^{-3} for highly energetic events is a magnitude that will be accepted by risk analysts and by the regulatory authorities. In other words, it appears that for a single relatively complex phenomenon in the nuclear business, "impossible" is identical to "inability to produce, manufacture, or observe the postulated phenomena,"[3] which allows one to take credit for a probability of 10^{-2} of the occurrence of the event before one has to accept the consequences.

In conclusion, the answer to the question of what is a realistic goal is certainly a probability no lower than 10^{-3} and is probably nearer 10^{-2}.

4. What LOA 3 Goal is Necessary to Produce a Reasonable Risk Curve? It is necessary to look at this question in relation to the rest of the risk curve. Using the four Lines of Assurance, Figure 4 is a schematic representation of our present research and development risk-assessment approach.

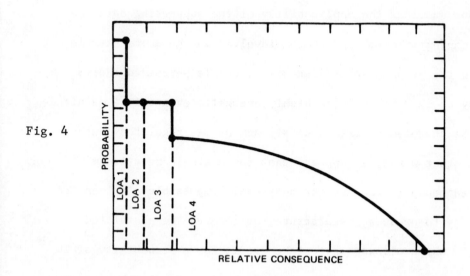

Fig. 4

This can be illustrated by an application to the Fast Flux Test

Facility (FFTF), for which no reliance was placed on Line of

Assurance 2 and little, if any, on Line of Assurance 4. Power or

flow transients with failure to scram and other low-probability

initiators were assumed to lead to whole-core meltdown, and risk

assessment was based on the ability to demonstrate accommodation of

accident debris within the primary system and/or the reactor cavity.

No attempt was made to evaluate the probability of violation of Line

of Assurance 3 for the FFTF in a quantitative manner. The smaller

FFTF core makes it easier to demonstrate accommodation of accident

energy and debris than will be the case for larger LMFBRs; but even

for FFTF one could argue that one could not achieve consensus on a

probability lower than 10^{-2} in a quantitative risk assessment.

The question one asks is: What can we do to improve the risk
curve for future LMFBRs? Again, one is influenced by the past.
There have been attempts to lower the probability of violation of
Line of Assurance 3 by subdividing the phenomena involved such that
energetic recriticality would require the combination of two or
three independent very low-probability events. These attempts have
had very limited success. The inability to prove negatives and rule
out conjecture continues to leave uncertainty. These attempts
have convinced many people that damaging recriticality energies are
impossible, but to date, at least, this cannot be accepted as proven.
One significant reason is the availability of conjectured bypass
paths, such as could be caused by an energetic FCI, that do not
require a forced progression through the path of low-probability
events.

An answer to the above can be based on the earlier suggestion
herein that, in establishing Line of Assurance 3, we do only the
work necessary to demonstrate a probability of violation of Line of
Assurance 3 of 10^{-2} to 10^{-3} and to accept the failure response as
input to LOA 4. We can then work elsewhere to lower the risk curve,
particularly in Lines of Assurance 2 and 4, where data can be more
easily developed and defended. For example, a scram system of a
different principle such as a self-actuated shutdown system within

the core can be demonstrated to lower the probability of "failure-
to-scram" events by 10^{-2} to 10^{-3}. Radiological source-term generation and attenuation information can demonstrate reduction in
consequences by orders of magnitude in defending Line of Assurance 4.

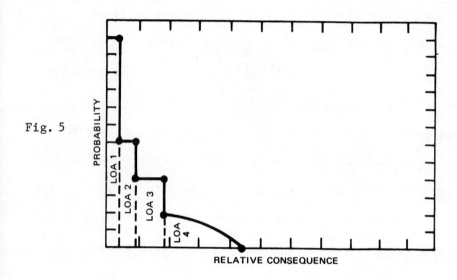

Fig. 5

RELATIVE CONSEQUENCE

Figure 5 illustrates a risk curve that assumes a successful
reduction in the probability of violation of Lines of Assurance 2
and 3 of 10^{-2} each and a substantial reduction in consequences in
Line of Assurance 4. The result is comparable to a reduction in
Line of Assurance 3 of 10^{-4} and is likely to be much more demonstrable.

In summary, it appears that the probability of events leading to
grossly disrupted cores can be made and defended to be at least as
low as 10^{-6} and probably as low as 10^{-8} per reactor operating year.
Also, unless events with probabilities lower than 10^{-2} to 10^{-3} are
postulated (such as for an FCI-induced recriticality event), the

core debris can be accommodated within some inner plant barrier.

Even for these extremely low-probability events, there will be no

substantial public consequences unless some other plant features

such as containment are postulated to fail. In a properly designed

plant, the containment can be protected, and should contribute

further reductions (10^{-2} to 10^{-3}) in the probability of severe

consequences. Thus, the sequences of events leading to severe public

consequences, given engineering independence of each LOA, are indicated

to be as low as 10^{-10} or even lower. Based on the comparable conse-

quences of LWR accidents, and the findings of the Reactor Safety

Study of relative risks to society, these results would seem adequate

for an LMFBR. The above described LOA approach requires only that

energetic recriticality be a low-probability event (i.e., $\sim 10^{-2}$

to 10^{-3}) following, for example, an accident involving a loss-of-

flow without scram.

    5. Implementation of the Concept. We believe we have shown

above that the CDA argument can be reduced to its component parts

and that if each of these have been shown to meet their goals as to

probability, the CDA argument will be made resolvable on a national

and international level.

Implementation of this concept presents both benefits and

challenges. We believe that risk-based R&D planning will give us

the following benefits:

- Identification of R&D packages with discrete cost and risk reduction potential.

- Identification of design features with known cost and risk reduction potential.

- The development of a full and integrated R&D information base for incorporation into risk assessments.

- The capability to trade off the cost of an R&D program or design modification against the potential reduction in risk.

- The capability to conduct an R&D program essentially independent of a specific LMFBR design and producing results at any prescribed level of uncertainty depending on the amount of R&D funding.

- A logical and understandable structure which will promote and strengthen communications internationally, which should aid in achieving an international consensus on LMFBR safety.

The challenges and problems ahead are seen to be:

- A need for education of researchers and designers in the concepts of probability and statistical test planning.

- The analysis and verification that the event decomposition contains no unknown bypass sequences, i.e., that the LOAs are constructed such that all accident sequences pass through all LOAs and that the set of initiators is all-inclusive.

. Development of a risk assessment structure which can be used

to guide the definition of LOAs and to aid researchers in

the formulation and planning of R&D test objectives.

. The development of methods for making tradeoff decisions

involving risk and cost of R&D and design features.

The program for developing the concept will be based on meeting

the challenges and solving the problems so that we can realize the

benefits.

As a beginning, we have prepared a Program Plan which identifies

first level products for each LOA and the necessary and sufficient

work that must be performed, expressed as second level products, to

solidly construct and demonstrate each LOA.

An example of a second level product is: "Demonstrate that the

probability of CDA initiation following a challenge to the Heat

Transfer System, Plant Protection System or reactor structures, can

be made to be less than 10^{-6}."

Sets of such goals will be assigned to four LOA Working Groups

for review and for definition of identifiable packages of R&D at a

detailed level. Each package must show a positive contribution to

meeting the requirements of an LOA, namely,

. Each LOA must be independent of other LOAs in an

engineering sense.

- Only work which is necessary and sufficient to meet the LOA product definition will be identified.

- Maximum use is to be made where applicable of the principles of redundancy, diversity and fail-safe inherent behavior characteristics within each LOA.

Risk assessment methodologists and statisticians will review and aid in structuring the R&D packages to assure consistency and thoroughness and to aid in test planning.

Cost of the testing program will be developed with cost minimized through the use of Bayesian test planning where possible. Current R&D work will be reoriented as necessary to meet the goals. As the work progresses, risk assessment methodologists will be developing methods for performing design/cost/availability tradeoff methods as well as design/R&D/cost/availability schemes. These methods will be used to optimize test planning and to attain the desired risk level at the minimum cost.

6. <u>Conclusion</u>. We believe that risk-based methods can be utilized to plan and manage an LMFBR safety R&D program so as to balance risk and development resources. We recognize that the approach entails a degree of technical risk. We are convinced, however, that the CDA issue needs a new approach and that one

based on risk assessment utilizing a positive demonstration of what

is known is most likely to succeed.

REFERENCES

[1] H. A. BETHE and J. H. TAIT, An Estimate of the Order of Magnitude
     of the Explosion When the Core of A Fast Reactor Collapses,
     UKAEA Report RHM(56)/113, 1956.

[2] B. J. GARRICK and S. KAPLAN, Reliability Technology and Nuclear
     Power, IEEE Transactions on Reliability, Special Issue -
     Nuclear System Reliability and Safety, Vol. R-25, No. 3,
     (August 1976).

[3] J. D. GRIFFITH, Reflections on the Recriticality Conference at
     Argonne National Laboratory, Nuclear Safety, Vol. 18, No. 1,
     (Jan-Feb 1977).

[4] R. FLORY, R. PAOLI, and R. MESLER, Molten Metal-Water Explosions,
     Chem. Eng. Prog., 65(1): pp. 50-54 (December 1969).

RELATIONSHIP BETWEEN RISK ASSESSMENT
AND RELIABILITY REQUIREMENTS

F. R. FARMER*

Abstract. Many accidents have occurred and have assisted the
development of many industries. Now some are called disasters, examples
are given. We cannot avoid all accidents and need to consider some
degree of improbability. Accepting limitations in knowledge and its use
the problem is how best to anticipate and reduce the risk of accident.
Research is required to provide background basic knowledge and to establish some data base.

Many assessments of systems, or pressure vessel behaviour or emergency cooling provisions proceed on a step by step basis - the logical
process variously called event or fault trees.

After some review of current objections to risk/consequent assessment the paper considers the current topic of risk acceptability and
leads into non acceptability as a more favoured route. The paper concludes with some discussion of individual and large scale societal
risks in industrial activities.

1. Industrial development based on experience of failures.

Accidents happen, people are killed, the environment is affected, and

this process will continue modified somewhat by the lessons of the past.

Hinton, responsible for the early and successful engineering achieve-

ments in atomic energy in the UK, said in 1957[1] :

"All other engineering technologies have advanced not on the basis

of their successes, but on the basis of their failures. The bridges

that have collapsed under load have added more to our knowledge of

bridge design than the bridges which have been successful; the

boilers which have blown up have added more to our knowledge of

boiler design than those which have been free from accident, and

---

*UK Atomic Energy Authority, Culcheth, Warrington, England. Visiting
Professor, Imperial College of Science and Technology, London
England.

the turbo-alternator rotors which have failed have taught us more
of turbo-alternator design than those which have continued in
satisfactory operation. Atomic energy, however, must forego this
advantage of progressing on the basis of knowledge gained by
failures."

A similar view has recently been stated in a report of the UK
Council for Science and Society[2]:

"A disaster quite often leads to legislative changes and also to
associated research. The amount of research done at the direct
instigation of the Merrison Inquiry[3] into box-girder bridges was
impressive, and has led to a number of improvements in design and
quality control. There would have been no inquiry and no sense of
urgency if the collapses had not taken place - if, for instance,
there had only been cracks and traffic limitations on a number of
bridges."

2.   Growing social reluctance to accept unusual accidents. We
are beginning to take a different view about accidents, their causes,
their "acceptability". Our western societies seem to react particularly
strongly against accidents which are unusual or contain new factors.

Atomic energy started with a bang and led most people to believe
that radiation induced carcinoma and mutagenesis was a new hazard.
There has been particular interest in England in the possibility that
two workers at Porton (UK) might have contracted Lassa fever. In Italy
there has been strong reaction following the accidental dispersion of
Dioxin at Seveso.

3.   What accident is called a disaster - a few case reviews.
Some accidents are called disasters - this does not depend on the

number killed or whether any were killed but rather whether the event
at one place, at one time, or of one type could have killed or hurt
many people:

"Windscale, Flixborough, Thalidomide, Ronan Point and Torrey
Canyon are now familiar words in our vocabulary; they epitomize
major technological disasters. In the wake of each of these,
there were questions about the technique or process responsible.
In some cases, improvements in safety procedures resulted
(Windscale); in others a technique was abandoned (Ronan Point);
sometimes policy changes were minimal (Torrey Canyon)."[2]

You may not be familiar with some of these accidents:

Flixborough. This accident occurred at the Flixborough Works of
Nypro (UK) Limited on 1 June 1974. The works were virtually demolished
by an explosion of warlike dimensions. Of those working on the site
at the time, 28 were killed and 36 others suffered injuries. If the
explosion had occurred on an ordinary working day, many more people
would have been on the site, and the number of casualties would have
been much greater. Property damage extended over a wide area, and a
preliminary survey showed that 1,821 houses and 167 shops and factories
had suffered to a greater or lesser degree. It was established by a
Court of Inquiry that the cause of the disaster was the ignition and
rapid acceleration of deflagration, possibly to the point of detonation,
of a massive vapour cloud formed by the escape of 200 tons of cyclo-
hexane under at least a pressure of 8.8 kg/cm^2 and a temperature of
155°C. It was subsequently estimated from the damage effects that the
equivalent of 15 to 30 tons of this exploded.

There were many lessons learnt from the accident which are
referred to in the Report[4]:

"It was clear on the evidence that no one concerned in the design
or construction of the plant envisaged the possibility of a major
disaster happening instantaneously. It is now apparent that such
a possibility exists where large amounts of potentially explosive
material are processed or stored. This possibility must therefore
be recognised when planning, designing, and constructing such
plants."

Ronan Point. This was one of a block of flats comprising 22 floors
of flats built in the Larsen Nielsen system resting on an in situ
concrete podium containing garages and a car deck. The immediate cause
of the disaster was a town gas explosion on 16 May 1968 in a flat on
the eighteenth floor. The explosion occurred when the tenant of the
flat struck a match to light her cooker. The explosion was not of
exceptional violence; the pressures produced were of the order of
3-12 lb/in^2; this is within the 'normal' range of domestic gas
explosions. An explosion of this force will cause local structural
damage to any form of domestic building; at Ronan Point the effect was
to blow out concrete panels forming part of the load-bearing flank wall
of the flat. The removal of this part of the load-bearing wall
precipitated the collapse of the south-east corner of the block above
the eighteenth floor; the weight of this part of the building as it fell
caused a collapse of the remainder of the south-east corner down to the
level of the in situ concrete podium of the block.

One of the conclusions reached by the Tribunal at the Public
Inquiry was [5]:

"Ronan Point was designed to comply with the British Standard Code
of Practice CP3: Chapter V on wind loading, but this is 15 years
old, and more recent research has shown that, during its lifetime,

a building of this height may have to withstand greater wind forces
than the Code of Practice envisages. The building in its present
form may suffer structural damage from high winds and this could
lead to progressive collapse."

And one of their Recommendations was that all blocks over six
storeys in height should be appraised by a structural engineer who
should consider (inter alia) whether they have been designed to resist
adequately the maximum wind loadings which they may experience.

4. What degree of improbability should be considered.
Unfortunately there is no guide given as to the maximum condition and
at this stage we touch on the nub of the problem in carrying out risk
assessment as to the degree of improbability to be considered.

You will all be familiar with the listing of man-made and natural
disasters in Rasmussen's report. The report considers the possible
death toll from tornadoes, earthquake, explosions, fires etc., and
indicates numbers from 100 to over 1,000 with estimated probability
from 10^{-2} to 10^{-4} per year[6]. For convenience I attach an abbreviated
list (Table 1).

You may or may not agree with the numbers, no doubt most will agree
that there is a small chance of accidents having severe consequences
and that for any specified accident of explosion, flood, release of
toxic substance, the consequences might vary over two decades or more.
Some disasters called "natural" or Acts of God, may now be seen to be
partly due to man's deficiencies, in lack of foresight or limitations
in technical competence. Some have occurred and will arise in the
future because we consciously choose not to design for highly improbable
events. We have codes of practice for ground accelerations, wind
speeds, height of flood etc., often selected on the basis that the

TABLE 1

RISK OF NATURAL OR MAN-MADE DISASTERS

	Estimated frequency per year	Number of deaths
Tornado	10^{-2}	>100
	$10^{-3}/10^{-4}$	$>1,000$
Earthquake	10^{-2}	$\approx 10,000$
Aircraft fatalities on ground	10^{-2}	100
	10^{-4}	$>1,000$
	$\approx 10^{-6}$	10,000
Explosions approx	10^{-1}	>100
	$\approx 10^{-2}$	$\approx 1,000$
Dams	10^{-1}	10
	10^{-2}	1,000
	10^{-3}	10,000
Fire	10^{-1}	100
	10^{-2}	1,000

specified condition might be exceeded once in 100 or 1,000 years,
although as indicated earlier no such guidance was given in the Ronan
Point recommendation.

A disaster might not imply lack of foresight, the event might fall
in the low risk range - which, before the event, is judged acceptable,
or incredible. After the event, some change might be demanded to reduce
future risk - it may or may not be possible to do so.

5.   The need to accept limitation in knowledge and its application
in accident initiation. Many accidents will reveal one or more of the
following characteristics:

(1) Lack of sufficient knowledge of the technology

(2) Complexity of management structure

(3) Man and his machines are not perfect, even when great care is
     exercised.

I list these as accident initiators in spite of man's best endeavour
- there are additionally many accidents which occur through poor management, poor design and poor discipline in operation.

When pursuing a difficult advance in technology - as in thick-
walled pressure vessel design and construction, there will be difference
in opinion, in interpretation of evidence, as in the 1960's in the
evaluation of pressure tests or Robertson Machine tests, leading to the
conclusion that pressure vessels would not fail if maintained at
NDT + 60°F. Lately there has been much more interest in crack growth
rates and improvement in inspection techniques.

Apart from the degree of understanding of various phenomena, there
is additionally the chance of failure in manufacture or inspection even
if every effort is made to achieve perfection. This is discussed in
the Marshall Study Group Report[7], for example, as regards efficiency

of crack detection by ultrasonic inspection methods:

"There is very little experimental information on the fraction of defects remaining undetected _after_ ultrasonic inspection has been carried out."

After careful investigation the Group suggested a best fit as curve Fig. 8.4 of their Report. From this curve it seems that the chance of failing to detect a crack of depth 1" is about 1 in 20, and a crack of 2" about 1 in 100.

"A rather similar survey carried out in the USA under the auspices of the PVRC produced results which when similarly interpreted are broadly consistent with our own."

6.   How to anticipate accidents. Reverting to an earlier theme, McCullough at the first Geneva Conference on the Peaceful Uses of Atomic Energy (1955) said[8]:

"So far, there have been essentially no reactor accidents leading to serious consequences. For this reason, statistical information about reactor accidents, although all favorable, does not suffice to give useful statistical information of the type needed by insurance companies, for example, in evaluating the nature of hazards. In other words, to determine what is an acceptable risk, a certain amount of judgment, detailed technical evaluation of a given reactor, and caution must be employed."

This statement implies that statistical information of a favourable type - that is in the absence of serious accidents - does not give us information necessary to evaluate the nature of hazards.

I maintain that it can do so - together with technical evaluation of associated research programmes. In fact, this need to extrapolate from lesser events in the prediction of major events is one of the

important lessons arising from the atomic energy programme and is now
being applied more widely in the field of major hazard analysis.

7.   Research to provide background data - rather than a demonstration. Some of you will recall the great interest in 1965, in devising
an experiment to show that even if a reactor depressurised, the result
would be tolerable. Since then the LOFT programme has been redirected
to give a better understanding of the relevant phenomena; to improve
the basis on which the results of serious accidents might be predicted.
I suggest that today we would not set up a programme to imitate major
accidents as their range would be limited to a selected number of
different types and the interpretation of such experiments is difficult
and seldom gives a unique solution.

8.   Step by step assessment (- perhaps a fault tree?). Hence,
our present assessment of the results of pipe fracture; of the need for
and effect of emergency cooling is itself a type of fault tree. It is
a stage by stage analysis of events in space and time feeding in the
best available information at each stage - at some stages the information has a distribution and various confidence levels. A like process
is appplied in the assessment of pressure vessels as in the Report
prepared by Marshall. In both examples, great zeal and money has been
applied to the search for information and the research programmes have
taken 10-20 years augmented by parallel programmes in many countries.

There has been a greater reluctance to search for corresponding
data base from which to assess the likely success or failure of components and plant items in complex plant (as for reactors).

9.   Objections to risk/consequence assessment. The ideas were
relatively new in 1965, we had little data and proposals to collect
data and use it in risk analysis were severely challenged and serious

attempts were made to stop this work by those who felt their judgment might be challenged. There have been some four major objections:

(1) "Data will be hard to find and will not be generally applicable":

> Will depend on type of use; quality of maintenance etc. In some areas there is sufficient data covering a range of use enabling experienced people to make a reasonable projection.

(2) "It is better to rely on engineering judgment"

> reductio ad absurdium - it is better to make judgment without information or those that seek information have no powers of judgment.

(3) "Having an appreciation of overall risk and consequences might encourage the design and operation of less safe plant than otherwise"

> In general, most reliability analysis with which I have been associated has revealed weaknesses in design or execution; subsequent changes have improved the safety of the plant.

(4) "We could not licence on the basis of our knowledge of risk"

> No one has said what risk is acceptable, therefore it would be better not to know. What a defeatist attitude.

10. Target of acceptability or non-acceptability. In conclusion, progress has been made in the development of techniques and the use of data of all sorts to give some order of magnitude assessment of risk and consequences and equally important to improve design, operation and maintenance to reduce failure of plant and systems.

There is still need for a target - the target may not be one of

acceptability, although this is now being widely debated - it may be a
target of non-acceptability. To some extent this is developing - partly
by comparison with other hazards without necessarily implying a constant
or uniform standard - and partly through a better and more honest recognition of the capabilities of man and machinery. Very low risk levels
are just not achievable. Let me conclude by giving my opinion.

I believe that we should aim to ensure that the risk to any one
person of death from radiation brought about through the nuclear
industry should not exceed 10^{-6} per person per year.

11. Consideration of large scale societal risk. The risk of
killing 10/100/1000 people in one accident is difficult to assess, and
difficult to discuss. It is not easy to accept the extrapolation from
limited evidence as by Bell[9] - deaths from aircraft crash or meteorites - or of Rasmussen or of Okrent presented to the Joint Committee on
Atomic Energy[10]. Okrent estimates the probability of a commercial
aircraft crashing directly into the grandstand of Hollywood Park whilst
it was occupied by a large crowd is about 10^{-5} per year; postulating
such a crash, he estimated the probable fatality to be in the vicinity
of 3,000-8,000, the maximum number of fatalities which might result was
estimated to be 30,000. He also came to the conclusion that the
failure of dams could lead to fatalities in the range 14,000-260,000
and the frequency of these events could be as high as 1 in 40 per year
for one affected by an earthquake to one in 30,000 per year.

It is difficult to come to terms with the possibility of very
severe consequences even if generally of very low probability. However,
major accidents will not change the overall risk of accidental death
over 10 years or so. In the USA, Rasmussen records accidental death
rate as 6×10^{-4} per year, i.e. about 100,000 fatal accidents/year in

the USA.   In Europe, roughly the same risk rate is estimated by Vinck[11].

How, then, to set a target for an event highly undesirable - one which might result in many deaths but cannot be set according to an individual risk rate?

If the target is not based on individual risk rate it might be based on the estimated likelihood of killing 10/100/1000 people. However, in order to estimate potential casualties, it is first necessary to set up an accident model, analyse it, and arrive at an estimated frequency (or range of frequencies).

I am currently interested in attempting to set an upper limit to the probability of a serious event, one which has the potential to kill many people, such as pressure vessel failure, a major reactor meltdown, or in the non-nuclear industry, the release of multi-tonne quantities of toxic or flammable gas or the equivalent.

We are gaining experience in assessing the ways in which accidents might occur; our confidence in these assessments will depend on the discipline applied to the exercise and the background of information within the particular industry or operation.

It is difficult to ascribe probabilities as low as $10^{-5}/10^{-6}$ per year; and to reach this level requires considerable repetition in systems and components. To ask for confident prediction at lower levels - as per 10^{-7} per event or year - presupposes a marked degree of randomness in the derivation of the number and very wide repetitive experience, as in the take-off and landing of aircraft.

At one time I sought to find a target - either by comparisons one with the other - or even as a cock-shy - that the chance of an accident having the potential to kill very many people should be less than some number - say $10^{-5}-10^{-7}$ per year.

I find it more and more difficult to move in this direction. Some
people say that no risk is acceptable. Yet we have to make decisions
about present and future industrial activities, the processing, storage,
transport of 1000 or 10,000's of tonnes of materials such as LPG, LNG,
HF, Cl, NH_3 etc. We do not know what is the current risk level; we are
attempting to assess this in some situations. There are many installations, or activities, in which an accident could bring >10,000 people
within the lethal range, yet most accidents will affect but few people.

I believe that accidents in new, complex activities can as yet in
many cases only be estimated down to the order of 10^{-4} per year, the
subsequent sequence from the accident to the variation in severity of
the consequence may spread over 2 decades or more - but not always so.

I can only ask for more systematic work to be continued covering
nuclear and non-nuclear hazards; for a continued and improved dialogue
regarding risk of major hazards and again ask for greater realism and
the rejection of the low number syndrome.

REFERENCES

(1)   C.HINTON, The Future for Nuclear Power, Axel Ax:Son, Johnson
        Lecture, Stockholm, 15 March 1975.

(2)   Superstar Technologies published by Barry Rose (Publishers) Ltd
        in association with the Council for Science and Society
        (SBN-85992-062-3).

(3)   Inquiry into the Bases of Design and Method of Erection of Steel
        Box Girder Bridges, Report of the Committee, Chairman
        Dr A W Merrison, (UK) HMSO, 1973

(4)   The Flixborough Disaster, Report of the Court of Inquiry to the
        Department of Employment, (UK) HMSO, June 1974.

(5)   Report of the Inquiry into the Collapse of Flats at Ronan Point,
        Canning Town, Presented to the Minister of Housing and
        Local Government by Mr.Hugh Griffiths, Professor Sir Alfred
        Pugsley and Sir Owen Saunders, (UK) HMSO, 1968.

(6) USNRC WASH-1400 Reactor Safety Study - An Assessment of Accident
     Risks in US Commercial Nuclear Power Plants, Washington DC,
     October 1975.

(7) An Assessment of the Integrity of PWR Pressure Vessels, Report of
     a Study Group under chairmanship of Dr.W.Marshall, UKAEA
     Report, 1 October 1976, pp.126-128.

(8) C. R. McCULLOUGH et al, The Safety of Nuclear Reactors, Proceedings
     of International Conference on the Peaceful Uses of Atomic
     Energy, Geneva, 1955, Vol. 13, p.79.

(9) G. D. BELL, Safety Criteria, Nuclear Engineering and Design, Vol. 13
     2, (1970).

(10) D. OKRENT, Investigation of Charges relating to Nuclear Reactor
     Safety, Hearings before the Joint Committee on Atomic Energy
     Congress of the United States, 94th Congress, Second Session,
     Vol. 2, Appendixes 12-19.

(11) W. VINCK and G. Van REIJEN, Quantitative Risk Assessment; the
     promised but not the sacred, ANS/ENS International Conference
     on World Nuclear Power, Washington DC, 14-19 November 1976.

ON THE SOCIAL ASPECTS OF RISK ASSESSMENT

HARRY J. OTWAY*, JOANNE LINNEROOTH** AND FRIEDRICH NIEHAUS*

Abstract. Plans for technological development have often been met by demands for a closer examination of the associated benefits and risks and the consideration of social values in public planning and decision processes. A theoretical framework for inter-disciplinary risk assessment studies is presented to aid the balancing of technical data with social values in decision making.

Methods for obtaining value measures are reviewed and an attitude-based method is developed in detail; this model allows identification of the relative importance of the technical, psychological and social factors which underlie attitudes and indicates which factors differentiate between social groups. Results of a pilot application to nuclear power are summarised. For these subjects, different attitudes between pro and con were primarily due to strongly differing beliefs about the benefits of nuclear power. Preliminary results are reported of an application of this model with a heterogeneous sample drawn from the general public. The cognitive limitations which affect rationality in intuitive decision making are summarised as background to introduce formal decision methodologies for the use of attitude data in public decision making.

1. Introduction. New technological systems have been developed to satisfy basic human needs for security and comfort. However, as technological innovations became larger and more complex, and thus capable of offering increasingly attractive benefits, their negative side-effects, or risks, began to emerge as unanticipated threats. Other subtle effects of technological development have been noted;

---

*International Atomic Energy Agency, Joint IAEA/IIASA Research Project, P.O. Box 590, A-1011 Vienna, Austria.

**International Institute for Applied Systems Analysis, Joint IAEA/ IIASA Research Project, P.O. Box 590, A-1011 Vienna, Austria.

they include complicated environmental interactions, feelings of dissatisfaction despite increasing standards of living and a sense of isolation from decisions affecting social welfare. The values implicitly accepted by technology-intensive societies have become increasingly subjected to challenge. The social response to nuclear energy must be viewed in the context of these general concerns (1).

This situation has provided incentives to investigate new procedures and methodologies to synthesize technical data with the needs and wishes of the many publics involved - the reconciliation of technical capabilities with social values. Risk assessment has come to describe studies oriented toward providing information on technological risks, and their social aspects, for use in decisions related to the management of risks. The intent of this paper is to suggest an interdisciplinary framework for risk assessment studies and to discuss methodologies that might be used within this framework to obtain indications of social value. The status of these methodological developments is summarised and results of a pilot application of an attitude-based method are given.

Figure 1 presents a theoretical risk assessment framework which was developed (2,3,4) as a collaborative effort by scientists from a number of disciplines; it is in essential agreement with the approaches used by other researchers (5,6). This framework does not intend to portray the interactions between technological and social systems in any "real world" sense; rather, it illustrates the relationships between the analyses, originating in various disciplines, which may form a risk assessment study. Risk assessment is divided here into three sub-topics: risk estimation, risk evaluation, and risk management.

2. Risk Estimation. The identification and quantification of the risks posed by the technological system under consideration is the most highly developed part of the risk assessment process and thus will not be discussed in detail.

A THEORETICAL FRAMEWORK FOR RISK ASSESSMENT STUDIES

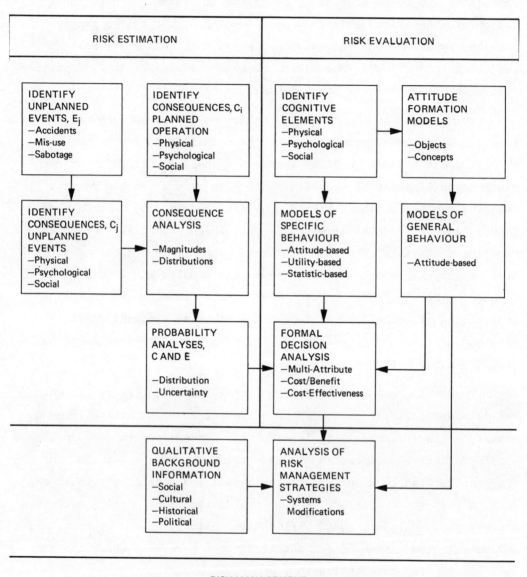

FIGURE 1

Risk estimation begins with the identification of the consequences of planned operations of the system. A sample result of a risk estimation study for projected planned operations of fossil-fuel energy production systems (7) is shown in Figure 2. Based upon a scenario specifying energy demand and the contributions of various energy sources, results are expressed in terms of changes in physical variables (e.g., average global atmospheric temperature) as a result of CO_2 emissions.

Unplanned events that might occur during normal operations have received considerable attention, and the methodologies and procedures are reasonably well understood. They include identification of possible unplanned events that might occur during operation, such as accidents, sabotage or mis-use; identification of their consequences; analyses of consequence magnitudes and their distributions in terms of time, space and social group; and finally, numerical analysis of the uncertainties of all events and consequences. Early risk estimation studies on unplanned events in nuclear power plants were largely carried out by individual scientists on an ad hoc basis (8,9,10,11). Well-financed, full-scale risk estimation studies were carried out later; the best known of these is that sponsored by the US Nuclear Regulatory Commission, the "Rasmussen Report" (12).

In addition to physical risks to health and environment, Figure 1 also includes mention of psychological and social levels of risk. The latter refer to the potential effects of perceived hazard upon the psychological well-being of individuals and the resulting risks to social values and relationships. The nature of some technologies might also imply the necessity of specific social structures and institutions and place requirements upon their stability. Quantitative methodologies for dealing with these "higher order" levels of risk are in the process of development.

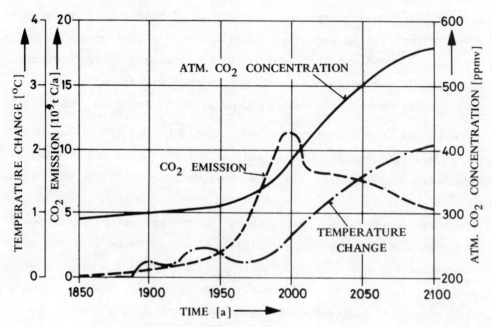

Simulation of the CO_2 burden to the atmosphere with regard
to an optimistic equilibrium strategy.

EXAMPLE OF A RISK ESTIMATION RESULT

PLANNED OPERATIONS

FIGURE 2

3. Risk Evaluation. The measurement of social values* and their reconciliation with technical risk estimates through the framework of formal decision-making methodologies is defined as risk evaluation.

The first four blocks in the risk evaluation part of Figure 1 refer to the social response to risk situations and its underlying determinants. Social response is not based only upon theoretical or statistical prediction of risk, but rather is multiply-determined through a variety of psychological functions such as perception, conditioning and learning. Figure 1 indicates methods for inferring response which are based upon attitudes, utility theory or statistical data.

3.1. Measures of Social Value: Methodologies. Utility-based methods have been used primarily for assessing the decision-makers' (or experts') expectation of "social utility" as a function of technical variables. Although some of these methodologies could, in principle, be extended to make utility measurements on a public survey basis, the technology for so doing does not exist at present; therefore, this method will not be discussed further here.

3.1.1. Methods Based Upon Statistical Data. One widely used approach to risk evaluation is simply to put estimates of risk "into perspective" by comparing them to statistical measures of other risks accepted by society. This method is capable of suggesting that a new risk might be too high. It cannot predict that a new risk, similar in magnitude to other existing risks, will be accepted. Comparison of different types of risks lacks meaning since each risk is characterised by many variables other than statistical expectation - e.g., the psychological effects of consent to exposure, control over outcome, the

---

*The "impossibility theorem" of Arrow (13) imples that measures of social value cannot be obtained. Edwards (14), although in agreement with the theorem as derived from its assumptions (e.g., ordinal utility, no interpersonal comparison of utilities), suggests that public values "can be elicited by some appropriate adaptation of the methods already in use to elicit individual values."

number of people at risk, delayed effects, etc. The importance of such factors will be shown later in the discussion of attitude formation.

The work of Starr (15) offered a broad philosophical basis for beginning risk assessment studies and was instrumental in calling attention to the importance of risk concepts in public decision making. National-level statistics were used as a basis for evaluating the risks and benefits from nine technologies or activities. Based upon this analysis, mathematical relationships were proposed between some determinants of risk acceptability: perceived benefit and acceptable levels of risk; the ratio of acceptable risk levels for voluntary and involuntary risk exposure; and the "psychological yard-stick" people use to judge the acceptable risk levels. This method implicitly assumes that risk levels resulting from past decisions were somehow optimal, thus providing an adequate basis for future decision making, and that those making the decisions had perfect knowledge of the data which were subsequently reflected in the statistical compilations. A subsequent study (16) found fault with the assumptions underlying the methodology and could not reproduce the numerical results using the same data base. The numerical results were found to be excessively sensitive to the assumptions required to extract these variables from national-level statistical data. Further, there is difficulty in proving cause-effect relationships between, for example, risk and benefit or participation in a risky activity and the actual risk level. At present, there is no evidence that analyses based upon statistical data could lead to useful rules for specifying risk acceptability or its determinants.

3.1.2. An Attitude-Based Method. An alternative to the statistics-based approach is to use attitude as a measure of value. Attitude may be defined simply as an evaluative judgement that one likes or dislikes some object or concept, that it is good or bad, that he feels favourable or unfavourable toward it. Attitudes are a useful measure of the values held with respect to some object or concept; the reason is simply that attitude creates a pre-disposition to behave in a consistent

manner toward the object in question. In other words, attitude pre-
disposes an individual to engage in a set of behaviours which, taken
together, are consistent with the attitude. So attitude is a measure of
the general behaviour that an individual will display toward the object
but it is not necessarily related to any specific behaviour (17). Until
rather recently, much attitude research was confused by the lack of a
clear definition of attitude and its underlying components. Often what
were reported as attitudes were, in fact, measures of opinion. Sometimes specific behaviours were used to infer attitude, thus giving rise
to the mistaken notion that there is no relationship between attitude
and behaviour. It has now become clear that beliefs, attitudes and behaviours are distinct variables, with different determinants, but with
stable and systematic relations among them. The relationships among beliefs, attitudes, intentions to perform a behaviour, and behaviour are
shown in Figure 3, developed by Fishbein and his colleagues (18).

Beliefs are the building blocks of attitudes. A belief is a person's subjective probability judgement that an object is characterised
by a certain attribute. People form a number of beliefs about the object on the basis of direct personal observations, information received
from outside sources or through inferential processes. Each of these
beliefs simply links the object with some attribute in the person's
mind. For example, one could directly observe that an automobile is big,
someone might provide the information that it has 150hp, but is economical to operate, and he might infer, from leather seats and wooden dash-
panel, the attribute that it is expensive. All of the beliefs an individual holds about an object are the informational basis for his attitude toward that object.

Each belief is weighted by the evaluation of the attribute in the
attitude formation process. Using the earlier example, "Automobile X is
economical to operate," the strength of this belief (which might also
be negative if one thinks this brand of automobile is uneconomical to
operate) is weighted by the evaluation of the attribute "economical to

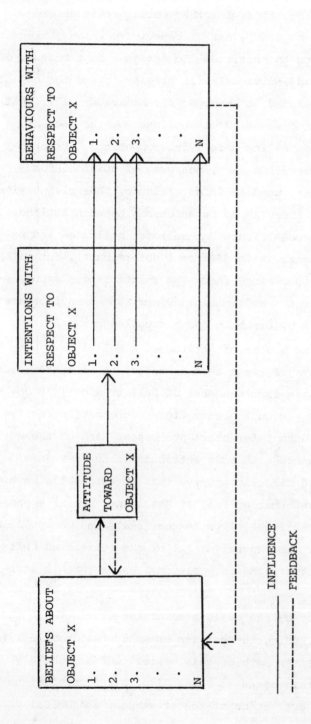

A MODEL OF ATTITUDE FORMATION
(AFTER FISHBEIN)

FIGURE 3

operate." The way in which beliefs and their evaluations combine to form attitude may be written mathematically as:

$$A_o = \sum_i^n b_i e_i$$

A_o is the attitude toward the object "o"

b_i are the beliefs which link the object to specific attributes

e_i are the evaluations of these attributes

n is the number of salient beliefs (usually 7 ± 2 in number)

The measure of attitude obtained from the equation is the sum of the eb products. To verify that this is indeed a measure of attitude, correlations can be made between the $\sum$ eb scores of the subjects and independent, direct measurements of the same attitude which can conveniently and reliably be made using the semantic differential method of Osgood et al. (19). The magnitude and statistical significance of this correlation coefficient provide a measure of the success of the model in estimating attitude and, in addition, ensure that the set of attributes used was adequate to describe the attitude object for the group tested. This test of validity is an important characteristic of the model.

The Fishbein Model has been successfully applied in areas such as attitudes toward minority groups, family planning, politicians; it has several interesting features which could make it useful in understanding the social response to a technology. For example, this model combines the belief and evaluative components of attitude in a form which preserves the distinction between them. The belief component (the b_i of the equation) represents knowledge or opinions about the attitude object while the evaluative component (the e_i) is a measure of affect or feeling.* The Fishbein Model allows not only the identification of the factors important in attitude formation, but also the respective

---

*The public opinion poll usually only measures the belief component of attitude. Unlike attitude, beliefs alone may be unrelated to behaviour. The poll also lacks the independent test to ensure that the attributes used adequately described the object for the subject group.

contributions of opinion and feelings to each factor. Further, by
aggregating the responses of individuals we can examine the response of
any social group and thus find out which factors differentiate between
groups, for example, groups pro or con nuclear power.

3.1.2.1. A Pilot Application of the Attitude Model. A pilot
application of the Fishbein model, to attitudes toward nuclear power,
was carried out in order to test its utility in the area of attitudes
toward technologies and their risks. A questionnaire was given to a
group of 30 people affiliated with a university institute engaged in
energy research. Details of the experimental design and complete
results may be found in Otway and Fishbein (20). The Spearman rank
order coefficient between the estimated and direct attitude scores was
0.66, statistically significant at a level of less that 0.1%, thus
demonstrating the validity of this application.

To better understand the factors differentiating between people
with favourable and unfavourable attitudes toward nuclear power, two
sub-groups were formed from the total sample. Using the direct attitude
measurement scores from the semantic differential as the criterion, the
ten subjects with the highest scores formed the "pro" group and those
with the ten lowest scores the "con" group. Figure 4 presents the mean
algebraic $\underline{eb}$ scores, the mean belief strengths $(\bar{b}_i)$, and the mean
evaluations $(\bar{e}_i)$ of each attribute, for the "pro" and "con" groups.
This table allows identification of those aspects which most clearly
differentiate between the two groups. The magnitude of the $\underline{eb}$ terms
represents their contributions to the overall attitudes.

For the pro group, the three attributes contributing most to
attitudes concerned benefits, i.e., providing good economic value, enhancing the quality of life, and providing benefits essential to
society. In contrast, the three attributes contributing most to the
attitude of the con group were risk-related, i.e., waste production,
the possibility of destructive mis-use of the technology, and the

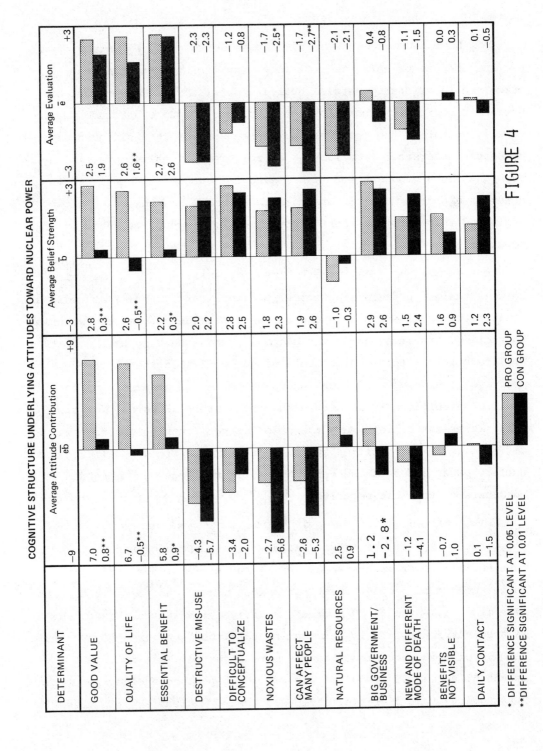

COGNITIVE STRUCTURE UNDERLYING ATTITUDES TOWARD NUCLEAR POWER

FIGURE 4

* DIFFERENCE SIGNIFICANT AT 0.05 LEVEL
**DIFFERENCE SIGNIFICANT AT 0.01 LEVEL

matter of catastrophic accidents.

For four attributes, the differences between the _eb_ values of the
pro and con groups were statistically significant. For example, the
perceived relationship between nuclear power and "big government or
business" contributed positively to the "pro" group's attitude, negatively to that of the "con" group. The reason for this difference can
be better understood from looking at beliefs and evaluations. It may be
seen that both groups strongly believed that nuclear power is in the
hands of big government or business. However, while the "pro" group
evaluated this attribute positively, the "con" group evaluated it
negatively.

There were three additional items for which the _eb_ differences
between the groups were statistically significant. These items were all
related to the benefits of nuclear power: providing benefits essential
to society, providing good economic value and enhancing the "quality
of life." In all three cases, both groups evaluated these attributes
positively, although the "con" group valued enhancement of the "quality
of life" significantly less than the "pro" group. However, for all
three items the beliefs were the major factor contributing to these
differences. More specifically, the "pro" group strongly believed that
nuclear power offers these benefits, while the "con" group tended to be
uncertain to somewhat negative.

There were no significant differences between the groups on the
_eb_ scores of any of the items related to risk. Both groups believed
that nuclear power is characterised by the attributes of affecting
large numbers of people, creating noxious wastes and possible destructive mis-use. Although both groups negatively evaluated these risk-
related attributes, it is interesting that the "con" group's evaluation
for two of them were significantly more negative. This indicates
essential agreement among the group with respect to nuclear power risks
but suggests that differing attitudes toward nuclear power may be

primarily determined by strongly differing beliefs about its benefits.*
The particular group used in this study was not representative, so it
should not be assumed that the results can be generalised to other
populations.

   3.1.2.2. A Further Test of the Attitude Model. Given the signif-·
icant results of the pilot study, the model was then tested with a
heterogeneous sample of 224 respondents residing in various parts of
Austria. By eliminating the separate section on risk attitudes, and
providing more detailed risk- and benefit-related attributes relevant
to the use of nuclear power, a set of 39 attributes was constructed.
Respondents were asked to indicate their beliefs that each of five
energy systems (i.e., nuclear, hydro-electric, solar, coal and oil)
were characterised by these attributes.

   Data collection has just been completed, so only preliminary
results related to attitudes toward nuclear power can be presented at
this time. Consistent with findings of the pilot study, it was possible
to predict respondents' attitudes toward nuclear power from the model.
The Pearson correlation coefficient between attitudes estimated from
the model and direct measures of these attitudes was 0.66 ($p < 0.001$,
df = 223).

   In general, these results agreed well with those of the pilot
study. To simplify interpretation, a factor analysis of the beliefs
underlying this attitude was conducted. It suggested four basic belief
factors: beliefs about "psychological" risks, about economic and technological benefits, about socio-political risks and about environmental
and physical risks. The psychological risk factor was characterised by

---

*In agreement with this result, many surveys on attitudes towards
smoking have found that smokers and non-smokers tend to agree on the
risks associated with smoking; significant differences are found in
their perception of the benefits.

beliefs relating to the use of nuclear power to risk-related attributes
of psychological significance. For example, the belief loading highest
on this factor was that using nuclear power will expose one to risks
without his consent. The belief with the second highest loading was
that, once exposed to these risks, the individual has no control over
them. The benefits factor was characterised by beliefs associating the
use of nuclear power with various benefits, such as increasing the
standard of living and leading to new forms of industrial development.

The third factor was labelled "beliefs about socio-political
risks." At first it seemed surprising that beliefs about the production
of noxious wastes and the transport of dangerous substances were
associated with socio-political risks. However, the storage and transport of nuclear wastes was viewed in relation to the need for physical
security measures and possible mis-use of the technology by terrorist
groups. As noted by Weinberg (22), the storage of long-lived radio-
isotopes places unprecedented requirements upon the stability of the
socio-political institutions charged with their care. Consistent with
this, these concerns were also seen as leading to dependencies upon
elite groups of technical experts and to the concentration of political
power in the hands of big industrial enterprises. The fourth factor was
characterised primarily by concerns about environmental damage, e.g.,
air and water pollution.

Although only preliminary, these results are of interest
because they suggest that people do not think of risks and benefits as
lying along a single bi-polar dimension. Rather, risks and benefits
appear to be viewed independently. Moreoever, in support of the hypo -
thesis proposed by Otway and Pahner (3), people do not seem to perceive
risks along a single dimension but instead they distinguish among their
physical-environmental impacts, the psychological characteristics of
the risk situation and their potential effects upon social and political
systems. The results will be reported in more detail in a forthcoming

paper (22).

3.2. The Decision Maker and Decision Methodologies. The final
integrative step in risk evaluation is an ordering of the alternatives
being considered. This may be viewed as the assimilation and balancing
of the complex technical data resulting from risk estimation analyses
with measures of the corresponding social values. The limitation of the
decision maker in handling the large quantities of probabilistic data
involved in many public decisions suggests the use of formal decision
methodologies to aid in this process.

3.2.1. Cognitive Limitations in Decision Making. Simon (23)
has proposed the theory of "bounded rationality" which asserts that,
in order to deal with the complexities of the world, cognitive limitations force the decision maker to intuitively construct a much simplified model. He then behaves rationally with respect to this simplified
model, but perhaps irrationally with respect to the real situation.
Psychological research on "cue utilisation" bears upon the limits of
rationality in decision making. People seem to be unable to use all the
information provided them in arriving at a decision and, even then, do
not know which information items have actually formed the basis for
their decisions. Further, there are limitations in the processing of
probabilistic information (24). Thus, formal decision methodologies
appear to be especially attractive for supporting complex, many-vari-
able decisions involving risk, which is probabilistic by definition.
These methods allow the decision maker to rationally assign weights to
information items in order to develop an ordered list of the options
under consideration.

Decision methods available include multi-attribute decision
analysis (25, 14), cost-benefit analysis (26), and cost-effectiveness
analysis. These methodologies are, in fact, closely related; cost-
benefit analysis may be derived from multi-objective analysis, cost-
effectiveness is a special case of cost-benefit analysis. Cost-benefit

analysis has been a popular method for making public policy decisions; it requires that all attributes of the decisions be expressed in common units - usually monetary.

A problem arises, however, if there is no observable market price for the attribute in question, such as is the case for most environmental concerns including the risks to the public's health and safety. The interest in assigning values to human life arises from the wish to evaluate changes in mortality risk for use with cost-benefit methodology. Linnerooth (27,28,29) surveyed the use of life value measures in public decision making and reviewed the theoretical models. It was concluded that a rigorous determination of life values is not possible because all methods are, in one way or another, dependent upon the income of those at risk. It was recommended that an arbitrary value, such as $300,000 per life, be chosen for use in cost-benefit analysis and this value be weighted by 3 factors representing the personal status of those exposed, third-party interest in life-saving, and psychological factors such as those mentioned earlier. The approaches which have been used to estimate life values, the values obtained and the limitations are summarised in Table I (30).

An alternative procedure to placing a monetary value on each of the impacts or attributes of the decision consequence is to evaluate them in terms of "utility." In the terminology of multiattribute decision analysis, the problem of valuing each of the decision consequences in terms of utility is referred to as assessing a multi-attributed utility function over the m attributes. The basic idea of multiattribute utility measurement is to elicit the value of each attribute in terms of the decision maker's preferences, one attribute at a time, and then to aggregate them using a suitable aggregation rule and weighting procedure.

Probably the most widely used, and certainly the simplest aggre-gation rule and weighting procedure, is the SMART (Simple Multiattribute Scaling Technique) procedure developed by Edwards (14), which involved taking a weighted linear average. A demonstration experiment (31) reported encouraging results in using attitude measures in Edwards' technique. In this experiment, a group of decision makers was given technical descriptions of six nuclear waste disposal sites as well as information on public attitudes toward these sites. The decision-making group was willing, and able, to use attitude measurement information in the decision process; public attitudes were an important factor in the decisions taken by the group.

4. Risk Management. Looking again at Figure 1, the section called Risk Management refers to the actions one might take, given the data on the technical system, its risks, and the corresponding risk evaluation information. The possibilities to resolve conflicts lie basically in changing the technological system, the social system or the decision process.

4.1. Technology Change. Risk estimation studies identify risks that might be too high and analyse ways to change the technology in order to reduce the risk. However, attitudes would be expected to be sensitive to changes in the physical characteristics of the technology primarily if the change could be directly observed (e.g., reduction in airport noise). This is not necessarily the case where evidence of the change is in the form of new information provided, e.g., an improved safety system in a nuclear plant. The description of a changed technology provides only one of many informational inputs and would not necessarily be reflected in the cognitive structure underlying attitude.

4.2. Social System Change. A survey of the social psychology literature on attitude change suggests that there is no quick and easy way to change people's attitudes. Research has failed to show any evidence of consistent and controlled attitude change. The regularity with which people conduct their daily lives and the persistence of customs, myths, ideals, and mores demonstrates the basic stability of attitudes

and their tendency to evolve rather than to change abruptly. This
research has found a large number of variables to be important in attitude change, thus, demonstrating its complexity. However, a few general
principles emerge from this literature which are of interest. First,
the credibility of the communicator is generally agreed to be an important variable. That is, if you want people to believe what you have to
say, then you yourself must be believable to them. Credible persons
are those known for expertness or prestige in the subject at hand, or
sometimes, on another subject. For the non-prestigeous, credibility is
established by always providing factual and balanced information.

Another variable agreed to be important is the discrepancy
between the message and the initial attitude. People have a "tolerance
band" about the position of their initial attitude in which they are
willing to accept and, at least, process information. This evidence
suggests that extreme messages, which fall outside this tolerance band,
tend to have an effect opposite to that intended. For example, a
strongly positive message may sound very appealing to an industry exec-
utive, but it might very well tend to change, in a negative direction,
the attitudes of those who are uncommitted. A few simple points might
be noted which could help make public communications about technological issues more effective. They are: (1) always be factual; (2) give
both sides of the story; and (3) avoid taking extreme positions.

In summary, attitude change does not seem to be a productive
area for risk management activities. Attitudes change, but usually
slowly. When abrupt changes do occur, the reasons are seldom known and
predictive capability is virtually non-existent.

4.3. The Decision Process. A promising area for risk management
is the use of decision methodologies to aid decision makers and to
provide a framework for broader public participation. This does not
mean that the decision process can be "mechanised;" however, the
decision-making group might be able to evaluate alternatives using a

decision model, and then, by an iterative procedure, make the model results and their holistic decision agree. This could provide a record of what variables were used in reaching a decision, what weights they were given and what values were assigned to them. This might help to make decision processes more understandable to those affected by them; direct participation may not be the real issue.

5. Concluding Remarks. The attitude model which was discussed is capable of answering many of the questions that have been raised about the effects of social and psychological factors, such as involuntary risk exposure, upon the acceptability of risks. It may be seen that these factors are indeed important, but they vary in their relative importance from one attitude object to another and from one social group to the next. This model also allows insights into the "irrationality" of the public in responding to risks without regard to the statistical estimates. Tables of figures have often been used to compare risk estimates with statistics on accepted risks in attempts to convince people that the new activity is "safe." That these "statistics" do not predict behaviour has been clearly shown in the case of nuclear energy. Attitude formation theory suggests that attitudes toward, for example, nuclear power, would be determined by from five to ten attributes. Statistical data on risks is only one informational input in the attitude formation process. To expect people's attitudes to be largely determined by estimates of risk represents a highly simplified behavioural model where one informational input, to the exclusion of other information and observations, determines behaviour. To expect people to behave according to such a simple model might be called hyper-rational and, thus, itself irrational.

More specific to nuclear power, it is characterised by all the psychological factors which have been found to increase the perception of risk. In addition, nuclear power also typifies many of the social concerns of centralisation of vital services, their control by big gov-

LIFE VALUATION FOR PURPOSES OF COST-BENEFIT ANALYSIS

(AFTER LINNEROOTH and OTWAY)

APPROACHES	VALUES	LIMITATIONS
(1) Implicit Value	$9,000 - $9,000,000	. Assumes Past Decisions Are Optimal
(2) Human Capital	$100,000 - $400,000	. Based Solely on Lifetime Income . Ignores Individual Preferences . Discriminates Against Unproductive Members of Society
(3) Insurance Premiums	Wide Range	. Does Not Take Into Account Individual's Interest in Protecting His Own Life
(4) Court Awards	$250,000	. Based on Lost Earnings
(5) Willingness to Pay	$180,000 - $1,000,000	. Difficult to Estimate . Depends on Risk Estimation

SUMMARY

All Measures Depend to Some Extent on the Lifetime Earning Potential
of the Individuals at Risk and Ignore Perception of Seriousness.

CONCLUSION

Cannot be Rigorously Determined. Choose Value (say $300,000)
Weigh According to: Personal Variables, Third Party Interests and
Psychological Factors.

TABLE I

ernment/industry, dependence upon technological elites and requirements
for long-term stability of social institutions. Thus nuclear power
contains, in one technology, virtually all of the issues which have
arisen in debates about technological development, is related to nuclear
weapons on both conscious and unconscious levels, and can be confronted
locally. This lends support to the suggestion (1) that nuclear power
is providing a forum to evaluate a wide range of social issues - that
nuclear power may be playing a symbolic role in a dialogue about the
shape and direction of a technologically-determined future.

The proposed framework for risk assessment studies allows an
inter-disciplinary approach to the formal consideration of social
values in public decision making. At present, the major contribution of
such studies is an improved understanding of the technical and social
systems being investigated and the acquisition of new insights into
their interactions; that is, the inter-disciplinary process involved
may be more valuable than the numerical results produced. Risk assessment may be only a transitory form in a developing analytical process,
but the importance of the issues being addressed suggests that research
along these general lines will continue.

REFERENCES

(1)   H.J. OTWAY, A review of research in the identification of factors
         influencing the social response to technological risks,
         IAEA/CN-36/4. Presented at the IAEA Conference on Nuclear
         Power and Its Fuel Cycle, 2-13 May, 1977, Salzburg, Austria.
         To be published in proceedings.

(2)   H.J. OTWAY, Risk assessment and societal choices, RM-75-2, Inter-
         national Institute for Applied Systems Analysis, Laxenburg,
         Austria, 1975.

(3)   H.J. OTWAY and P.D. PAHNER, Risk assessment, Futures 8, 2 (1976),
         pp. 122-134.

(4)   H.J. OTWAY, The status of risk assessment. Presented at the 10th
         International TNO Conference on Risk Analysis: Industry,
         Government and Society, 24-25 February, 1977, Rotterdam,
         Netherlands. To be published in proceedings.

(5) W.A. ROWE, An "Anatomy" of risk, United States Environmental Pro-
        tection Agency, Washington, D.C., 1975.

(6) R.W. KATES, Risk assessment of environmental hazards, SCOPE report
        8, Scientific Committee on Problems of the Environment,
        Paris, France, 1976.

(7) F. NIEHAUS, A non-linear eight level tandem model to calculate
        the future CO_2 and C-14 burden to the atmosphere, RM-76-35,
        International Institute for Applied Systems Analysis,
        Laxenburg, Austria, 1976.

(8) F.R. FARMER, Reactor safety and siting: a proposed risk criterion,
        Nuclear Safety, 8 (1967), pp. 539.

(9) J.R. BEATTIE, Risks to the population and the individual from
        iodine releases, Nuclear Safety, 8 (1967), pp. 573.

(10) H.J. OTWAY and R.C. ERDMANN, Reactor siting and design from a risk
        viewpoint, Nuclear Engineering and Design, 13 (1970),
        pp. 365.

(11) H.J. OTWAY et al., A risk estimate for an urban-sited reactor,
        Nuclear Technology, 12 (1971), pp. 173.

(12)   Reactor safety study: an assessment of accident risks in
        U.S. commercial nuclear power plants, WASH-1400
        (NUREG 75/014), United States Nuclear Regulatory Commission,
        Washington, D.C., 1975.

(13) K. ARROW, Social Choices and Individual Values, Wiley, New York,
        1951.

(14) W. EDWARDS, How to use multiattribute measurement for social
        decision making, Technical Report 001597-1-T, Social
        Science Research Institute, University of California,
        Los Angeles, 1975.

(15) C. STARR, Social benefits vs. technological risks, Science, 165
        (1969), pp. 1232-38.

(16) H.J. OTWAY and J.J. COHEN, Revealed preferences: comments on the
        Starr benefit-risk relationships, RM-75-5, International
        Institute for Applied Systems Analysis, Laxenburg, Austria,
        1975.

(17) M. FISHBEIN, Attitude and the prediction of behaviour, in Readings
        in Attitude Theory and Measurement (M. Fishbein, ed.),
        Wiley, New York, 1967.

(18) M. FISHBEIN and I. AZJEN, Belief, Attitude, Intention and
        Behavior: An Introduction to Theory and Research, Addison-
        Wesley, Reading, Massachusetts, 1975.

(19) C.E. OSGOOD et al., The Measurement of Meaning, University of
     Illinois, Urbana, 1957.

(20) H.J. OTWAY and M. FISHBEIN, The determinants of attitude
     formation: an application to nuclear power, RM-76-80,
     International Institute for Applied Systems Analysis,
     Laxenburg, Austria, 1976.

(21) H.J. OTWAY and M. FISHBEIN, Public attitudes and decision making.
     Presented at the 6th Research Conference on Subjective
     Probability, Utility and Decision Making, September 1977,
     Warsaw, Poland. To be published in proceedings.

(22) A. WEINBERG, Social institutions and nuclear energy, Science, 177
     (1972), pp. 27-34.

(23) H.A. SIMON, The Models of Man, Wiley, New York, 1957.

(24) P. SLOVIC, B. FISCHHOFF, S. LICHTENSTEIN, Cognitive processes
     and social-risk taking, in Cognitive and Social Behavior
     (J.S. Carroll, W.J. Payne, eds.), Lawrence Erdlbaum Assoc.,
     Potomac, Maryland, 1976.

(25) H. RAIFFA, Decision Analysis, Addison-Wesley, Reading, Massa-
     chusetts, 1968.

(26) E.J. MISHAN, Cost-Benefit Analysis, Allen and Unwin, London,
     England, 1971.

(27) J. LINNEROOTH, The evaluation of life-saving: a survey, RR-75-21,
     International Institute for Applied Systems Analysis,
     Laxenburg, Austria, 1975.

(28) J. LINNEROOTH, A critique of recent modelling efforts to determine
     the value of human life, RM-75-67, International Institute
     for Applied Systems Analysis, Laxenburg, Austria, 1975.

(29) J. LINNEROOTH, Methods for evaluating mortality risk, Futures,
     8 (1976).

(30) J. LINNEROOTH and H.J. OTWAY, Methodologies for the evaluation of
     mortality risk: the implied social objectives, RM-77-XX,
     International Institute for Applied Systems Analysis,
     Laxenburg, Austria, 1977 (in press).

(31) H.J. OTWAY and W. EDWARDS, Application of a simple multiattribute
     rating technique to evaluation of nuclear waste disposal
     sites: a demonstration, RM-77-31, International Institute
     for Applied Systems Analysis, Laxenburg, Austria, 1977.

Topic 2
EVALUATIONS AND APPLICATIONS

DISCUSSION BY THE EDITORS

Topic 2
Evaluations and Applications

This section is concerned with evaluations and applications of reliability and risk assessment technology. The first two papers are specifically concerned with evaluating consequences of nuclear system accident chains. The other four papers deal with probabilistic and safety evaluations.

The first paper in this section concerns the methods used to evaluate consequences during the preparation of the Reactor Safety Study (RSS) [1]. The authors, I. B. Wall, S. S. Yaniv, R. M. Blond, P. E. McGrath, H. W. Church, and J. R. Wayland discuss the manner in which the consequence calculations are performed, the results of RSS in summary, and the individual consequences with emphasis on how the consequences depend on the model parameters.

The second paper, by D. E. Simpson, is concerned also with accident consequence assessment. Specifically, deterministic evaluations of the Fast Flux Test Facility and the Clinch River Breeder Reactor Plant are discussed. Extensive analysis results are presented. Also, assessment methodology and computer programs are discussed.

V. Joksimovic, W. J. Houghton and D. E. Emon present the results of a recent High Temperature Gas Reactor risk assessment study in the next

paper. The methodology used in the study, called the Accident Initiation and Progression Analysis Study, is similar to the methodology
used in the Reactor Safety Study [1]. The paper presents the probabilistic aspects of the AIPA study rather than the consequence aspects.

In the fourth paper, M. E. Lapides discusses evaluation of the
productivity of nuclear power generation systems. This work, that has
been carried out by the Electric Power Research Institute, highlights
the potential value of reliability analyses as an aid to efforts during
the procurement and operational phases of the plant. The paper presents
results, summarizes the present status of the work, and presents future
requirements.

The fifth paper, by R. Billinton, R. N. Allen and M. F. DeOliveira
is concerned with analysis of nuclear power station auxiliary systems.
The details of modeling these systems for qualitative and quantitative
results are presented. The paper emphasizes the comparison of adequacy
of various configurations in terms of their impact on the unit output.

In the final paper in this section, A. Carnino and J. Dubau present
a method for determining reactor accident sequences. An example is
given that involves analysis of a pressurized water reactor plant. The
method called "Barrier" analysis is compatible with the methods used
during the preparation of the Reactor Safety Study [1] and with the
cause consequence analysis method [2].

REFERENCES

[1] U. S. Nuclear Regulatory Commission, Reactor Safety Study,

     WASH-1400, October 1975.

[2] D. Nielsen, "Use of Cause Consequence Charts in Practical Systems

     Analysis," Reliability and Fault Tree Analysis, SIAM.

     Philadelphia, 1975, pp 849-877.

OVERVIEW OF THE REACTOR SAFETY STUDY CONSEQUENCE MODEL

IAN B. WALL*, SHLOMO S.YANIV*, ROGER M. BLOND*, PETER E. MCGRATH[+],
HUGH W. CHURCH[+] AND J. R. WAYLAND[+]

Abstract. This paper describes the calculation of potential nuclear reactor accident consequences as performed for the Reactor Safety Study, an Assessment of Accident Risks in U.S. Commercial Nuclear Power Plants (WASH-1400) [1]. The objective of the study was to realistically assess the risk to society from commercial nuclear power plant accidents. The engineering analysis of the plants, which is described in detail in the Reactor Safety Study, provides an estimate of the probability versus magnitude of the release of radioactive material. The consequence model, which is the subject of this paper, describes the postulated accident after the release of radioactive material from the containment. Separate models trace the released radioactive material through the environment and assess its impact on man.

1. Introduction. The specific purpose of the Reactor Safety

Study (WASH-1400) [1] was to realistically assess the risk to society

from potential accidents in commercial nuclear power plants. To

perform the assessment, event trees and fault trees were used to

estimate the probability of release of radioactive material to the

human environment. To estimate the magnitude of the release, phenom-

enological models were used to describe the liberation of radioactive

materials from the fuel and their transport to the exterior of the

reactor containment building. The assessment of the accident after

release of the radioactive material from the containment is performed

with the consequence model which is the subject of this paper.

*U.S. Nuclear Regulatory Commission, Washington, D.C.
+Sandia Laboratories, Albuquerque, New Mexico.

The first section of this paper provides a description of the consequence calculations from breach of containment and release of radioactive material to predicted consequences (early fatalities, property damage, etc.). The second section of this paper provides a discussion of the individual consequences. The purpose is to indicate how the consequences depend on the many parameters of the model. This discussion also shows the relative degree of uncertainties in the consequences to the modeling considerations and selection of input data.

For a complete description of the consequence model and computed results, the reader is referred to Appendix VI of WASH-1400 [1].

2. Description and Function of Consequence Model. A schematic outline of data and models of the consequence calculation is shown in Figure 1. The following general description gives an idea of the makeup

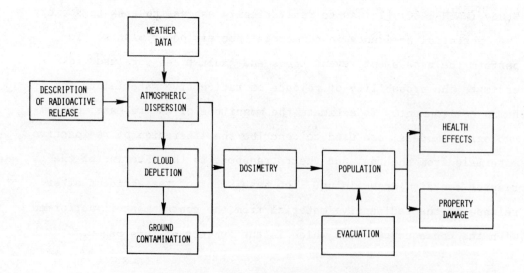

Figure 1.   SCHEMATIC OUTLINE OF CONSEQUENCE MODEL

of the data, the manner in which it is used, the output of the various individual models, and a general flow of the accident progression (consequence calculations).

The results of the consequence model is a set of complementary cumulative distribution functions (ccdf)* for specific consequences (e.g., early fatalities, property damage). Basically, these ccdfs are generated by calculating the magnitude of each consequence for each combination of postulated accident release magnitude, weather and population, and then, after ranking these magnitudes, by plotting the aggregate probability of all combinations which exceed a specified magnitude versus that magnitude.

2.1. Accident Description. The initiating point of the consequence calculation is the specification of the postulated accident in terms of the quantity of radioactive material that could be released to the environment, amount of energy associated with the release, the duration of the release, time of release after accident initiation, warning time for evacuation, elevation of release, and probability of the accident occurrence.

The range of postulated radioactive releases is characterized by 9 PWR and 5 BWR release categories as stated in Table 1. The data in this table represent the basic input to the consequence model and are generated by the engineering analysis of the PWR and BWR reactors as analyzed in the Reactor Safety Study.

---

*A ccdf is a plot of the probability of equalling or exceeding a specified value versus the specified value.

Table 1. SUMMARY OF RELEASE CATEGORIES REPRESENTING HYPOTHETICAL ACCIDENTS

Release Category	Probability (reactor-yr^{-1})	Time of Release (hr)	Duration of Release (hr)	Warning Time for Evacuation (hr)	Elevation of Release (meters)[g]	Energy Release (10^6 Btu/hr)[e]	Fraction of Core Inventory Released[a]						
							Xe-Kr	I[b]	Cs-Rb	Te-Sb	Ba-Sr	Ru[c]	La[d]
PWR 1	9×10^{-7}[e]	2.5	0.5	1.0	25	20 and 520[e]	0.9	0.7	0.4	0.4	0.05	0.4	3×10^{-3}
PWR 2	8×10^{-6}	2.5	0.5	1.0	0	170	0.9	0.7	0.5	0.3	0.06	0.02	4×10^{-3}
PWR 3	4×10^{-6}	5.0	1.5	2.0	0	6	0.8	0.2	0.2	0.3	0.02	0.03	3×10^{-3}
PWR 4	5×10^{-7}	2.0	3.0	2.0	0	1	0.6	0.09	0.04	0.03	5×10^{-3}	3×10^{-3}	4×10^{-4}
PWR 5	7×10^{-7}	2.0	4.0	1.0	0	0.3	0.3	0.03	9×10^{-3}	5×10^{-3}	1×10^{-3}	6×10^{-4}	7×10^{-5}
PWR 6	6×10^{-6}	12.0	10.0	1.0	0	N/A	0.3	8×10^{-4}	8×10^{-4}	1×10^{-3}	9×10^{-5}	7×10^{-5}	1×10^{-5}
PWR 7	4×10^{-5}	10.0	10.0	1.0	0	N/A	6×10^{-3}	2×10^{-5}	1×10^{-5}	2×10^{-5}	1×10^{-6}	1×10^{-6}	2×10^{-7}
PWR 8	4×10^{-5}	0.5	0.5	N/A[f]	0	N/A	2×10^{-3}	1×10^{-4}	5×10^{-4}	1×10^{-6}	1×10^{-8}	0	0
PWR 9	4×10^{-4}	0.5	0.5	N/A	0	N/A	3×10^{-6}	1×10^{-7}	6×10^{-7}	1×10^{-9}	1×10^{-11}	0	0
BWR 1	1×10^{-6}	2.0	0.5	1.5	25	130	1.0	0.40	0.40	0.70	0.05	0.5	5×10^{-3}
BWR 2	6×10^{-6}	30.0	3.0	2.0	0	30	1.0	0.90	0.50	0.30	0.10	0.03	4×10^{-3}
BWR 3	2×10^{-5}	30.0	3.0	2.0	25	20	1.0	0.10	0.10	0.03	0.01	0.02	4×10^{-3}
BWR 4	2×10^{-6}	5.0	2.0	2.0	25	N/A	0.6	8×10^{-4}	5×10^{-3}	4×10^{-3}	6×10^{-4}	6×10^{-4}	1×10^{-4}
BWR 5	1×10^{-4}	3.5	5.0	N/A	150	N/A	5×10^{-4}	6×10^{-11}	4×10^{-9}	8×10^{-12}	8×10^{-14}	0	0

(a) Background on the isotope groups and release mechanisms is presented in the Reactor Safety Study, Appendix VII.
(b) Organic iodine is combined with elemental iodines in the consequence calculations. Any error is negligible since its release fraction is relatively small for all large release categories.
(c) Includes Ru, Rh, Co, Mo, Tc.
(d) Includes Y, La, Zr, Nb, Ce, Pr, Nd, Np, Pu, Am, Cm.
(e) Accident sequences within PWR 1 category have two distinct energy releases that affect consequences. PWR 1 category is subdivided into PWR 1A with a probability of 4×10^{-7} per reactor-year and 20×10^6 Btu/hr and PWR 1B with a probability of 5×10^{-7} per reactor-year and 520×10^6 Btu/hr.
(f) Not applicable.
(g) A 10 meter elevation is used in place of zero representing the mid-point of a potential containment break. Any impact on the results would be slight and conservative.

2.2. Atmospheric Dispersion and Weather Data. The atmospheric
dispersion of the released radioactive material is estimated with a
Gaussian dispersion model. The model is used to calculate ground
level air concentrations, and subsequently ground concentrations of
radioactive material to great distances from the reactor. The Gaussian
model is utilized in a fashion different from usual applications in
that the model includes specification of thermal stability, wind
speed, and precipitation occurrence on an hourly basis. The calculated
plume may change in characteristics every hour of travel time. The
model assumes no temporal variations in wind direction. This simplifying assumption is justified on the grounds that the spatial representation of population is relatively simple* and that only single-station
meteorology is available at most reactor sites. By propagating the
unidirectional plume in all directions from the reactor, each geographical location is potentially exposed.**

The weather data have been collected from six sites which were
judged to be representative of all reactor sites with respect to
variability of climatic or topographic features. Each of the sites has
one year of complete hourly recorded data, i.e., there were 8,760 readings of thermal stability, wind speed and precipitation occurrence.

---

*Representation of the population is simple in the sense that the data
 do not account for diurnal movements and that the population is
 assumed uniform within a spatial mesh having an azimuthal resolution
 of 22-1/2%.
**For an individual site, the relative probabilities of the wind blowing
 in each direction are represented by the wind rose. In the calculations for the 100 reactors at 68 sites, the consequence model utilized
 a uniform probability distribution of wind direction. This simplifying assumption is justified for the calculation of aggregate risk
 from many reactor sites [2].

Since the atmospheric dispersion of the radioactive material
depends on the weather over a period of many hours, and since there is
a large (almost infinite) number of combinations of hourly weather
sequences, the year's worth of data is sampled in such a way that the
true frequency distribution of accident consequences would be closely
approximated. This simulation is performed by systematically selecting
(stratified sampling) during a year the time at which a postulated
accident might be initiated. The atmospheric dispersion of the radioactive material, therefore, is described by utilizing the hourly
sequences of weather data following the selected accident starting
time. A 4-day sampling interval was found to cover the predominant
weather cycles and to provide an acceptably small variance on the
computed consequences. Diurnal cycles are accommodated by a 13-hour
shift in the starting hour.

This procedure of sampling from actual weather data allows one to
calculate more realistically the dependence of the computed consequences
on the type and sequence of weather conditions. It provides a more
realistic estimate of the probability distribution function of consequences than the commonly used procedure of associating the consequence estimates from a time-invariant atmospheric dispersion model
with the frequency distribution of weather conditions [3].

For thermally hot plumes the centerline is modified to account
for buoyant plume rise. The vertical growth of the plume is constrained
to the space under the mixing-layer depth, i.e., no penetration of the
mixing-layer is presumed.

2.3. Cloud Depletion and Ground Contamination. As the plume of radioactive material travels outward from the reactor, competing mechanisms remove the airborne material. In addition to radioactive decay, the radioactive material is removed by deposition processes, e.g., impaction on obstacles (dry deposition), and by precipitation scavenging (wet deposition).

These deposition mechanisms cannot be specified precisely. There are significant dependencies of removal rates on, among other things, precipitation type and rate, particle density and size distribution, surface characteristics of the ground, and weather conditions. For simplicity, the dry deposition velocity (ratio of the deposition flux to the air concentration at a particular distance from the surface) is assumed to be constant and characteristic of 1 micron diameter aerosol which is thought to be a conservative selection. Wet deposition occurs simultaneously with dry deposition when precipitation occurs. Wet deposition is modeled by a simple exponential removal rate. When precipitation occurrence is specified by the weather data, it is assumed to occur uniformly within time and throughout the spatial interval. The removal rate is a function of the thermal stability. The noble gases are assumed to be insoluble and nonreactive, and therefore are not removed by either dry or wet deposition.

The ground concentration is calculated from the air concentration and the deposition rate. The material deposited on the ground is subtracted from the airborne material.

2.4. Air and Ground Concentrations of Radioactive Materials. The previous discussion describes how the consequence model calculates the air and ground concentrations of radioactive materials at various

distances downwind after a postulated accident. It is useful to
reiterate briefly on the calculational flow which leads to estimates
of air and ground contamination without some of the confusing mechanistic details. For a postulated accident, the consequence model selects
a starting hour for the accident from one year of weather data. The
dispersion model utilizes the weather data given at the starting time
of the accident to describe the plume development for the first hour.
The distance of travel from the reactor within that hour is determined
by the wind speed. The plume development is changed each hour after
the release to reflect the changing wind speed, thermal stability,
mixing-layer height, and precipitation occurrence. In an outward
direction from the reactor, the air and ground concentrations of
radioactive material are specified on a fixed polar grid. Air and
ground concentrations are assumed uniform over the interval and they
reflect atmospheric conditions existing at the center point of the
interval. The released radioactive material is followed in the above
manner, hour after hour, until the plume becomes essentially depleted
or until 500 miles is reached. If there is any remaining airborne
material (except noble gases) at the end of the calculation, it is
deposited uniformly in an interval of 500 to 2,000 miles within which
a uniform population density of 78 people per square mile is assumed.

Therefore, from the one selected accident starting time, spatial
distributions of air and ground concentrations of the released radioactive material are estimated. By selecting another starting time,
another spatial distribution of air and ground concentrations is
estimated for the same postulated release. This procedure is repeated

often enough to approximate closely a frequency distribution of air
and ground contamination isopleths following a postulated accident.

   2.5. Dosimetry. The potential radiation dose to individuals and
to populations is calculated from the previously described air and
ground concentrations of radioactive material by using suitable dosimetric models. For this purpose, it is convenient to categorize the
exposure pathways as those associated with the passing cloud and those
associated with ground contamination.

   The airborne radioactive material results in radiation doses
through external radiation from the plume and radiation from inhaled
radioactive material. To receive the external radiation, the individual
must either be immersed in the plume or in its general vicinity. The
model relates the air concentration of radioactive material to an
external dose to various body organs, e.g., bone marrow, GI-tract.

   The radiation dose from inhaled radioactive material is proportional to the exposure to airborne concentration of radionuclides at
roughly 2 meters above the ground, and to the individual's breathing
rate. The dosimetric model describes the time dependent movement of
the radioactive material within the body. The model is essentially
the ICRP Task Group Lung Model [4], modified to calculate a single
short term exposure as opposed to long term inhalation and with some
changes in the model parameters to reflect newer data. The deposition
site of the inhaled material within the respiratory tract depends on
the aerosol size (assumed to be 1 micron). The movement of a specific
radionuclide after its deposition is determined by its chemical form
and particle size and shape. Particle size is the major factor in
determining the fraction of inhaled material retained in the deep
lung. Solubility determines the retention time of the particles in

the deep lung.   The dose from inhaled radioactive material is calculated
over varying time periods up to 50 years.

The radioactive material deposited on the ground results in
radiation doses to individuals through basically three pathways.   The
radiation dose from direct external irradiation due to deposited
material and from inhalation of resuspended radioactive material
starts immediately upon depositing on the ground.   The third pathway
is the ingestion of deposited radioactive material through food and
water.   The ingestion of the radioactive material is a result of its
depositing directly on vegetation which is consumed by man or by
animals furnishing food for man.   The more indirect ingestion pathways
involve the uptake of ground deposited radioactive material through
the roots of vegetation.   The dosimetric models for ingested radioactive material are similar to the models for inhaled material except
that material is directly deposited in the gastrointestinal tract.

The quantity of ingested radioactive material is calculated in
the consequence model with a relatively simple environmental model
which considers the soil-grass-cow-milk pathway and an "other" pathway
in which all other possible pathways are combined.   The soil-milk
pathway model was derived from the large amount of experimental data
that exists.   The results of this model, which agreed well with field
results, were correlated with the many years of data collected from
the nuclear weapon fallout studies.   This correlation yielded results
for other possible environmental pathways to man.   The availability of
these ingestion exposure models was determined from specific data on
the agricultural characteristics for each reactor site.

The radioactive material deposited on the ground is subject to weathering. The material may become bound to large soil particles and thereby be less susceptible to resuspension or to uptake by vegetation. In addition, the material may be leached downward below root zones. The downward movement also provides additional shielding to man from emitted penetrating radiation. Therefore, the dose rate to man will decrease faster with time than that accountable by radioactive decay.

2.6. Population Distributions. The population distributions within 16 radial sections around each of the 68 reactor sites on which the first 100 LWRs are located were obtained from 1970 U.S. Census Bureau data. In order to reduce the number of calculations necessary to estimate risk for 100 reactors, each reactor site was assigned to one of six composite sites on the basis of comparable meteorology. The makeup of these six composite sites is stated in Table 2.

Table 2.   NUMBER OF REACTORS ASSIGNED TO THE COMPOSITE SITES

Composite Site Characteristics	Number of Sites	Number of Reactors BWR	PWR
Atlantic coastal site	10	5	9
Large river valley in northeast	10	6	8
Great Lakes shore	4	3	2
Southeast river valley influenced by Bermuda High	17	7	23
Central midwest plain	23	13	18
Pacific coastal site	4	0	6
	68	34	66

The population sectors for each composite site were generated
from those of the assigned sites in the following manner which was
designed to correctly represent both the average and the peak population
sectors; the first site listed in Table 2 is discussed as an example.
Fourteen reactors at 10 actual sites are assigned to this composite
site. The actual population around each of these 14 reactors was
described by sixteen 22.5° sectors. These 224 sectors were then ranked
from highest to lowest population based on the cumulative population
within 50 miles of the reactor. These 224 ranked population sectors
were used to generate 16 representative population sectors in the manner
indicated in Table 3. For instance, the highest and second highest

Table 3.   CONSTRUCTION OF A COMPOSITE REACTOR SITE

Sector of Composite Site	Sector From Ranked Listing	Conditional Probability of Sector Being Exposed
1	1	1/224
2	2	1/224
3	3,4 (a)	2/224
4	5,6 (a)	2/224
5	Average of next 6	6/224
6	Average of next 6	6/224
7	Average of next 12	12/224
8	Average of next 22	22/224
9	Average of next 22	22/224
10	Average of next 23	23/224
11	Average of next 22	22/224
12	Average of next 22	22/224
13	Average of next 20	20/224
14	Average of next 20	20/224
15	Average of next 21	21/224
16	Average of next 22	22/224

(a) Two reactors at one site.

ranked sectors (of the 224) were assigned to sectors 1 and 2, respectively, of the composite site. The third sector of the composite site
had a radial population distribution that is the average of the population distributions in the third and fourth sectors of the 224. The
population distributions for all 6 composite sites resulting from the
above procedure are characterized in Table 4. The most significant
assumption implied by the above procedure is a uniform frequency
distribution for wind direction. A wind rose for a typical individual
site shows a maximum difference in frequency of about a factor of
three. When assessing the risk for 100 reactors, this variation is
small compared to other modeling approximations [2].

Since airborne and ground concentrations of radioactive material
and hence radiation doses are assumed to be uniform within a given
annular section of a sector, the radiation dose to an individual or to
a population is determined simply by their assigned annular sector.

2.7. Evacuation. Population evacuation was incorporated in the
consequence model. Evacuation should be distinguished from relocation.
"Evacuation" denotes an expeditious movement of people to avoid exposure
to the passing cloud. "Relocation" denotes a post-accident response
to reduce exposure from long term ground contamination.

The evacuation model is based on an analysis of evacuation data
assembled by the U.S. Environmental Protection Agency [5]. This
statistical analysis, which is reported in Appendix VI of WASH-1400 [1],
found that evacuations occur as a uniform mass of people all moving
with the same speed and that the effective evacuation speed is log-
normally distributed. The effective evacuation speed is defined as
the total distance travelled divided by time since warning. The

Table 4.   CUMULATIVE POPULATION VERSUS DISTANCE FOR COMPOSITE SITES

		5	10	20	50	200	500
Atlantic coastal	(a)	9.8×10^3	3.7×10^4	1.4×10^5	1.8×10^6	1.6×10^7	4.5×10^7
	(b)	2.1×10^3	7.6×10^3	1.0×10^5	2.1×10^6	1.6×10^7	2.3×10^7
Large river valley in northeast	(a)	2.1×10^4	8.2×10^4	4.5×10^5	5.3×10^6	3.9×10^7	8.0×10^7
	(b)	3.1×10^3	1.1×10^4	9.9×10^4	8.1×10^6	8.6×10^6	8.6×10^6
Great Lake shore	(a)	2.7×10^3	3.7×10^4	1.9×10^5	9.6×10^5	9.1×10^6	8.0×10^7
	(b)	0	7.5×10^3	3.5×10^5	4.7×10^5	1.9×10^6	1.1×10^7
Southeast river valley influenced by Bermuda High	(a)	3.2×10^3	2.7×10^4	1.5×10^5	7.5×10^5	1.1×10^7	6.1×10^7
	(b)	4.3×10^3	5.4×10^3	2.1×10^5	7.4×10^5	9.5×10^5	1.8×10^6
Central midwest plain	(a)	9.8×10^3	5.1×10^4	2.2×10^5	2.4×10^6	1.8×10^7	6.9×10^7
	(b)	1.5×10^3	1.6×10^4	1.2×10^5	3.3×10^6	4.5×10^6	5.1×10^6
Pacific coastal	(a)	2.4×10^3	2.2×10^4	1.2×10^5	1.8×10^6	1.1×10^7	1.9×10^7
	(b)	1.9×10^3	2.0×10^4	5.1×10^4	1.7×10^6	6.7×10^6	1.2×10^7

(a) Population within 16 sectors weighted by the conditional probability of a sector being exposed (see Table 3).
(b) Population, unweighted, in peak sector as defined by cumulative population within 50 miles.

distributions are very skewed towards very low speeds. In the consequence model, the distribution is approximated by three discrete
speeds. The population within 25 miles is assumed to move radially
outward from the reactor with a 30% probability of having an effective
speed of 0.0 mph, a 40% probability of 1.2 mph, and a 30% probability
of 7.0 mph. The time available for each individual's evacuation is
the sum of the warning time and the time required for the radioactive
plume to reach the individual.

It is assumed in the evacuation model that the population movement
is always radially outward from the reactor, i.e., there is no cross-
wind movement. During the evacuation, the population is assumed to be
unshielded from exposure to airborne radioactive material both exter-
nally and through inhalation and to be shielded from exposure to
ground contamination due to surface roughness and the automobile. If
the evacuating population is overtaken by the cloud of radioactive
material, it is assumed that people will have moved outside of the
contaminated area within a 4-hour period.

Beyond 25 miles, the people are assumed to be relocated within 7
days if the chronic dose due to exposure to ground deposited radionuclides exceeds a specified value. However, if the dose accumulated
within the first 7 days due to exposure to contaminated ground exceeds
200 rads, then the people are assumed to be relocated within 1 day.

2.8. Health Effects. Three categories of potential health
effects are calculated: early and continuing somatic effects, late
somatic effects (cancers), and genetic effects. Early and continuing
somatic effects manifest themselves within a year of exposure. In
contrast, latent cancers would be observed 2 to 40 years after exposure

and genetic effects in succeeding generations. Late somatic and
genetic effects stemming from a major release of radioactive material
would manifest themselves as an increase in the spontaneous incidence
of cancer or genetic effects in the exposed population. Since early
and continuing somatic effects are usually observed after large, acute
doses of radiation (e.g., whole-body doses of 100 rads), they would be
limited to persons within 50 miles or so of the reactor even for the
largest conceivable release. Conversely, late somatic and genetic
effects may result from very low doses albeit with low incidence.
Consequently, these effects may occur at long distances from the
reactor.

   2.8.1. Early and Continuing Somatic Effects. Early and continuing
mortalities may result from radiation damage to the bone marrow, lung
or gastrointestinal tract with radiation damage to the bone marrow
being the most important contributor. Under the conditions likely to
exist as a result of reactor accidents, radiation damage to the lung
or to the gastrointestinal tract is not likely to be lethal unless
accompanied by bone marrow damage.

   The medical advisors to the Reactor Safety Study proposed three
bone marrow dose-mortality criteria [6], depending on the degree of
medical therapy given to the exposed individual. These curves are
reproduced in Figure 2 and are denoted by A, B and C for minimal,
supportive, and heroic therapy, respectively. Mortality criteria are
often stated in terms of the dose that would be lethal to 50% of the
exposed population within 60 days (denoted by $LD_{50/60}$). In Figure 2,
the $LD_{50/60}$ may be read on the abscissa opposite the 50% value on the
ordinate.

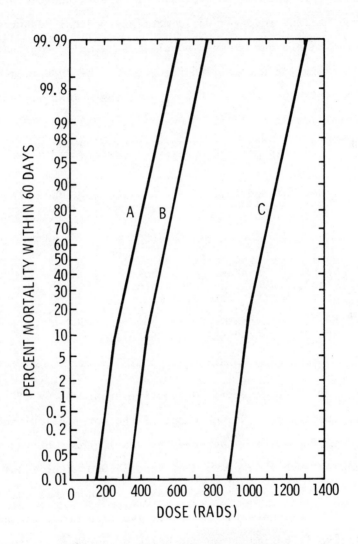

Figure 2. ESTIMATED DOSE-RESPONSE CURVES FOR
MORTALITY IN 60 DAYS WITH MINIMAL TREATMENT (CURVE A),
SUPPORTIVE TREATMENT (CURVE B), AND
HEROIC TREATMENT (CURVE C)

The Reactor Safety Study judged that, in the event of a serious reactor accident in the United States, the Federal and State governments and the utility involved would mobilize medical resources throughout the nation to aid the exposed population; a major constraint would be the availability of specialized resources. The type of medical therapy, the estimated value of $LD_{50/60}$ and the number of people who could potentially be treated are summarized in Table 5.

Table 5.   EFFECT OF MEDICAL THERAPY ON THE BONE
MARROW DOSE MORTALITY CRITERIA

	Medical Therapy	$LD_{50/60}$ rads	No. of People
Minimal	None	340	(large number)
Supportive	Barrier nursing, copius anti-biotics, transfusions of whole blood, packed cells and platelets. Start within 20 days	510	2500-5000
Heroic	Supportive treatment plus bone marrow transplantation. Start within 10 days	1050	50-150

The critical dose range for supportive treatment is 350 to 550 rads to the bone marrow. In the event of the worst calculated accident (a probability of about 10^{-9} per reactor-year), the expected number of people receiving such a dose would be about 5,000. For less severe accidents, the number would be much smaller. Since this number is consistent with the constraint on supportive treatment, the number of early fatalities is estimated on the basis of curve B in Figure 2.

The study also estimated the number of prenatal deaths and cases of early morbidities including hypothyroidism, temporary sterility, congenital malformations, growth retardations, cataracts and prodromal vomiting [1].

2.8.2. Late Somatic Effects. Late somatic effects are limited to latent cancer fatalities plus benign and cancerous thyroid nodules. The following discussion is limited to latent cancer fatalities since the methodology for all these effects is similar and it is the most significant effect.

Following the irradiation of a large number of people, there is a latent period during which no increase in cancer incidence is detectable. After this period, the radiation-induced cancers appear at an approximately uniform rate for a period of years, which is termed the plateau. The risk of latent cancers is normally stated either in terms of the incidence rate during the plateau period (cases per million exposed population per year per rem) or in terms of the expected number of cases (cases per million man-rem). The latter value is merely the integral under the curve, or the incidence rate times the plateau period.

As a starting point, the study uses the estimates stated in a report issued by the National Academy of Sciences on the Biological Effects of Ionizing Radiation (the BEIR report) [7]. The BEIR report estimates risks on both an absolute and relative basis and by using 30-year and lifetime plateaus. For the reasons given in Appendix VI of WASH-1400 [1], the study accepts the absolute basis and a 30-year plateau as being more appropriate for the evaluation of reactor risks.

The BEIR report relied heavily on the ongoing study of the
Japanese atomic bomb survivors, who received very high dose rate
exposure of gamma, beta and neutron (high-LET)* irradiation. Furthermore, the dose magnitudes were estimated to range from 10 to over 300
rem. Those survivors receiving less than 10 rem were used as a control
population group for the BEIR estimates. The doses from a reactor
accident would be almost exclusively due to low-LET radiation (i.e.,
no neutrons and less than 1% due to alpha radiation). Except for a
few individuals who might be irradiated by the passing cloud very
close to the reactor, the dose rates to the whole body would be less
than 1 rem per day which, with respect to latent cancer induction, is
a low dose rate. For all these reasons, the exposures resulting from
a reactor accident would be different from the exposures on which the
BEIR report bases its estimates with respect to quality of radiation,
dose rate and dose magnitude. The risk estimates generated in the
BEIR report are based on a linear extrapolation from the aforementioned
data to zero doses without any threshold dose (i.e., a dose magnitude
below which there would be zero induction of cancer). Both the BEIR
and United Nations [8] reports caution that this linear hypothesis is
likely to overestimate the risks for low doses and/or low dose rates
of low-LET radiation and that, in cases of low exposure, it cannot be
ruled out that the risk may actually be zero.

---

*Linear energy transfer (LET) is a measure of the rate of energy loss
along the track of an ionizing particle. High-LET radiation includes
alpha particles and neutrons.

Since the objective of the Reactor Safety Study is to make as realistic an assessment of risks as is possible and to place bounds on the uncertainty, the study makes three estimates of the number of latent cancers from a reactor accident. The upper bound is based on the BEIR estimates with some small changes reflecting recent data. For the central (most realistic) estimate, the upper bound is modified by dose-effectiveness factors which are stated in Table 6.

Table 6. DOSE-EFFECTIVENESS FACTORS

Total Dose (rem)	Dose Rate (rem per day)		
	< 1	1-10	> 10
< 10	0.2	0.2	0.2
10 - 25	0.2	0.4	0.4
25 - 300	0.2	0.4	1.0

These factors, which are based on recent experimental data for animals, reduce the expected incidence of latent cancers for small doses and/or low dose rates. In the opinion of the study, these central estimates would represent a more realistic assessment of latent cancer fatalities arising from a reactor accident; the study's medical advisors were of the unanimous opinion that these dose effectiveness factors still probably overestimate the true risk. The overall pattern of data shows no observable difference from an unirradiated control population for persons receiving either an acute dose of less than 25 rem or a chronic dose of less than 1 rem per day to the whole body. As an approximate indication of a possible nonzero lower bound, the study estimates the population dose received by individuals in excess of a threshold and applies the incidence rate used for the central estimate.

The BEIR report estimates the incidence of radiation-induced latent cancer fatalities for individual organs and summarizes the overall effect in terms of whole-body radiation. The latter approach was appropriate since the BEIR report was primarily concerned with external radiation to the whole body. In the event of a reactor accident, inhalation of radioactive material from the passing cloud may result in a nonuniform dose distribution in the body; certain organs (e.g., lung) may receive much higher doses than others. In order to accommodate this nonuniform dose distribution, the doses and the expected radiogenic latent cancer deaths are calculated for individual organs and summed to determine the overall risk. For reference purposes, the whole-body values are also calculated.

There is a detailed discussion in Appendix VI of WASH-1400 of the changes to the risk coefficients and plateaus recommended in the BEIR report; the changes are very small. The impacts of the dose effectiveness factors and of the organ-by-organ dose calculation will be discussed in Section 3.2.

2.9. Property Damage. Property damage following a postulated reactor accident is not of the same nature as that resulting from other potential catastrophic events, i.e., there is no physical damage to the property offsite. The property damage arises as a result of contamination by radioactive material and the possible radiation dose that could be received if the property were utilized in its intended manner. The restriction in the property use results in economic loss.

The components of property damage, as assumed and modeled in the consequence model, are evacuation costs, loss of agricultural products, decontamination costs and population relocation costs.

2.9.1. Evacuation Costs. The evacuation model designed for low probability high consequence accidents assumed that all persons within a 5-mile circumference of the reactor and within a 45° arc out to 25 miles centered on the prevailing wind direction at the time of the accident would participate in evacuation. An analysis of EPA evacuation data [5] provided an estimate of the costs, corrected for inflation, of evacuating large numbers of persons and furnishing or paying for temporary food and shelter.

2.9.2. Interdiction Model. The radioactive contamination of a large area may result in the contamination of milk produced by cattle grazing on contaminated pastures, in the external contamination of crops and/or excessive radiation doses to man. In such events, the milk and crops may be impounded and/or the people relocated for a period of time. All of these actions are termed "interdiction."

The interdiction model is based upon maximum acceptable doses in the unlikely event of a reactor accident. The dose criteria used by the Reactor Safety Study are based upon the recommendations of the U.S. Federal Radiation Council [9] and the British Medical Research Council [10].

The dose criteria are translated into corresponding contamination levels (Ci/m^2) of different radionuclides by the dosimetric models described in Section 2.5 and an environmental model which incorporates the grass-cow-man or soil-root-crop-man pathways. The milk interdiction level is the most restrictive so that the area over which milk would be impounded would be the largest. Conversely, the people interdiction level is the least restrictive so that the area from which people would be relocated would be the smallest. The "weathering" of deposited

radionuclides is properly incorporated so that as the cumulative
lifetime dose decreases, the interdiction distance slowly moves towards
the reactor.

The cost of agricultural losses is calculated on the basis of
current prices and average land fractions devoted to agricultural use.
Relocation costs are composed of the loss of income during the period
of relocation (90 days), moving costs (to a distance of 100 miles),
and the economic loss of the physical property. Property value is
amortized over 30 years. For relocation periods greater than 10 years,
the total value of the property is assessed.

2.9.3. Decontamination Model. Decontamination is defined as the
cleanup and removal of radionuclides. The possible decontamination
methods include physical removal of the radionuclides, stabilization
of the radionuclides in place, and environment management. The particular procedure utilized in a given case would depend on many factors,
including (1) the type of surface contaminated, (2) the external
environment to which the surface is exposed, (3) the possible hazards
to man, (4) the costs involved, (5) the degree of decontamination
required, and (6) the consequences of the decontamination operation.

A measure of effectiveness of decontamination operations is the
decontamination factor DF, which is defined as the original contaminant
concentration (in curies per square meter) divided by the contaminant
concentration after decontamination.

The decontamination model is illustrated in Figure 3. Without
decontamination, the interdiction criterion (L) translates to a distance
R_1. After a maximum decontamination factor of 20, the land area
between R_1 and R_2 will become available for reoccupation. Subsequent

weathering of the radionuclides will reopen the land area between R_2 and R_3. In fact, the model is slightly more sophisticated since two decontamination factors of 2 and 20 are incorporated; the decontamination costs for the former are significantly lower than for the latter. These costs were estimated on a per capita basis so that for the population within a spatial interval requiring decontamination, the total decontamination costs could be easily calculated in the consequence model.

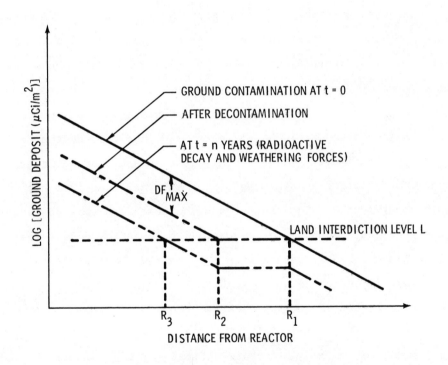

Figure 3.   ILLUSTRATIVE DECONTAMINATION MODEL FOR
            GROUND LEVEL RELEASE

3. Discussion of Accident Consequences. This section of the
paper provides a discussion of the individual estimated consequences,
i.e., early fatalities, latent cancer fatalities, and property damage.
The purpose of this section is to provide the reader with an indication
of how these individual consequences depend on the various models in
the consequence model and their input data.

3.1. Early Fatalities. Before exploring the sensitive parameters,
it is instructive to consider the cause of early deaths and the principle radionuclides involved. Doses to three organs are considered for
the calculation of early fatalities. On the basis of clinical and
experimental data described in the Reactor Safety Study, it is evident
that the dominant mechanism for early death is the radiation damage to
the bone marrow. Figure 4, which is for a BWR 1 release, shows the
relative doses at 0.5 miles from the reactor to the bone marrow from
the major exposure modes associated with the passing cloud and the
relative contributions from the different radionuclides to the doses.
It is evident in Figure 4 that for this release and distance the
external dose from the passing cloud, the external dose from the
contaminated ground within 4 hours, and the dose from inhaled radionuclides contribute equally to the overall bone marrow dose. The
principle radionuclides resulting in bone marrow dose are Te-132,
I-132, I-133, I-135, Sr-89 and Ba-140.*

The calculation of early fatalities is one of the most sensitive
results of the consequence model. The reason for this sensitivity is
due to a number of important factors. First, clearly the number of
early fatalities is sensitive to the assumed value for $LD_{50/60}$.

---

*The dose contributions are for radionuclides either in the passing
cloud, deposited on the ground or inhaled and include any contributions
from their radioactive daughters.

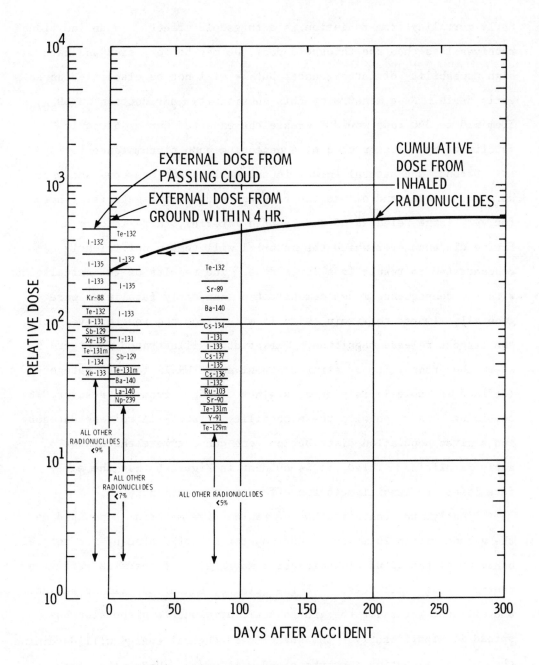

Figure 4. RELATIVE DOSES TO BONE MARROW AT
0.5 MILES FROM REACTOR

Early mortality from radiation is a threshold effect. If an individual receives less than the threshold dose, he may be very sick and have a high probability of latent cancer but he will not be classified as an early death. As a measure of this sensitivity, a reduction in $LD_{50/60}$ from 510 to 340 rads would increase the expected number of early fatalities by a factor of 3 or 4 depending upon circumstances.

Second, the natural forces in the environment are constantly at work to disperse and dilute the released radioactive material. Even for very large releases of radioactive material, there is only a finite distance over which the material will remain sufficiently concentrated to result in a large dose. The results of the calculations with the consequence model demonstrated that early fatalities were generally limited to within 10 to 15 miles from the reactor even for the largest release magnitude. This fact is illustrated in Figure 5 where the probability of early death given a PWR-1A release category (defined in Table 1) is plotted against distance from the reactor. By combining this graph with the probability of a PWR-1A release category and a given population distribution, one can estimate the number of early fatalities. Third, it is evident in Figure 5 that the early fatalities are sensitive to the effective evacuation speed.

Finally, the concentration of radioactive material deposited on the ground within 20 miles of the reactor is influenced by the initial behavior of the cloud of radioactive material. If there is sufficient thermal energy associated with the release, the cloud will be buoyant and will not intersect the ground until atmospheric dispersion has spread it significantly. The quantity of thermal energy will determine the initial elevation that the cloud will reach. Under the same atmospheric conditions, the higher the initial elevation of the cloud

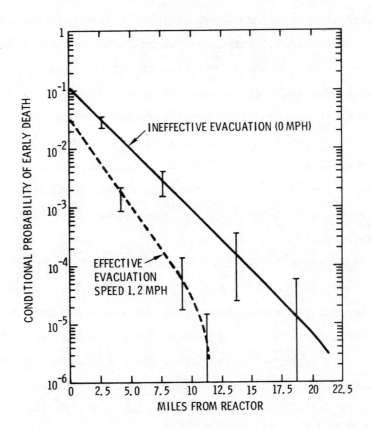

Figure 5. CONDITIONAL PROBABILITY OF EARLY DEATH AS A FUNCTION OF DISTANCE FROM REACTOR FOR THREE EFFECTIVE EVACUATION SPEEDS GIVEN A PWR-1A RELEASE

(a) Approximately, absolute mortality probabilities are 10^{-6} per reactor year times quoted values

(b) The error bars denote the variation in the mean values for the six meteorological data sets

(c) For effective evacuation speeds of 4.7 and 7 mph, the conditional probability of early death is effectively zero within 25 miles.

the greater the distance from the reactor that the cloud will intersect
the ground. When the cloud intersects the ground at increasing distance,
the air concentration at ground level will be decreasingly smaller.
It is for this reason that, generally, thermal energy of the release
has an important effect on the calculation of early fatalities. The
exception occurs when radioactive material is deposited on the ground
by precipitation. Since precipitation scavenging removes airborne
material from all parts of the cloud, the important parameter is the
total quantity of airborne material and not the ground level air
concentration. For this reason, the maximum number of early fatalities
is fairly independent of release energy (see Figure VI 13-25 of Appendix VI of WASH-1400 [1]).

3.2. Latent Cancer Fatalities. The calculation of latent cancer
fatalities is performed on the basis of population doses. Since no
threshold is assumed for cancer induction from radiation exposure, it
is reasonable to expect that the number of latent cancer fatalities
would not be very sensitive to the various assumptions made in the
consequence model. The single parameter of primary importance is the
total quantity of radioactive material released in the accident.

Table 7 states the average effective incidence per million man-
rem to the whole-body for different calculational methods. Tables 8
and 9 state the percentages of latent cancer fatalities attributable
to each exposure pathway calculated from both organ-by-organ doses and
whole-body doses. Tables 7 and 8 are different from their counterparts,
Tables VI 13-3 and VI 13-4, in Appendix VI of WASH-1400 [1]. First,
during preparation of this paper, a programming error in the consequence
model was discovered whereby dose effectiveness factors were being
erroneously applied to the calculation of breast cancer. Correction

Table 7. EFFECTIVE INCIDENCE OF RADIATION-INDUCED
LATENT CANCER FATALITIES PER MILLION MAN-REM
OF WHOLE-BODY RADIATION

Method	Release Category	Upper Bound (BEIR)	Central Estimate	Lower Bound Threshold 10 rem	25 rem
Whole Body		122	48	46	31
Sum of Individual Organs	PWR-1	208	107	103	87
	PWR-2	143	75	73	62

of this error (i.e., no dose effectiveness factors for breast cancer)
increases the average effective incidence of latent cancer calculated
on an organ-by-organ basis by a few percent (Table 7) and enhances the
relative importance of breast cancer (Table 8). Second, the calculational method was improved for Tables 8 and 9 compared to Table VI 13-3.
For example, the original calculations for the distribution were based
upon an actual population distribution which created a misleading
result whereas those for Tables 8 and 9 used a uniform population
distribution. Further, the contribution of each exposure pathway in
Table VI 13-3 was estimated by setting its conversion factors to zero
which method is only approximate since the central estimate is based
upon a piece-wise linear model; this approximation was circumvented
for Tables 8 and 9. Third, Tables VI 13-3 and VI 13-4 were based upon
a BWR-1 (PWR-1) release category. The results of calculations based
upon a PWR-2 release category have been added to Table 7 and in
Table 9. The very low probability ($< 10^{-8}$/reactor-year), high latent
cancer fatality estimates are dominated by the PWR-1 and BWR-1 release
categories but the other release categories (PWR-2 through PWR-9 and
BWR-2 through BWR-5) dominate the higher probability ($> 10^{-8}$/reactor-
year) end of the spectrum. The PWR-1 and BWR-1 release categories

Table 8. CONTRIBUTION OF DIFFERENT EXPOSURE PATHWAYS[a] TO LATENT CANCER FATALITIES FOR A PWR-1 RELEASE CATEGORY[a]

Percentages

	Leukemia	Lung	Breast	Bone	GI Tract [b]	All Other [c]	TOTAL	WHOLE BODY [d]
External Cloud	.1	.1	.3	.1	.1	.1	1	1
Inhalation from Cloud	.3	22	.4	.1	.5	.2	24	5
External Ground (< 7 days)	2	2	5	.9	.7	2	13	18
External Ground (> 7 days)	8	5	19	2	2	7	43	64
Inhalation of Resuspended Contamination	.2	13	.2	.3	.2	.2	14	4
Ingestion of Contaminated Foods	1	.6	2	.6	.5	.7	5	8
SUBTOTALS	12	43	27	4	4	10	100	100

[a] Except thyroid cancer which is calculated separately as discussed in Appendix VI of WASH-1400 [1]

[b] The gastrointestinal tract includes stomach and the rest of alimentary canal

[c] "All other" denotes all cancers except those specified in the table

[d] Whole-body values are proportional to 50-year whole-body man-rem

Table 9. CONTRIBUTION OF DIFFERENT EXPOSURE PATHWAYS TO LATENT CANCER FATALITIES FOR A PWR-2 RELEASE CATEGORY[a]

Percentages

	Leukemia	Lung	Breast	Bone	GI Tract	All Other	Total	Whole Body
External Cloud	.2	.1	.5	.1	.1	.1	1	1
Inhalation from Cloud	.5	4	.7	.2	.4	.2	6	3
External Ground (< 7 days)	3	2	7	.7	.9	3	16	16
External Ground (> 7 days)	12	8	28	3	4	11	66	68
Inhalation of Resuspended Contamination	.2	1	.2	.4	.2	.1	3	2
Ingestion of Contaminated Foods	2	1	3	1	1	1	9	10
SUBTOTALS	18	16	39	5	6	16	100	100

[a] All footnotes to Table 8 apply

have a relatively very large ruthenium content compared to the other release categories. This high Ru-106 content results in a preferential exposure of the lung due to inhalation from the passing cloud and from resuspended contamination and in the dominance of lung cancer in the latent fatalities as shown in Table 8. For the other release categories, typified by PWR-2, the dominant exposure pathway is from ground

contamination which gives a more uniform dose to the different organs
and is comparable to a whole-body dose. The more prevelant single
cancers are estimated to be breast, leukemia and lung. The more
uniform dose associated with the PWR-2 release category also results
in a less significant difference between the organ-by-organ and the
whole-body dose calculations for PWR-2 than for the PWR-1 release
category as shown in Table 7. Finally, it is also evident in Table 7
that the use of threshold doses of 10 or 25 rem would not greatly
reduce the estimates of latent cancer fatalities. This low sensitivity
is due to the already low effectiveness of small doses in the central
estimate.

Figure 6 shows the average probability of late death versus
distance from the reactor given a PWR-2 release. (This figure is
different from its counterpart, Figure VI 13-26, in Appendix VI of
WASH-1400 [1] which was based upon a PWR-1A release.) On the average,
the probability of a latent cancer fatality per individual will be
roughly proportional to the airborne concentration of radioactive
material with distance from the reactor. The difference between the
"total" and "chronic" probabilities in Figure 6 is the conditional
probability of late death due to radiation exposure within the first
week; the discontinuity at 25 miles is caused by evacuation. The
chronic risk of latent cancer fatality per individual is roughly
constant between 5 to 80 miles since decontamination and/or interdiction
actions will limit an individual's dose to approximately 25 rem.
Within a radius of 5 miles, the average chronic risk is reduced since
this area will be permanently interdicted under extreme accident
conditions. Beyond 80 miles, the chronic risk falls rapidly with
distance since the airborne concentration decreases. By combining

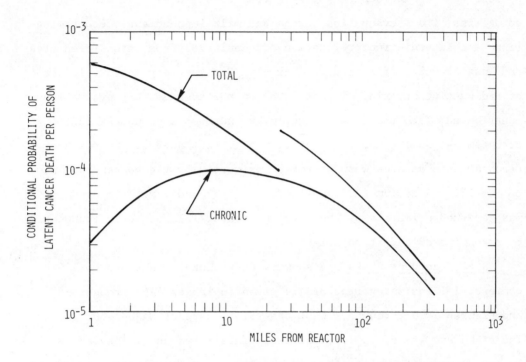

Figure 6.   CONDITIONAL PROBABILITY OF LATENT CANCER DEATH GIVEN
A PWR-2 RELEASE (APPROXIMATELY, ABSOLUTE MORTALITY
PROBABILITIES ARE 10^{-6} PER REACTOR YEAR TIMES STATED ONES)

this graph with the probability of a PWR-2 release and a given population distribution, one can estimate the number of late fatalities.
Since the population is predominantly located beyond 25 miles, the
number of late fatalities is roughly proportional to the population
living between 25 and 100 miles from the reactor.

3.3. Property Damage. The estimation of costs associated with a nuclear reactor accident involves a consideration of costs in four categories, i.e., evacuation, crops and milk impoundment, decontamination, and interdiction (relocation of population from interdicted area and loss of value of interdicted area). For the very large releases of radioactive material, it was found on the average that evacuation contributed 9% to the total costs, milk and crop impoundment 13%, decontamination 19%, and interdiction 59%. The need to interdict or decontaminate an area was determined primarily by the concentration of cesium-134 and cesium-137 deposited on the ground. Milk impoundment was governed primarily by the level of iodine-131, and crop impoundment by strontium-89 and strontium-90.

3.4. Accident Risks. A summary of accident risks due to 100 commercial light water reactors is given in Tables 5-6 through 5-8 and Figures 5-10 through 5-16 of the Main Report of WASH-1400 [1]. The risks are stated in terms of early fatalities and illnesses, latent cancer fatalities, thyroid nodules, genetic effects, and property damage. Correction of the programming error noted in Section 3.2 increases the average number of latent cancer fatalities stated in the Reactor Safety Study* by roughly 15%. This change in the latent cancer fatalities varies from a small increase at higher probability, low consequence events to about 25% increase at low probability, high consequence events. Therefore, the maximum calculated latent cancer fatalities increase from 1500 to about 1900/year per reactor-year.

---

*Latent cancer fatalities are stated in Tables 5-5, 5-6, and 5-8 and Figures 5-5 and 5-12 of the Main Report, and in Figure VI 13-33 of Appendix VI.

The reader was cautioned (page 13-26) in Appendix VI of WASH-1400 "that a given trial is unlikely to maximize each consequence. That is, one trial may result in maximum early fatalities but another may result in maximum property damage." As a result of additional work, it can now be stated more emphatically that, in the probability range of 10^{-8} to 10^{-9} per reactor-year, there is a very low probability of any early fatalities occurring simultaneously with a large number of latent cancer fatalities. Thus, it is incorrect to suggest that the total risk is represented by adding the stated early and latent cancer fatalities for two reasons. First, for a given probability, e.g., 10^{-9}/reactor-year, the events resulting in the corresponding early and latent cancer fatalities are disjoint. Second, early and latent cancer deaths have different characteristics; the former would occur within a short time period after exposure and involve an average of about 40 years life-shortening for each fatality, whereas most of the latter would occur within the time period 10 to 40 years after exposure and involve an average life-shortening of about 10 years for each cancer fatality [11].

The contributions of the various postulated release categories to the overall risk are stated in Table 10. It is evident in this table that the PWR-1 through PWR-3 and BWR-1 through BWR-3 release categories are the major contributors to accident risks. The release categories not involving a core meltdown, PWR-8, PWR-9 and BWR-5, are negligible contributors to the public risk.

Table 10.   CONSEQUENCES OF INDIVIDUAL RELEASE CATEGORIES

Type	Accident Probability Per Reactor Year	Risk(a)			$\approx 10^{-9}$(b)		
		Early Fatality	Latent Fatalities	Damage $	Early Fatality	Latent Fatalities(c)	Damage ($X10^9$)
PWR-1A	4×10^{-7}	4.4×10^{-6}	6.2×10^{-4}	420	500	500	10
PWR-1B	5×10^{-7}	1.0×10^{-5}	5.5×10^{-4}	290	500	660	10
PWR-2	8×10^{-6}	1.7×10^{-5}	6.1×10^{-3}	6700	1000	625	10
PWR-3	4×10^{-6}	9.8×10^{-6}	5.2×10^{-3}	2700	300	625	10
PWR-4	5×10^{-7}	0	2.1×10^{-4}	87	0	200	2
PWR-5	7×10^{-6}	0	9.7×10^{-5}	53	0	83	0.5
PWR-6	6×10^{-5}	0	1.2×10^{-4}	320	0	12	0.1
PWR-7	4×10^{-5}	0	1.3×10^{-5}	2100	0	2	0.07
PWR-8	4×10^{-5}	0	3.0×10^{-4}	34	0	7	0.01
PWR-9	4×10^{-4}	0	5.1×10^{-6}		0	0	0.0
BWR-1	1×10^{-6}	4.3×10^{-6}	8.7×10^{-4}	440	500	660	10
BWR-2	6×10^{-6}	3.2×10^{-7}	3.2×10^{-3}	2900	30	825	10
BWR-3	2×10^{-5}	0	5.8×10^{-3}	2900	0	625	10
BWR-4	2×10^{-6}	0	7.8×10^{-5}	21	0	41	0.1
BWR-5	1×10^{-4}	0	1.8×10^{-7}	0	0	0	0

(a) Risk is defined as probability of release category times the integral of the product of probability and consequence magnitude. The risk values are weighted by the fraction of the particular light water reactor type in the 100 reactor population, i.e., 0.66 for PWR and 0.34 for BWR. The average consequence can be obtained by dividing the risk by the accident probability per reactor year.

(b) These values indicate the approximate consequence for a 10^{-9} per reactor-year probability. The overall consequences as shown in Appendix VI of WASH-1400 [1] were obtained by the summation of probabilities for a given consequence magnitude. These values cannot be reproduced by the summation of the consequences at a given probability (e.g., 10^{-9}/reactor-year).

(c) 10^{-9} latent fatalities are per year values.

ACKNOWLEDGEMENTS

A work of the magnitude of the Reactor Safety Study consequence model represents the contributions of numerous people. The authors wish to acknowledge, in addition to their outstanding leadership of the overall study, the major contributions to the consequence model of Norman C. Rasmussen and Saul Levine. The authors received the support of a very talented group of medical advisors: Victor P. Bond, James F. Crow, Marvin Goldman, Leonard D. Hamilton, George G. Hutchison, Clarence C. Lushbaugh, Roger O. McClellan, Harry R. Maxon, Samuel C. Morris, James V. Neel, Dean R. Parker, William L. Russell, Eugene L. Saenger, Leonard A. Sagan, Maurice F. Sullivan, Niel Wald, and Joseph A. Watson. A complete list of acknowledgements is contained in Appendix VI of WASH-1400 [1].

REFERENCES

[1] Reactor Safety Study, "An Assessment of Accident Risk in U.S. Commercial Nuclear Power Plants," U.S. Nuclear Regulatory Commission, WASH-1400, NUREG-75/014 (1975).

[2] Sprung, J. L. and G. P. Steck, "Correlations between wind direction probability and population at nuclear reactor sites" (Sandia Laboratories report to be published).

[3] McGrath, P. E., Ericson, D. M., and Wall, I. B., "The Reactor Safety Study (WASH-1400) and its Implication for Radiological Emergency Response Planning," IAEA International Symposium on Handling of Radiation Accidents, February 28 - March 4, 1977, Vienna, Austria.

[4] ICRP Task Group on Lung Dynamics, "Deposition and Retention Models for Internal Dosimetry of the Human Respiratory Tract," Health Physics 12, 173-207 (1966).

[5] Hans, J. M., Jr., and Sell, T.C., Evacuation Risk - An Evaluation, U.S. Environmental Protection Agency, EPA-520/6-74-002 (1974).

[6] Wald, N. and Watson, J. A., "Medical Modification of Human Acute Radiation Injury," Fourth International Conference of the International Radiation Protection Association, April 24-30, 1977, Paris.

[7] The Effects of Exposure to Low Levels of Ionizing Radiation (BEIR Report), Advisory Committee on the Biological Effects of Ionizing Radiation, Division of Medical Sciences, National Academy of Sciences, National Research Council, Washington, D.C. (1972).

[8] United Nations Scientific Committee on Atomic Radiation (UNSCEAR), Ionizing Radiation: Levels and Effects, Report E 72 IX 18, Vols. I and II, United Nations, New York (1972).

[9] Federal Radiation Council, Background Material for the Development of Protective Action Guides for Strontium-89, Strontium-90 and Cesium-132, FRC Staff Report No. 7 (1965).

[10] Medical Research Council, Criteria for Controlling Radiation Doses to the Public After Accidental Escapes of Radioactive Material, Her Majesty's Stationery Office, London (1975).

[11] Davis, H. T. et al., "Estimates of life-shortening due to latent cancer from potential nuclear reactor accidents" (Sandia Laboratories report to be published).

LMFBR ACCIDENT CONSEQUENCES ASSESSMENT

D. E. SIMPSON*

Abstract. Deterministic analyses of hypothetical LMFBR accidents
have been applied in safety evaluations of FFTF and CRBRP. These analyses involve the application of complex inter-related computational
techniques to assess a wide range of circumstances more extreme than
any which are realistically foreseen. Such analyses are well suited to
be employed in combination with probabilistic techniques for risk
assessment.

I. Introduction. The first important risk assessment for nuclear

reactors in the United States was the Reactor Safety Study [1] directed

by Dr. Norman Rasmussen and published in October 1975. This study

addressed accident risks associated with commercial light-water reactor

power plants in the United States. The study has stimulated increased

interest in the assessment of risks and in the associated probabilistic

analysis techniques. The major departures of the reactor safety study

from the safety evaluation techniques which have traditionally been

employed in reactor licensing were its explicit consideration of the

probabilities of accidents, and its consideration of accidents of such

low probability that they had not been specifically included in

licensing reviews.

─────────────────────

*Hanford Engineering Development Laboratory, Richland, Washington,
operated by Westinghouse Hanford Company, a subsidiary of Westinghouse
Electric Corporation, for the United States Energy Research and Development Administration under contract EY-76-C-14-2170.

The experience in safety analysis for liquid metal fast breeder
reactors is, of course, less extensive than that of light-water reactors,
and there have been no risk assessments of liquid metal fast breeder
reactors on the scale of the Reactor Safety Study. However, in recent
years, there have been extensive safety analyses performed for the Fast
Flux Test Facility (FFTF) [2] and the Clinch River Breeder Reactor
Plant (CRBRP) [3].

These safety analyses have included identification and evaluation
of conventional types of accidents considered for light-water reactors
and also have considered safety margins for extremely improbable, hypothetical core disruptive accidents (HCDA). Those studies have included
analysis of postulated failure of post-accident cooling, including melt-
through of the core from the reactor vessel. Analytical models and computer codes have been developed and applied to calculate the inherent
response of the reactor and plant systems to extreme circumstances far
beyond design conditions.

II. Summary. The LMFBR, like the LWR, is designed so that plant
faults and malfunctions can be sustained without excessive plant damage
or release of radioactive material. However, the evaluation of risk
associated with low probability, high consequences events includes
consideration of circumstances which are thoroughly protected against
by design, but which, if they should occur, might involve substantial
core disruption and the release of significant amounts of energy. Such

postulated events have been termed (hypothetical) core disruptive
accidents (HCDA, or CDA).

LMFBR core disruptive accident initial phases are analyzed with
complex computer codes which calculate reactor power and thermal transients, and the corresponding reactivity effects of core physical
changes, up to the point of significant core disruption. Additional
calculations are made to evaluate core meltdown or hydrodynamic disruption, and to define the energy release and the core conditions resulting from dispersal of core material.

The mechanical effects of a given energy release are analyzed by
structural dynamics codes, supplemented by analytical simulation of the
specific reactor system and in many instances, by scale model experiments.

Calculation of the expulsion or escape of sodium and radioactive
materials from the reactor vessel into the containment is an essential
step in the consequences assessment, and one which currently has
limited mechanistic or experimental basis. Accordingly, it is current
practice to incorporate a wide range of parametric variation at this
stage, or to identify clearly conservative assumptions.

Several computer codes are used in appropriate combinations for
analysis of containment conditions and radioactivity escape. These
codes include analysis of chemical reactions; heat transfer; atmosphere
pressure and leakage; and the suspension, transport and deposition of
radioactive materials in the containment atmosphere.

Calculations of off-site consequences are carried out by analysis
of the meteorological transport and dispersion of radioactive material
released from the reactor building. Conventional calculations can be
made using reactor licensing methods, or for risk assessment, probabilistic analyses of the effects of the calculated release can be made,
similar to the analyses of the Reactor Safety Study.

III. Discussion.

A. Accident Consideration in LMFBR Design. The LMFBR design provides defense-in-depth comparable to that of light-water reactors in
the United States. Design considerations and safety assessments
include identification and evaluation of accidents ranging from relatively insignificant plant malfunctions up to severe accidents with the
potential for significant release of radioactivity. A recent comparison of typical light-water reactor accidents and accidents identified
by the Nuclear Regulatory Commission (NRC) for evaluation for the CRBRP
is shown in Table I, taken from the CRBRP Final Environmental
Statement [4]. Accident classes 1 through 8 are included in the plant
design basis. The plant is designed specifically to sustain any of
these accidents without severe adverse impact upon the public; specifically without exceeding the radiation exposure guidelines of 10 CFR 100.
Class 9, Hypothetical Events, are not specifically included within the
plant design basis. However, for the CRBRP the NRC staff determined
that the plant should include capabilities to reduce the risks associated with the spectrum of events in this category. Core melt and core

TABLE I

CLASSIFICATION OF POSTULATED ACCIDENTS AND OCCURRENCES

NO. OF CLASS	DESCRIPTION	EXAMPLES (9/1/71 LWR GUIDANCE)	CRBRP EXAMPLES—GENERAL
1	Trivial Incidents	Small spills Small leaks outside containment	Single seal failures, minor sodium leaks
2	Misc. Small Releases Outside Containment	Spills Leaks and pipe breaks	IHTS valve, seal leaks, condensate storage tank valve leak Turbine Trip/Steam Venting
3	Radwaste System Failures	Equipment failure Serious malfunction or human error	RAPS/CAPS valve leaks RAPS surge tank failure cover gas diversion to CAPS Liquid Tank leaks
4	Events that release radioactivity into the primary system	Fuel failures during normal operation. Transients outside expected range of variables	Loss of hydraulic holddown Sudden core radial movement Maloperation of Reactor Plant Controller
5	Events that release radioactivity into the secondary system	Class 4 & Heat Exchanger Leak	Class 4 & Heat Exchanger Leak*
6	Refueling accidents inside containment	Drop fuel element Drop heavy object onto fuel Mechanical malfunction or loss of cooling in transfer tube	Inadvertent floor valve opening Leak in CCP in EVTM Drop of fuel element Crane impact on head
7	Accidents to spent fuel outside containment	Drop fuel element Drop heavy object onto fuel Drop shielding cask—loss of cooling to cask Transportation incident on site	Shipping cask drop EVST/FHC system leaks Loss of forced cooling to EVST
8	Accident initiation events considered in design-basis evaluation in the Safety Analysis Report	Reactivity transient Rupture of primary piping Flow decrease-Steamline break	S-G leaks Steamline break Primary Na storage tank failures Cold trap leaks Rupture of primary piping
9	Hypothetical sequences of failures more severe than Class 8	Successive failures of multiple barriers normally provided and maintained	Successive failures of multiple barriers normally provided and maintained.**

*The CRBRP has a closed cycle secondary heat transport system which separates the primary coolant from the power conversion system. Class 4 failures and coincident heat exchanger leaks therefore do not result in a significant release to the environment.

**Class 9 accidents are not included in the design basis of the plant protective system and engineered safety features. However, the staff has determined that the plant should include capabilities to reduce the risks associated with a spectrum of events in this category (see Sec. 7.1.1).

IHTS = Intermediate Heat Transfer System
RAPS = Radioactive argon processing system (purifies contaminated core gas)
CAPS = Cell atmosphere processing system
EVST = Ex-vessel storage tank (in spent fuel)
FHC = Fuel handling cell
SG   = Steam generator
EVTM = Ex-vessel transfer machine (for fuel handling)

disruptive accidents are considered to be in Class 9 because design features are incorporated to assure that such accidents are of very low probability (no more than one chance in a million per year) [4].

For the FFTF, the safety review included analysis of similar postulated accident sequences. During the FFTF construction authorization review, a hypothetical core disruptive accident with the release of theoretically available work energy of 150 MW-sec was identified as a basis for evaluation of containment safety margins. It was further established that the consequences of even more energetic accidents would be evaluated and the possible incorporation of "fallback" design features would be considered. One of these "fallback" features was related to the ex-vessel core retention in case of reactor vessel melt-through. The evaluation of greater accident energetics and the consideration of potential reactor melt-through effects led to extensive calculations of accident consequences considering possibilities much more severe than any which were predicted mechanistically [5].

B. Hypothetical Accident Consequences Analysis Basis. Hypothetical accidents are analyzed to assess safety margins or to estimate risks associated with extremely improbable events. The term "hypothetical" implies that such an event requires circumstances which are carefully designed against or which are inconsistent with technical understanding of applicable phenomena. Design for prevention of such an accident is illustrated in Figure 1. Multiple malfunctions or an

event exceeding the capacity of engineered protective systems must

occur for the core disruptive stage (HCDA) to be reached.

Historically, considerable emphasis in safety evaluation for fast

reactors has been focused on the energetics of a core disruptive nuclear

excursion. The consideration of such an accident arises from charac-

teristic differences between the fast reactor and the LWR. The liquid

metal cooled fast reactor has no stored mechanical energy in the cool-

ant, as does the pressurized system of the LWR, and therefore, the

possibility of a primary coolant boundary break is less important. On

the other hand, the fast reactor core is not arrayed in its most reac-

tive configuration. Therefore, mechanical rearrangements, such as

those which can be conceived of in case of fuel melting, could result

in increased reactivity and a power excursion which would be terminated

only by the dispersal of the core. Originally, highly simplified

analyses were performed to scope the possible energetics of a core

excursion [6]. Subsequent emphasis has been placed on understanding

the pertinent reactor phenomenology and on development of computer

codes for analysis of accident characteristics and energetics. Gener-

ally, two classes of accidents are considered: those in which normal

cooling to the core is lost and those in which power increase in the

reactor is postulated; in each case, assuming an accompanying total

failure of the plant protective system. A third class of event can be

included: that of an initial reactivity insertion at such a rate or

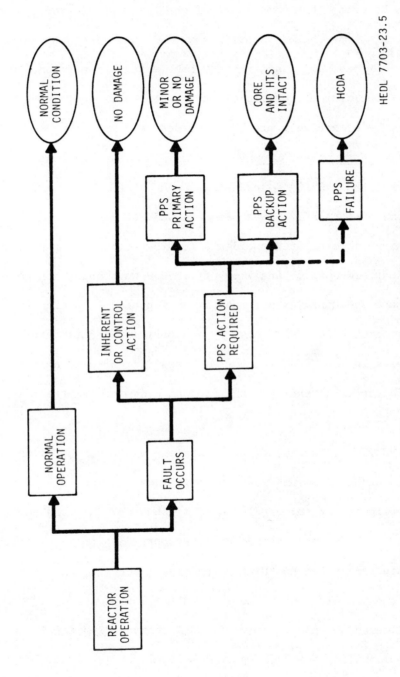

HEDL 7703-23.5

FIGURE 1. Reactor Operation Event Tree – Accident Prevention Summary

magnitude as to exceed the capability of the plant protective system.
Examples of this type include control rod ejection, core support structure failure, or passage of a large gas bubble selectively through the
positive void coefficient portion of the core. Such events have been
considered to be effectively precluded by design.

C. Core Transient and Energetics Analysis. Two major series of
computer codes, the SAS [7][8] code developed by ANL and the MELT [9]
[10] code developed by HEDL, have been created for analysis of the
characteristics of postulated accidents leading to core disruption.
These codes perform coupled analyses of reactor neutronics, thermal
hydraulics, and the dynamics and reactivity feedback of coolant and
fuel motion. Both codes rely heavily on experimental data developed in
transient tests in the TREAT reactor as bases for calculating fuel
failure characteristics and fuel and coolant dynamics under accident
conditions.

For the FFTF, evaluation of the transient overpower HCDA concluded
that the most probable result would be relatively benign, with the
transient terminated by the expulsion of molten fuel into the coolant
channel after fuel failures caused by overpower. Hydraulic sweepout of
fuel would counteract the postulated positive reactivity insertion,
terminating the transient with the core intact and coolable. While
this is considered the most likely result, uncertainties exist and
there is a possibility that molten fuel or fuel debris would plug cooling channels to the extent that gross core meltdown would occur. In

that case, subsequent circumstances would not be significantly different from some of those investigated in the loss of flow HCDA studies.

In the FFTF, the loss of flow without response of the plant protective system would lead (if no protective action by the operator were
taken in the 10-15 seconds available) to coolant boiling followed by
dryout, clad melting and fuel melting. For relatively small reactors,
including FFTF and CRBR, the net reactivity feedback of these effects
would be negative or sufficiently small that no reactivity excursion
would be produced. In larger LMFBR's, the possibility of positive
reactivity feedback due to coolant boiling exists. This could lead to
a nuclear excursion with reactivity insertion rates determined by coolant voiding dynamics. For the smaller reactors, with no nuclear excursion, the melting of fuel could lead to non-energetic core dispersal
into a subcritical array due to the forces of fission gases or sodium
or steel vapor generated in the core. It has been postulated, alternatively, that molten fuel could slump under gravity, leading to an
increase in reactivity. This reactivity addition rate can be calculated to be of the order of 10-20 $/sec, leading to a mild hydrodynamic
disassembly, with the core being dispersed by fuel vapor pressures of
the order of tens of atmospheres.

It is possible that neither of the above dispersal sequences would
take place, perhaps due to prior plugging of coolant channels by molten
and re-solidified cladding material. In such an event, a generalized

molten core condition would be reached which has been termed a "transition phase" [11][12]. The transition phase involves a liquid-vapor
system in which the residual reactor power plus decay heat generates
vapor tending to maintain the core material in an expanded configuration. This condition would persist until melting of surrounding steel
structure could produce an opening for expulsion of the core material.
Transition phase characteristics have been investigated phenomenologi-
cally, with idealized assumptions. These studies indicate the stability of the boiled-up condition. More detailed analytical techniques
are being developed for investigation of transient effects and geometri-
cal interactions. The FUMO [13] code is being developed at HEDL, for
use alone or as a sub-routine of SAS; the SIMMER [14] code is under
development at LASL for whole core analyses inclusive of the transition
phase.

From this brief discussion, it is clear that there are a number of
possible core dispersal modes for the loss-of-flow HCDA. The termina-
tion of the accident is predicted to occur with expulsion of most of
the core as a two-phase mixture into the sodium in the reactor vessel.
This core dispersal is predicted to be essentially nonenergetic or only
mildly energetic, except for the possible large-core voiding excursion.
Nevertheless, consideration of pessimistic possibilities of fuel slump-
ing, molten pool void collapse, or dispersed fuel reentry are such that
energetic core disassembly has not been totally discounted, although

its probability seems to be much less than the probability of non-
energetic meltdown and dispersal.

The occurrence of prompt criticality in any accident sequence
generally requires introduction of a different analytical technique and
computer code. Most widely used is the VENUS [15] code. VENUS performs a coupled neutronics-hydrodynamics analysis, approximating the
core as zones of homogeneous fluid, with material motion calculated due
to pressure gradients and neutronic feedbacks due to temperature and
material displacement.

For relatively low reactivity insertion rates such as 10-20 $/sec,
excursion energy is low compared to the mechanical strength of reactor
vessel components. Rates of this order are characteristic of postulated gravity collapse of molten fuel, or of sodium voiding feedback in
a large LMFBR. Rates of the order of 100 $/sec are considered pessi-
mistic allowances for extremely improbable recriticality conditions.

The potential core accident terminating modes are illustrated in
Figure 2, which is an event tree for the HCDA up to the point of core
dispersal. The end-conditions can be grouped in three classes. The
two more likely of these are nonenergetic, and the third, least likely,
presents a range of energetics with decreasing probability as energy
increases.

Calculation of the energy available to cause damage to the reactor
system is based on the thermodynamics of materials capable of being
vaporized by the thermal energy of the accident. Generally, the

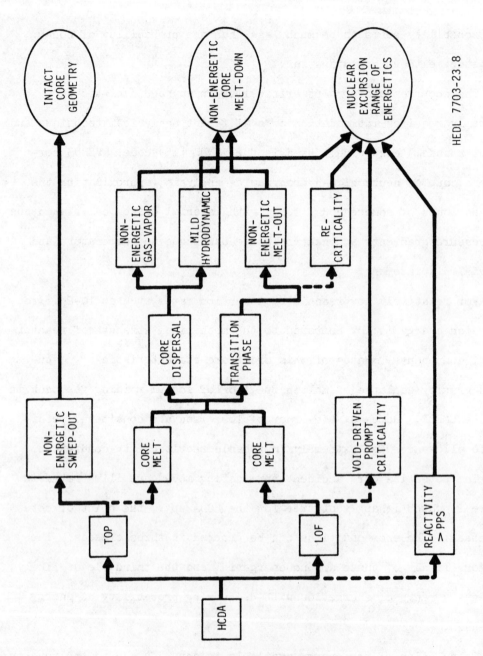

FIGURE 2. Possible Hypothetical Accident Paths

HEDL 7703-23.8

thermodynamically available work from expansion of fuel vapor is considered to represent a reasonable estimate of damaging work potential [16].
The possibility of rapid heat transfer to sodium and release of additional work energy through a sodium vapor expansion has been widely
evaluated. For mixed oxide fueled fast reactors, the possibility of an
extensive molten fuel-coolant interaction with substantially increased
work energy has essentially been discounted [17].

D. Mechanical Consequences. An energetic reactor accident would
be capable of doing damage to the surrounding system because of the
generation of pressurized vapor at or near the core location. The
pressure would be felt on the reactor vessel internals and would be
transmitted through the fluid to the boundaries of the system. The
downward force against the core support structure would be transmitted
to the reactor vessel support system. The radially propagating pressure wave would reach the wall of the reactor vessel and, if sufficiently high, could cause deformation of the vessel wall. If there is
a free sodium surface, which is characteristic of sodium cooled
reactors, a sodium slug would be accelerated upward. Impact of this
slug on the reactor vessel head would transfer momentum to the head,
with the excess slug energy being dissipated radially by deformation of
the upper vessel wall. The upward motion of the vessel head would
generally be resisted by its supporting structure and holddown system.
Effects of an energetic accident on the remainder of the heat transport

system would result from the displacement of the reactor vessel, and from pressure waves propagated through the system. These conditions are illustrated by Figure 3.

A mechanistic analysis of a reactor system can be initiated using the REXCO [18] code to compute deformation and force histories in the vessel. REXCO is a hydrodynamic structural response code developed at ANL particularly for LMFBR accident analysis applications; other hydrodynamic codes exist with similar capabilities. The code ICECO [19] has been developed at ANL for similar applications. The HCDA energy source is input in the form of pressure-volume data. The output from REXCO includes vessel wall strains, pressure history at HTS nozzles, and force histories on the vessel head and core support plate.

Additional mechanical analyses are required to complete the mechanical consequences assessment. For example, for FFTF, dynamic movement of the vessel was computed with a special purpose computer simulation of the vessel, head, support structure and connecting bolt and suspension systems. Input data were REXCO force histories on the vessel head and vessel support structure. Computed displacements of the reactor vessel were used to evaluate clearances between vessel nozzles and the guard vessel.

Primary HTS analysis includes the transmission of pressure pulses from the reactor vessel inlet and outlet nozzles, using computer codes such as NAHAMMER [20], HAMOC [21], ICEPEL [22] and SUPWAN [23].

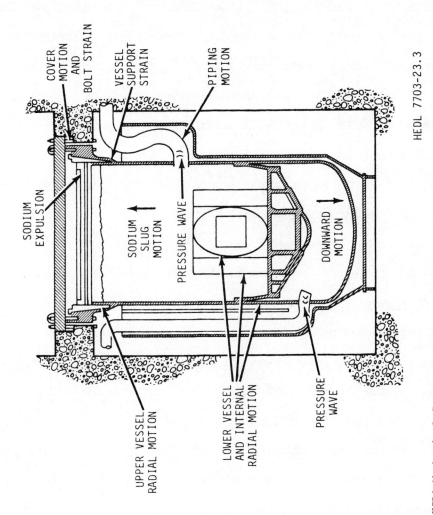

HEDL 7703-23.3

FIGURE 3. HCDA Mechanical Consequences

The mechanical response analysis of the FFTF has shown that a reactor vessel and heat transport system designed for normal LMFBR service is capable of absorbing significant amounts of energy without rupture, provided clearances are available in those locations where displacements or deformations occur.

Detailed analyses also provided predictions of failure modes and the extent of system boundary failure, especially in reactor head-mounted components, at various accident energy levels. For example, at a 150 MW-sec energy level, no failure sufficient for sodium expulsion was predicted. Reactor head and system structural integrity, with possible development of leakage through damaged seals, was predicted to be maintained for energy levels up to about 350 MW-sec equivalent [5].

E. Core Disposition. Deterministic analysis of the disposition of the dispersed core material is one of the most difficult aspects of evaluation of core disruptive accident consequences. It is an important consideration because the initial disposition of the core will determine its coolability, so long as heat removal capability by sodium circulation exists. For FFTF, analysis [24][25] led to the conclusion that most of the core material would be expelled upward into the reactor vessel outlet sodium plenum where it would be mixed with sodium, fragmented and deposited in particulate debris beds on available horizontal surfaces. This assessment is based upon several considerations. Hydraulic pressure drop, mechanical and thermal conditions surrounding the core would result in preferential penetration upward, rather than downward or laterally. Pressures in the core, whether from fission gas, sodium vapor, or even fuel vapor, would expel the core material as a jet through the first available opening. The jet of fuel material would penetrate well into the sodium, where mixing of the expelled fuel and sodium would take place.

Considerable experimental data have been developed which show that molten UO_2 fuel, upon contacting sodium, fragments into particles. Experimental data have also established the coolability of such particle beds up to a thickness of the order of 5-10 cm, depending on decay heat level [26][27]. Assessment of the coolability of the core debris, then, depends upon the available surfaces, the amount of debris, and the availability of sodium heat removal.

In the FFTF, it was concluded that the entire core could be cooled on horizontal surfaces in the upper plenum, and that heat would then be removed from the fuel material to the sodium and subsequently be removed from the sodium by the main heat transport system. The limited amount of fuel which might not be expelled from the core would be coolable either as fragment beds or as molten fuel within the reactor core support structure. For larger reactors, the ratio of available cooling surface to core quantity is substantially less and the debris may not be coolable; in this case analyses can be carried out to estimate the amount of time which would be required for the resulting debris to melt its way downward and through the reactor vessel.

Considering both the mechanical consequences of the HCDA and the
core disposition, alternate accident paths subsequent to an HCDA, up to
disposition of the core debris in the reactor vessel, or release from
the vessel, are illustrated in Figure 4. Five potential conditions are
identified; the two conditions with debris expelled are most unlikely,
requiring accident energetics sufficient to rupture the reactor coolant
boundary and expel core material. The other events are predictable
based on the technology summarized here. Hand calculations are
required for core disposition; deterministic computer codes have not
been developed for this application.

F. Containment Transients and Radiological Analysis. Deterministic analyses as discussed above, with allowance for branch points to
reflect alternate possibilities, provide for definition of reactor and
heat transport system conditions which could result from postulated
HCDA's. Analysis of subsequent behavior requires calculation of
several time-varying parameters in a closely coupled manner. Computational techniques treat a number of applicable sequences and parameters
separately. Application of these techniques has been coupled for an
integrated analysis, as illustrated in Figure 5.

(1) Containment and Radiological Source Terms. Calculation of
containment transient conditions and radioactive material transport and
release from the plant is dependent upon the extent, nature and timing
of release of sodium and core materials from the reactor system.

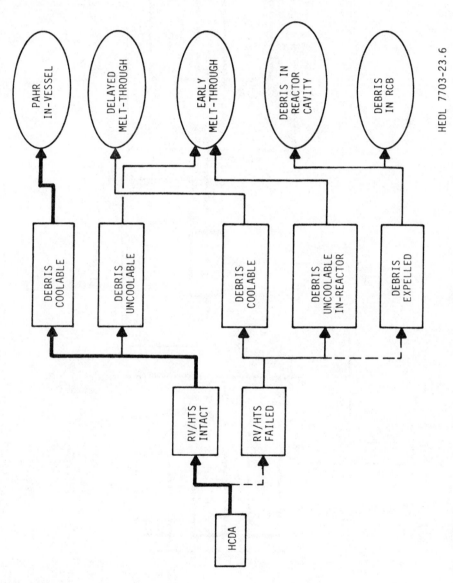

HEDL 7703-23.6

FIGURE 4. Possible Post-HCDA Core Dispositions

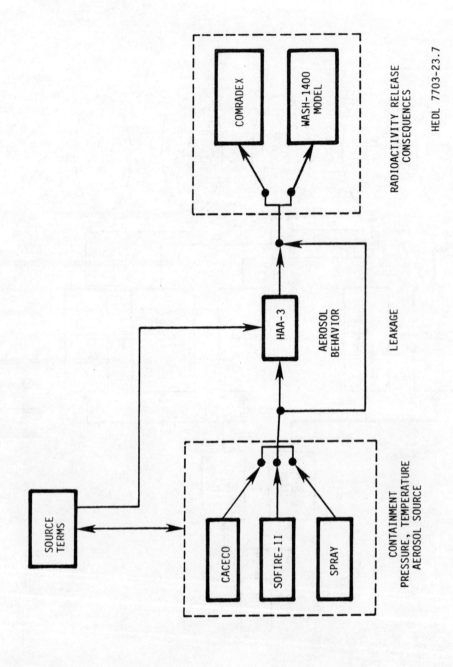

FIGURE 5. Integration of computations for Containment Transients and Radiological Consequences

Release of sodium could occur either at the time of the initial accident or, if applicable, at the time of core melt-through from the vessel. The initial release is calculated on the basis of leakpaths predicted by the mechanical consequences analysis and sodium locations and pressures from the accident characteristics analysis. For FFTF, the maximum calculated sodium expulsion for energetics up to the head integrity limit was about 240 lb (110 kg), corresponding to an assumption of failure of reactor head seals with a resulting open area of a few square inches.

In case of reactor vessel/guard vessel melt-through, the entire reactor vessel sodium inventory would of course be released into the reactor cavity.

The sodium release is important to subsequent analysis because of its potential for chemical reaction. Release and distribution of radioactive materials is important for calculation of decay heating effects and radioactivity transport.

A fraction of the core radioactive inventory can be released initially from the reactor vessel if leakage paths are produced. The magnitude of this fraction depends upon the extent of the leakpaths and upon processes which tend to retain the materials within the sodium system. Noble gas fission products, of course, become free to escape in any case of core disruption; generally 100% release at the time of the accident is assumed. Processes tending to retain other materials

within the reactor vessel include condensation, turbulent mixing, dissolution, and particulate agglomeration and deposition.

Deterministic analytical techniques for calculation of radioactive
material release are not well-developed. To date, little credit has
been taken for source term depletion by processes within the reactor
vessel, and conservative assumptions or parametric sensitivity studies
have been made to evaluate release consequences.

The most important uncertainties exist in the source terms for
radioiodine and plutonium. The iodine uncertainty is of less concern
in the sense that a release of at least one percent of the iodine is
generally assumed, and there seems to be no possibility that this could
be optimistic by more than, say, one order of magnitude. Source term
assumptions as high as 25% have been applied, however, by NRC in CRBRP
evaluation [4].

The plutonium release uncertainty is larger and is dependent upon
the accident postulated. In spite of the phenomena which would tend to
condense any fuel vapor and scrub any fuel particles out of the gas
space so that fuel would remain predominantly within the sodium in the
reactor vessel, there is a potential that some fraction of the fuel
vaporized or fragmented into fine particles could be suspended as an
aerosol above the sodium surface, and subsequently leaked or expelled.
The source terms postulated range from one-hundredth of one percent to
as much as ten percent of the core inventory.

Core materials retained in the reactor vessel at the time of the initial accident would distribute according to their physical and chemical condition. Theoretical studies have identified those radioactive materials which would tend to be dissolved in sodium or associated with fuel or steel core debris [27]. Generally, the relatively volatile fission products would dissolve in sodium. HEDL analyses consider the elements As, Se, Br, Cd, Rb, I and Cs as soluble and volatile. Volatilization of these elements becomes a time-dependent radioactive material source in any case in which sodium heatup, especially boiling, occurs.

In case sodium boils dry, subsequent heating of core debris without overlying sodium can be a consideration. Calculations indicate that any additional volatilization of radioactive material would not be significant. Sparging of non-volatile materials is a consideration, as discussed in the Reactor Safety Study for LWR's [1].

(2) Containment Transients. Analysis of temperature, pressure, and leakage or release of material from the containment requires analyses of thermal conditions, chemical reactions, and mass transport in the containment.

Computer codes such as the SPRAY [28] code and the SOFIRE [29] code are capable of performing sodium burning and heat transfer calculations for temperature and pressure from the release of sodium into the containment; the additional heating due to fission products can also be included in SOFIRE. SPRAY, as its name implies, is designed

to calculate sodium burning in a spray configuration. SOFIRE calculates burning in a pool geometry. Burn rate results from either code
can provide input for sodium-oxide aerosol calculations.

For conditions in which in-vessel PAHR is not effective, long term
transient analyses are more complex. Two classes of conditions without
in-vessel PAHR can be postulated. First, the core may be coolable in
the vessel, but with failure of the HTS to remove decay heat. Second,
the core configuration may be uncoolable with resultant melt-through.
The computer code, CACECO [22], has been developed at HEDL for calculation of temperature, pressure and chemical reaction transients under
either of these conditions. The CACECO code evolved originally from
the water reactor containment code CONTEMPT [31] and uses the CONTEMPT
methodology for its basic heat transfer calculations. However, CACECO
has been developed to incorporate chemical reaction and vapor phase
mass transport terms appropriate for consideration of LMFBR transients,
in particular chemical reactions involving sodium and its reaction products. Major phenomena which must be taken into account in analyzing
containment transients are the following:

- Radioactive decay heat.

- Heating of concrete and resultant release of water vapor and
gases from concrete.

- Chemical reactions of sodium with air and water.

- Chemical reaction of sodium with concrete and its evolved gases.

• Generation of hydrogen as result of sodium water reactions, and subsequent chemical reactions.

• The interaction of fuel and steel debris with concrete. Consideration of these factors is outlined below:

Decay Heat. Fission product decay heating is a significant heat input and may be the largest source of heat, depending on chemical reactions. This heat source is distributed in the sodium, fuel debris, or atmosphere in accordance with the physical state of the various elements.

Water and Gas Release From Concrete. The heating of concrete causes release of free water and water of hydration, dependent upon the temperature above prior operating temperature. Preliminary experimental data have been developed for the characteristics of water release from concrete [32]. The effects of the released moisture can be several. Release of the moisture can be a means for heat transfer from the reactor cavity region by transport of water vapor and recondensation in the containment. The moisture is also a potential reactant with sodium including contact of the vapors in the containment building atmosphere. The moisture also tends to pressurize the space between the reactor cavity concrete and the steel liners. Heating of limestone concrete to high temperatures will also cause release of CO_2 gas. Experimental data on CO_2 release characteristics are being developed in reactor safety R&D programs.

Sodium Reaction with Air and Water. Sodium reacts with air to form sodium oxide, and with water to form sodium oxide and hydrogen. Sodium oxide will react with water to form sodium hydroxide.

Sodium Reaction with Concrete and Evolved Gases. Sodium will react chemically with water and CO_2 evolved from concrete, and also with cement and aggregate materials in the concrete. Small scale tests have established initial sodium-concrete chemical reaction rates [33]. These tests indicate a significant reduction in reaction rate with time, which may indicate that the buildup of reaction products impedes the reaction. It is likely that the presence of a steel liner, even if penetrated, would significantly impede the reaction of sodium and concrete. Additional R&D in this area is being carried out to establish definitively the rates and effects of sodium-concrete reactions.

Hydrogen Generation and Reaction. The reaction of sodium with water in the presence of air generally results in the formation of sodium oxides and/or hydroxides without the production of free hydrogen. The hydrogen produced in the sodium-water reaction tends to react immediately with oxygen available in the air. However, in a low-oxygen atmosphere the release of free hydrogen from the sodium-water reaction has been observed. Under certain conditions of temperature and constituent concentrations, mixing of hydrogen with air in the presence of sodium oxide particles (sodium smoke) results in hydrogen reaction without building up large concentrations [34].

Fuel-Concrete Reactions. Theoretical studies have indicated that
in the absence of sodium or another heat removal mechanism, fuel in
contact with concrete would cause melting of concrete and penetration
of the molten fuel into the concrete [35]. These materials are
mutually soluble, so dilution of the molten fuel with the molten concrete would reduce the specific heat at the same time as increasing the
heat transfer surface; such a system would be expected to come into
equilibrium with a limited penetration of concrete by the molten fuel.
The ANL code, GROWS [36], and the LASL code, AYER [37], have been used
for this analysis.

The CACECO code accommodates all of the above considerations
except the last. CACECO computes these processes in up to four interconnected spaces within the containment structure, with any of the
spaces having the potential for atmosphere mass transfer to other
spaces or to outside. The code is based on experimentally determined
data, where such data are available.

(3) Radioactivity Transport and Release. Radioactive materials
are considered in four groups for purposes of computation of transport
and deposition in the containment building: noble gases, halogens,
volatile solids, and nonvolatile fission products and fuel materials.
The noble gases, of course, are most readily released from the fuel and
would mix in the containment atmosphere, where they would leak or be
released at a rate fixed by the containment leakage or release rate.

Halogens and other volatile solids are generally analyzed as being attached to sodium oxide aerosol particles. Transport and deposition of these materials is therefore governed by the behavior of sodium-oxide aerosol. The aerosol behavior is computed by codes of the HAA series; the HAA-3B code [38] is in use at HEDL. These codes compute the aerosol dynamics of particle agglomeration and deposition within the building atmosphere. The HAA-3B code has been coupled with the CACECO code in integrated analysis at HEDL, using sodium vapor and sodium-oxide generation rates determined by CACECO as source input into HAA-3B calculations. In this way, long term aerosol concentrations are computed consistent with the CACECO pressure, temperature, and chemical composition calculations.

Both the CACECO and the HAA-3B computations are coupled by the use of integrated output and input formats with the COMRADEX [39] code for calculation of radioactivity release and off-site radiation exposures [40].

Vaporization of sodium provides a long term source of volatile radioactive materials, with the vaporization rates of the various species established by fractionation coefficients relative to the sodium vaporization rates. Volatile species released to the containment are deposited in accordance with the computed aerosol behavior. Activity suspended at any time is leaked or released at the same relative rate as the atmosphere leakage. Filtration or atmosphere cleanup may be applicable, if engineered systems are provided. Deposition in leakpaths has not generally been assumed, but is an option which has some experimental support [41].

(4) Atmosphere Dispersion. Standard licensing evaluation calculations are made to compute off-site potential radiation exposures corresponding to specific accident sequences and their calculated radioactivity releases. COMRADEX is one popular code for this purpose. Alternatively, integrated consequences can be calculated, with probabilistic incorporation of meteorological conditions, by techniques such as those used for this purpose in the Reactor Safety Study.

(5) Example. Recent calculations at HEDL have been performed to assess the consequences of an FFTF HCDA with the arbitrary additional condition that in-vessel PAHR would not be effective [40]. The study was intended to determine existing safety margins for such conditions and also to evaluate possible additional margins associated with operation of certain engineered equipment for containment atmosphere cleanup or control. For these purposes, a variety of alternative cases were analyzed. These cases are identified in the event trees of Figures 6 and 7.

Two different conditions representing failure of in-vessel PAHR were considered. For one condition, the core was assumed to remain within the reactor but failure of the heat transport system to cool the sodium from the reactor vessel was postulated (in-vessel cases). For

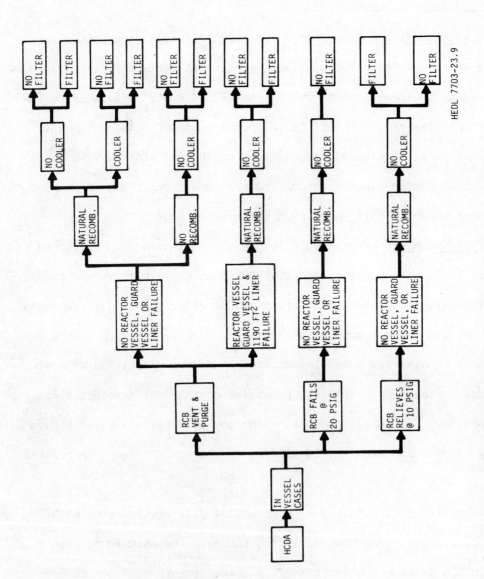

HEDL 7703-23.9

FIGURE 6.   Alternate Deterministic Analysis Cases – Example of Safety Margin Assessment
(In-Vessel Core Disposition)

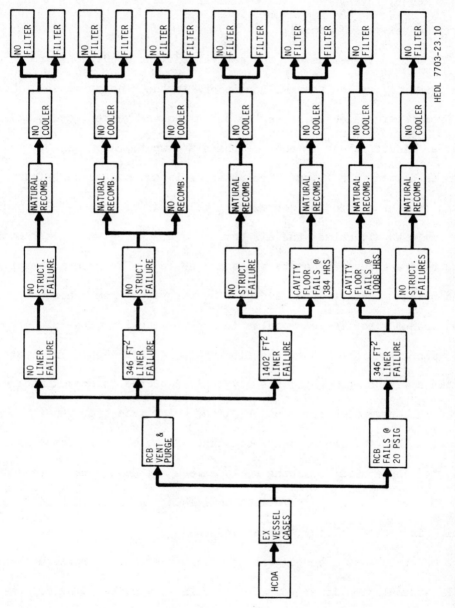

FIGURE 7. Alternate Deterministic Analysis Cases - Example of Safety Margins Assessment (Ex-Vessel Core Disposition)

the second, a postulated melt-through and release of the core and the
reactor vessel inventory of the sodium into the reactor cavity was
postulated (ex-vessel cases).

For the in-vessel cases, the early transient conditions are determined primarily by the thermal conditions in the reactor vessel. Decay
heat from the fuel debris heats the sodium in the vessel, and heat
losses are insufficient to balance the heat input. Accordingly, the
sodium is heated to boiling. Sodium vapor is assumed to be released
through leakpaths into the containment building atmosphere. Reaction
of sodium with air raises the air temperature. Water vapor is released
as the concrete reactor cavity walls are heated. The general heating
of the building adds to the release of water vapor from concrete. Continued sodium vaporization results in sodium reaction with both air and
water vapor, with the latter reaction producing hydrogen. The hydrogen
recombines with oxygen until the oxygen is depleted to approximately
half of its original concentration, after which free hydrogen is
released within the containment building. If no control action is
taken, the containment building would exceed its design pressure and
presumably would eventually fail at some location. If, instead, the
building is vented at, say, 10 psi and subsequently purged with fresh
air to maintain $<4\% $ H_2 concentration, failure due to overpressure and
also H_2 accumulation is prevented. In either case, building depres-
surization would be followed by an increased rate of sodium boiling,
with its accompanying contribution to radioactivity release.

Eventually, the sodium would all boil away leaving the core debris within the reactor vessel; this material would still have sufficient heat generation to melt down through the reactor vessel and into the concrete at the bottom of the reactor building. The maximum temperature reached by the fuel is estimated to be in the range of the melting points of steel and concrete. A stable core debris configuration would be reached after penetration of several feet into the concrete and perhaps the earth.

For the ex-vessel cases, transient conditions are greatly affected by the conditions within the steel-lined reactor cavity. Integrity of the liner or the extent of liner failures are important with respect to analysis of the containment transients. The heating of concrete by heat transfer from the spilled sodium causes release of water vapor. The moisture is vented to prevent pressurizing the space between the steel liners and concrete, and is thus a factor in heating and pressurizing the containment building. Because of the great heat absorption in the building structure, sodium boiling may not occur. Cavity liner failures, if postulated, add chemical reaction energy and increase the tendency for sodium to boil. Liner failures also result in H_2 generation in the sodium pool in the cavity. The H_2 tends to recombine when entering the air space accompanied by sodium vapor.

For all cases in this FFTF study, the potential site boundary radiation exposures were significantly below 10 CFR 100 guideline values. Although it is not necessarily a general result, the cases

investigated in this study for the ex-vessel situation were less
severe in terms of off-site exposures than the in-vessel cases.

IV. Conclusion. Extensive and complex analyses are required in
order to carry out deterministic computation of postulated LMFBR hypothetical accident scenarios. Analytical techniques are available and
are being further developed for such analyses. Many branch points
exist in the analysis process where the more severe path, even if less
likely, is not excluded. Nevertheless, large safety margins exist for
current fast reactor designs and concepts. The situation for further
evaluation of these margins is well suited to risk assessment techniques wherein the alternate deterministic consequences are assessed
as to their quantitative probability.

REFERENCES

[1] Reactor Safety Study - An Assessment of Accident Risks in U.S.
        Commercial Nuclear Power Plants, U.S. Nuclear Regulatory
        Commission, WASH-1400 (NUREG-75/014), October 1975.

[2] FFTF Final Safety Analysis Report, HEDL-TI-75001, Hanford
        Engineering Development Laboratory, December 1975.

[3] Clinch River Breeder Reactor Project - Preliminary Safety
        Analysis Report, Project Management Corporation.

[4] Final Environmental Statement - Clinch River Breeder Reactor
        Plant, U.S. Nuclear Regulatory Commission, February 1977,
        Docket No. 50-537 (NUREG-0139).

[5] D. E. SIMPSON, Resolution of Key Safety-Related Issues in FFTF
        Regulatory Review, ANS/ENS International Conference on Fast
        Reactor Safety and Related Physics, Chicago, October 1976.

[6] H. A. BETHE & J. H. TAIT, An Estimate of the Order of Magnitude
     of the Explosion when the Core of a Fast Reactor Collapses,
     RHM (56)113, Atomic Energy Research Establishment, 1956.

[7] M. G. STEVENSON, et al., Current Status and Experimental Basis of
     the SAS LMFBR Accident Analysis Code Systems, Proceedings of
     Conference on Fast Reactor Safety , Beverly Hills,
     California, April 1974, Report CONF-740401-P3, Vol. 3,
     pp. 1303-1321.

[8] F. E. DUNN, et al., The SAS-3A LMFBR Accident Analysis Computer
     Codes, ANL/RAS 75-17, Argonne National Laboratory, Argonne,
     Illinois, June 1975.

[9] C. H. LEWIS & N. P. WILBURN, MELT-IIIA: An Improved Neutronics,
     Thermal Hydraulics Modeling Code for Fast Reactor Safety
     Analysis, HEDL-TME 76-73, Hanford Engineering Development
     Laboratory, December 1976.

[10] A. E. WALTAR, A. PADILLA, JR. & R. J. SHIELDS, MELT-II: A Two-
     Dimensional Neutronics-Heat Transfer Code for Fast Reactor
     Safety Analysis, HEDL-TME 72-43, Hanford Engineering
     Development Laboratory, April 1972.

[11] R. W. OSTENSEN, R. J. HENNINGER, J. F. JACKSON, The Transition
     Phase in LMFBR Hypothetical Accidents, ANS/ENS International
     Conference on Fast Reactor Safety and Related Physics,
     Chicago, October 1976.

[12] H. K. FAUSKE, Boiling Fuel-Steel Pool Characteristics in LMFBR
     HCDA Analysis, Trans. Amer. Nucl. Soc., 22:386, November
     1975.

[13] A. E. WALTAR, J. MURAOKA & F. J. MARTIN, A Computational Model
     for Analyzing Postulated LMFBR Accident Fuel Boil-up Condi-
     tions, ANS/ENS International Conference on Fast Reactor
     Safety and Related Physics, Chicago, October 1976.

[14] L. L. SMITH, et al., SIMMER-I, An LMFBR Disrupted Core Analysis
     Code, ANS/ENS International Conference on Fast Reactor
     Safety and Related Physics, Chicago, October 1976.

[15] J. F. JACKSON & R. B. NICHOLSON, VENUS-II: An LMFBR Disassembly
     Program, ANL-7559, Argonne National Laboratory, September
     1972.

[16] D. H. CHO, et al., Work Potential Resulting from a Voided-Core
      Disassembly, Trans. Amer. Nucl. Soc., 18:220.

[17] H. K. FAUSKE, The Role of Energetic Mixed-Oxide-Fuel-Sodium
      Thermal Interactions in Liquid Metal Fast Breeder Reactor
      Safety, Proceedings of Third CSNI Specialist Meeting on
      Sodium Fuel Interactions in Fast Reactors, Tokyo, March 22-
      26, 1976.

[18] Y. W. CHANG & J. GVILDYS, REXCO-HEP: A Two-Dimensional Computer
      Code for Calculating the Primary System Response in Fast
      Reactors, ANL-75-19, Argonne National Laboratory, June 1975.

[19] C. Y. WANG, ICECO - An Implicit Eulerian Method for Calculating
      Fluid Transients in Fast Reactor Containment, ANL-75-81,
      Argonne National Laboratory, December 1975.

[20] W. L. CHEN, D. H. THOMPSON & Y. W. SHIN, NAHAMMER, A Computer
      Program for Analysis of One-Dimensional Pressure-Pulse
      Propagation in a Closed Fluid System, ANL-8059, Argonne
      National Laboratory, May 1974.

[21] H. G. JOHNSON, HAMOC, A Computer Program for Fluid Hammer
      Analysis, HEDL-TME 75-119, Hanford Engineering Development
      Laboratory, December 1975.

[22] M. T. A. MONEIM, ICEPEL, A Two-Dimensional Computer Program for
      the Transient Analysis of a Pipe-Elbow Loop, ANL-75-35,
      Argonne National Laboratory, July 1975.

[23] U.S. Energy Research and Development Administration, Washington,
      D.C.: A Compendium of Computer Codes for the Safety
      Analysis of LMFBR's, ERDA-31, June 1975.

[24] D. H. CHO, et al., Work Potential Resulting from a Mechanical
      Disassembly of the Voided FFTF Core, ANL/RAS 74-19, Argonne
      National Laboratory, Argonne, Illinois, August 1974.

[25] D. H. CHO, M. EPSTEIN & H. K. FAUSKE, Work Potential Resulting
      from a Voided-Core Disassembly, Trans. Amer. Nucl. Soc.,
      18:220, June 1974.

[26] J. D. GABOR, et al., Studies and Experiments on Heat Removal from
      Fuel Debris in Sodium, Proceedings of Conference on Fast
      Reactor Safety, Beverly Hills, California, Report CONF-
      740401-P2, Vol. 2, pp. 823-844, April 1974.

[27] L. BAKER, JR., et al., Post-Accident Heat Removal Technology,
      ANL/RAS 74-12, Argonne National Laboratory, July 1974.

[28] P. R. SHIRE, SPRAY Code Users Report, HEDL-TME 76-94, Hanford
      Engineering Development Laboratory, March 1977.

[29] P. BEIRIGER, et al., SOFIRE-II Users Manual for the LBL-CDC-7600
      Computer, N707T1130045, Atomics International, June 1976.

[30] R. D. PEAK & D. D. STEPNEWSKI, Computational Features of the
      CACECO Containment Analysis Code, Trans. Amer. Nucl. Soc.,
      21:274, 1975.

[31] L. C. RICHARDSON, et al., CONTEMPT, A Computer Program for Pre-
      dicting the Containment Pressure-Temperature Response of a
      Loss-of-Coolant Accident, USAEC Report, Idaho Operations
      Office, IDO-17220, June 1967.

[32] J. D. McCORMACK & A. K. POSTMA, Water and Gas Release from Heated
      Concrete, HEDL-SA-117, Hanford Engineering Development
      Laboratory, September 1976.

[33] J. A. HASSBERGER, et al., Sodium-Concrete Reaction Tests, HEDL-
      TME 74-36, Hanford Engineering Development Laboratory.

[34] G. R. ARMSTRONG & R. W. WIERMAN, Hydrogen Formation and Control
      Under Postulated LMFBR Accident Conditions, HEDL-SA-119,
      Hanford Engineering Development Laboratory, September 1976.

[35] G. JANSEN & D. D. STEPNEWSKI, Fast Reactor Fuel Interactions with
      Floor Material After a Hypothetical Core Meltdown, Nucl.
      Technol., 17:85-95, January 1973.

[36] R. KUMAR, et al., Ex-Vessel Considerations in Post-Accident Heat
      Removal, ANL/RAS 74-29, Argonne National Laboratory, October
      1974.

[37] R. G. LAWTON, The AYER Heat Conduction Computer Program LA-5613-
      MS, May 1974.

[38] R. S. HUBNER, E. U. VAUGHAN & L. BAURMASH, HAA-3 User Report,
      AI-AEC-13038, Atomics International, March 1973.

[39] G. W. SPANGLER, et al., Descriptions of the COMRADEX Code,
      AI-67-TDR-108, 1967.

[40] S. BANKERT, et al., Containment Margins in FFTF for Postulated
      Failure of In-Vessel Post-Accident Heat Removal, HEDL-TME
      77-18, Hanford Engineering Development Laboratory, April
      1977.

[41] C. T. NELSON & N. P. JOHNSON, Aerosol Leakage Through Capillaries,
      Trans. Amer. Nucl. Soc., 22:451-452, November 1975.

HTGR RISK ASSESSMENT STUDY

V. JOKSIMOVIC,* W. J. HOUGHTON* AND D. E. EMON†

Abstract. The Accident Initiation and Progression Analysis
(AIPA) Study being performed for the high-temperature gas-cooled
reactor (HTGR) utilizes methods of probabilistic risk assessment similar to those used in the Reactor Safety Study (WASH-1400). Analytical methods are employed for the treatment of common cause failures
involving human errors and other dependencies in redundant systems.
The original objective of the AIPA study was to provide guidance for
safety research and development programs for HTGRs. Core heatup
events appear to be relatively important though the risk from such
accidents in the HTGR is found to be low. Plateout and fission
product transport within the prestressed concrete reactor vessel
(PCRV) during core heatup rank high as important subjects for safety
research and development to confirm the degree of retention of fission
products. Conclusions about the relative and, in particular, absolute
safety of HTGRs are being reached with care. Results of the study to
date provide a reasonable and strong indication that the HTGR has
excellent relative and promising absolute safety characteristics
because of its inherent safety features that reduce the potential for
fission product release.

1. Introduction. In February 1974, a study funded by ERDA was

conducted by General Atomic Company (GA) using probabilistic risk

assessment techniques to attempt to determine priorities in ERDA-

funded safety research for the high-temperature gas-cooled reactor

(HTGR) programs. Early development of the methodology benefited

extensively from the work of prior researchers, particularly the

*General Atomic Company, P.O. Box 81608, San Diego, California 92138.
†U.S. Energy Research and Development Administration. This research
was supported by the U.S. Energy Research and Development Administration under Contract EY-76-C-03-0167, Project Agreement 51.

original conceptions presented by Farmer of the United Kingdom [1].
In addition, it benefited from the early studies done on probabilistic
assessment of nuclear power plants, one in the United States under the
direction of Mulvihill [2], and one in the United Kingdom by
Joksimovic [3]. Other work within the United States was also fundamental to providing a proper breadth of background, in particular the
presentation made by Starr [4] on benefit-cost studies and the testi-
mony provided by Rasmussen to the Congress on plans for the Reactor
Safety Study [5].

Since the GA study is based on gas-cooled reactors and HTGRs in
particular, it benefited from prior histories of power plants such as
MAGNOX reactors in the United Kingdom. The study also included the
helium-cooled, coated-particle-fuel experience relating in particular
to the retention of fission products, obtained through operation of
Peach Bottom Unit 1 in Pennsylvania and the Dragon Project in the
United Kingdom. The plant chosen for the study is the reference
General Atomic design as of the summer of 1975. This plant resembles
the Fulton Nuclear Power Project design for Philadelphia Electric Company, which was successfully reviewed by the Nuclear Regulatory Commission and ACRS for construction permit purposes prior to cancella-
tion.

The title of the program is the Accident Initiation and Progression Analysis (AIPA) Study. A seven-volume Status Report was

published in 1975-76, and in January 1977 an eighth volume was published responding to 175 comments from reviewers and providing some new results [6]. The probabilistic techniques employed were similar to those used in the Reactor Safety Study [7]. Some considerations associated with modeling of consequences in the AIPA study are presented in a paper to appear in Nuclear Safety [8] later this year.

2. Methodology and Preliminary Analysis. The methodology employed is displayed in Fig. 1. The overall method is to select initiating events on as broad and rational a basis as possible and then to develop event trees that form the basis for projecting the course of events in the plant during an accident. The probabilities of proceeding down one branch or another in the event tree are determined with the aid of various reliability techniques and, in the most important cases, with fault trees. The consequences of significant accident sequences are then estimated in terms of the effect on the health and safety of the public; generally, this is in terms of radiological dose in rem. The risk in the study is defined as the frequency of the event multiplied by the consequences of the event in rem or man-rem. More recently, a health effect and property damage concept has been introduced, and this concept is briefly discussed later in this paper.

The selection of initiating events is a screening process. First, all the sources of radioactivity in the power plant are identified and

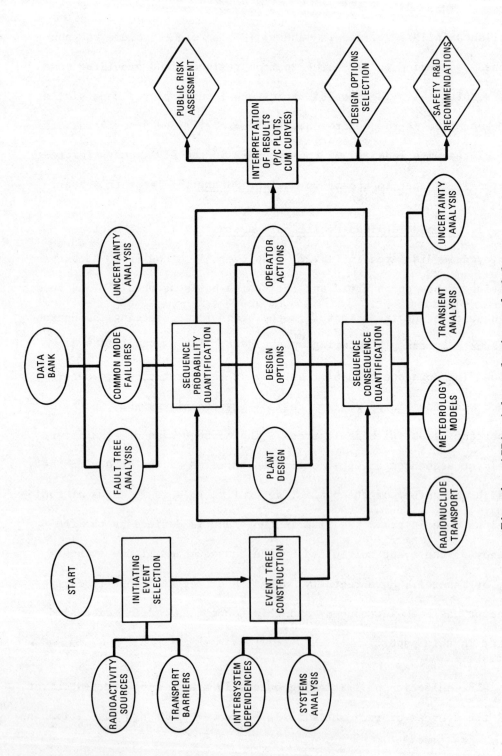

Fig. 1. AIPA risk analysis methodology

listed. Then the barriers to the release of radioactivity to the

public are determined, including any stray pipe or valve that could

provide an apparent short-circuit in the passage of radioactivity to

the public. These configurations are then examined, one by one, to

determine the event that appears to have a combination of a relatively

high probability and a high enough fission product inventory to pro-

vide potentially a relatively high risk. At least one event is chosen

for a possibility of releasing each identified source of radioactivity

in the plant. These courses and the initiating events for accidents

chosen are listed in Table 1.

Table 1. Accident classes, preliminary analysis

Classes of radioactive sources	Representative initiating events
Secondary coolant activity	Loss of offsite power Steam line break
Primary coolant activity	Rapid PCRV depressurization Slow PCRV depressurization Rupture disk/relief valve failure Helium instrument line break outside containment PCRV purge header break outside containment Reheater tube leak Steam ingress to primary coolant
Core	Loss of offsite power Safe shutdown earthquake PCRV closure failure Hydrolysis from steam ingress
Spent fuel	Drop of fuel shipping container
Gaseous waste	Gas waste tank rupture
Liquid waste	Liquid waste tank rupture
Solid waste	Drop of high temperature filter adsorber shipping container
Irradiated hardware	Drop of spent control rod shipping container
Neutron source	Sheared in two by heavy object
Reload fuel	Drop of fuel element

A preliminary analysis was conducted to quantify various possible sequences suggested by the first examination. This analysis yielded an initial set of event trees and preliminary evaluations of the probabilities of the sequences in these trees. It also included rough estimates of the consequences for the various sequences. The relatively higher risk sequences were plotted as shown in Fig. 2. It should be recognized that in preliminary analyses, very low probabilities have significantly less accuracy than those at higher values.

Information summarized in Fig. 2 can be used to determine which sequences are of interest. The criterion is that a relatively high risk event is of greater interest for more detailed study. The highest risk event is a reheater leak and, in this particular case, it is a very infrequent but large reheater leak in which all equipment works as designed so that the dose is kept low but the total result is that the risk is relatively high. The next sequence having a high risk in Fig. 2 starts with the representative transient initiating event, which consists of a loss of offsite power. This sequence has two particularly significant steps. One is the loss of the main turbine generator and the other is multiple failures in the core auxiliary cooling system. This sequence results in a core overheating at a fairly low probability. Core damage, however, releases enough fission products to constitute one of the highest doses. These two initiating events, reheater leak and loss of offsite power, were the primary

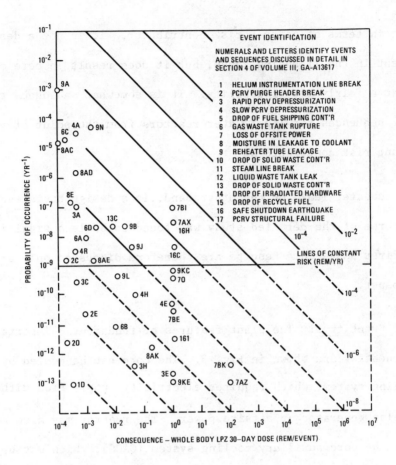

Fig. 2. Preliminary risk summary plot

events chosen for more detailed analysis. A third event chosen (see Fig. 2) is an earthquake, a safe shutdown earthquake in particular, which is accompanied by some small probability of resulting core over-heating.

3. Detailed analysis.

3.1. Initiating events. The representative transient event leading to the multiple failures stated above turned out to be the most

important in terms of being a risk contributor. It is not a design
basis event in the licensing sense, but it does result in core overheating at a fairly low probability. It is somewhat analogous to the
core melt sequences for light water reactors from the point of view
of dominant risk contributors.

The reheater leak, on the other hand, is a design basis event but
in the course of the detailed study was found to be less significant
than at first assessed. Results are therefore discussed only briefly
in this paper.

3.2. Event tree. The plant features of fundamental importance to
core overheating are shown in Fig. 3. The core can be cooled by the
main cooling system, which requires electricity supply from either the
main turbine generator or offsite power. The reactor may also be
cooled by the core auxiliary cooling system (CACS), which may be
powered by either offsite power or diesel generators.

The FY-75 event tree for this event is shown in Fig. 4. A more
elaborate tree was subsequently developed in FY-77. The plant is
designed for the main turbine to remain on line, thereby supplying
house load so that the main cooling loops maintain hot standby during
the outage. In case main loop operation is not retained, the reactor
is tripped and the CACS is started. The CACS will normally operate
until the main loops are restored. If the CACS fails, there is a

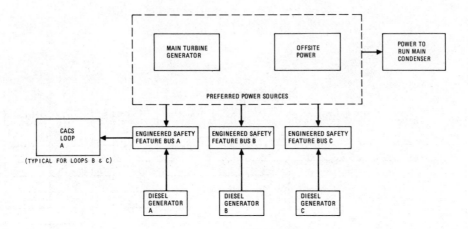

Fig. 3. Possible flow paths of electric power to core
auxiliary cooling system

probability that it can be repaired or the offsite power and main

loops can be restored prior to excessive damage to the plant.   The

containment isolation system is actuated and the filter system oper-

ates.   From the analyses performed, it was determined that the BD

sequence was the one leading to the highest risk, and further discus-

sion will therefore concentrate on that branch.

One of the significant events in branch BD is the failure of the

core auxiliary cooling system to start.   In this branch, the offsite

power has been lost and the main turbine generator is no longer oper-

ating so that the power for this cooling system is derived from diesel

generators.   Since this is a redundant cooling system, common cause

failure considerations, often referred to as common mode failures,

are essential.

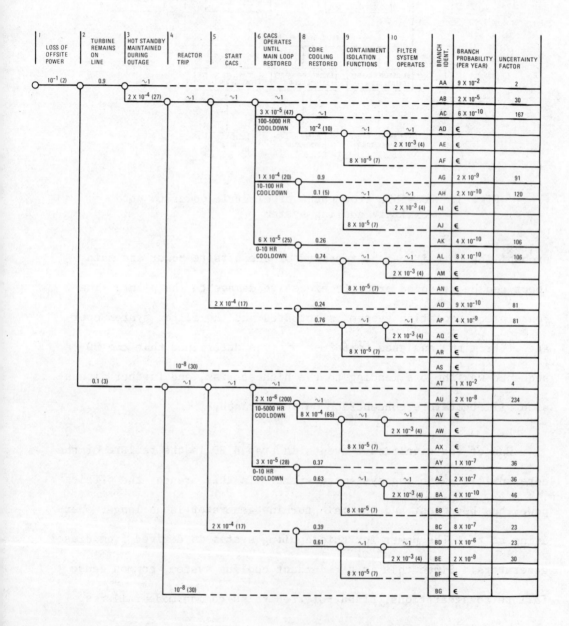

Fig. 4. Event tree for the transient, loss of offsite power

3.3. Common cause failures. The beta factor method for dealing with assessment of common cause failures was originated by Fleming [9] and developed in this study. Its initial derivation is shown in Fig. 5 for failures on demand. The beta factor is defined as the conditional probability that the system fails if one of the redundant subsystems fails. This formulation is convenient because to a first approximation the common cause failure of an entire system with redundancy is beta times the probability of failure of the single redundant subsystem. Beta factor equations were also derived using the Markov process.

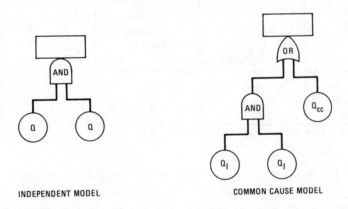

INDEPENDENT MODEL   COMMON CAUSE MODEL

SUBSYSTEM FAILURE PROBABILITY ON DEMAND $= Q = Q_{INDEPENDENT} + Q_{COMMON\ CAUSE}$

β = CONDITIONAL PROBABILITY THAT THE SYSTEM FAILS BY COMMON CAUSE IF ONE REDUNDANT SUBSYSTEM FAILS

$$Q_{cc} = \beta \cdot Q$$

SYSTEM FAILURE PROBABILITY $= Q_I^2 + Q_{cc}$

$$\approx \beta Q \qquad \text{WHEN } Q_{cc} \gg Q_I^2$$

Fig. 5.   Common cause beta factor method

It is, of course, preferable to use data where possible in order to obtain a beta factor. In the case of branch BD in the event tree, startup of the diesels is an event for which data exist (see Tables 2 and 3). The data are taken from the AEC report of June 1974 for nuclear diesel experience [10]. Of the 49 failures in the sample, there were four failures of redundant diesels from a common cause. This result yields a beta factor of approximately 20%. As shown in Table 3, the cause of this common error was in many cases a design error, which is to say a human error.

Table 2. Estimates of common cause failure parameters
for diesel generators[a]

Failure	No. of failures in sample	No. of common cause failures	No. of causal failures	Beta
Failure to start	49	4	0	0.08
Failure to run	28	2	1	0.07

(a) Data taken from Ref. 11.

Table 3. Relative contributions of causes to common
cause failure in diesel generators (%)[a]

Design error	Fabrication/ manufacturing error	Storage/ shipping error	Human operator error	Environmental cause
50	0	17	17	16

(a) Data taken from Ref. 11.

3.4. Fault tree and probability data. The pruned fault tree for the startup of the CACS is shown in Fig. 6 to illustrate the fact that it includes the common cause failures of the CACS in a way compatible with the beta factor method.

When this fault tree is solved for the failure probability, it is found that the diesel generator common cause failure probability contributes about two-thirds of the total failure probability. The diesel generator failure-to-start probability has been adjusted downward from the American nuclear experience data [10], consistent with a Battelle Northwest assessment [11], because the HTGR core heats up slowly enough to allow the operator additional time to try to start

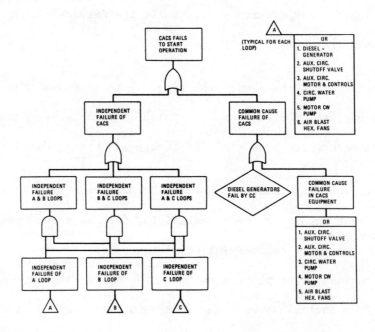

Fig. 6.   Fault tree for CACS startup after LOSP and turbine trip

the diesels in case they do not start automatically at the first signal. The two components that are next in importance are the auxiliary
circulator motor and the helium shutoff valve. The probability that
the circulator motor fails to start is derived from data in Ref. 12.
The helium shutoff valve is a check valve for which the failure rate
is estimated based on test data for the valve and general industrial
experience.

Another junction point on the BD sequence of interest is the 10%
probability that the main turbine generator does not remain on line.
This is consistent with British data and is a reflection of the fact
that the British power plants are designed to keep the turbines on
line following loss of offsite power, while the American light water
reactor plants are not.

The last item on the event tree (Fig. 4) to be considered is the
probability of loss of offsite power of 10^{-1}. This is based on the
AEC abnormal occurrence data [13] for five years ending March 1974.

The data described above for branch BD are summarized in Table 4,
which shows that input for the probabilities for the event tree comes
from a wide variety of sources including nonnuclear power system data,
gas-cooled nuclear power plant data from the United Kingdom, diesel
experience from nuclear power plants throughout the United States,
motor experience from the United States industrial base, and an assessment of a unique HTGR component.

Table 4. Sources of critical probability data for core
          overheating sequence BD

Event	Probability characteristic [a]	Original data source	Reference
Loss of offsite power	0.1/yr	Predominantly non-nuclear power grids	Abnormal Occurrence Reports (Ref. 14)
Main turbine-generator trip	0.1/demand	Nuclear power plants in United Kingdom	--
Failure of diesel to start in 20 min	2×10^{-3}/demand β = 7.5%	Nuclear power plants in United States	AEC and BNWL reports (Refs. 11 and 12)
Failure of auxiliary circulator motor to start	3×10^{-4}/demand β = 4%	Industrial base	WASH-1400 (Ref. 13)
Failure of auxiliary circulator shutoff valve to open	3×10^{-4}/demand β = 7%	Subcomponent experience and tests of the valve	--

(a) β = common cause failures/total failures

3.5. Consequence considerations. The consequence assessments are
discussed here only briefly in relation to the fission product transport occurring in branch BD. Given that forced circulation cannot be
reestablished, core temperatures only gradually increase because of
the large heat capacity of the graphite moderator and reflector, as
seen in Fig. 7. As the core heats up, primary system pressure rises,
with a pressure vessel relief valve lifting after about 1-1/2 hr.
Eventually, the relief valve fails open due to temperature increase,
and the primary coolant blows down into the containment. The containment isolation system automatically isolates, and only a slow controlled leakage of the containment gases to the environment occurs.
The fission products in the fuel elements are not released at this
time because the fuel particle coatings have not failed.

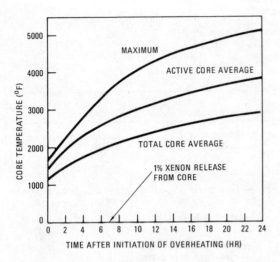

Fig. 7. Core overheating temperatures

As the core temperatures continue to rise, the fuel particle coatings fail, releasing fission product gases and vapors into the PCRV. For example, release of Xe-133 from the core reaches 1% after 7 hr. Nuclides other than noble gases are partially adsorbed on the metallic surfaces of the primary coolant system. The fractional release of these nuclides from the PCRV is not well known. Experiments and computer modeling are under way but until these become sufficiently developed, interim release fractions are being used [14], and these have large uncertainty bands. For example, the median PCRV release fraction for iodine is currently taken to be 2% and the 95th percentile is 30%. Possible reentrainment of these nuclides will also be clarified by ongoing experiments.

The deposition of nuclides other than noble gases in the containment is significant because of plateout and fallout and, in the case of sequence BD, because of the operation of the containment cleanup filter system. Some of these nuclides still leak out of the containment to the environment, resulting in a radiation exposure.

3.6. Other events.   Other initiating events (reheater leak and earthquakes) were studied with similar techniques.   The studies yielded results including the highest risk branches, which are plotted in the summary risk plot of Fig. 8.   The large reheater leak is now estimated to cause a rather low dose, but a new small reheater leak appears in a relatively high risk position.   The earthquake branches are shown relatively low but with large uncertainties.   They are the same in concept as the previous earthquake points in that the probability includes both the probability of a very large earthquake and the probability that the equipment will not survive and will therefore allow core overheating to occur.   The $\alpha = 1.5$ on the higher risk earthquake point refers to the fact that the acceleration for that earthquake is 1-1/2 times the acceleration for the SSE. We see, indeed, that branch BD for the representative transient consisting of loss of offsite power provides the highest risk

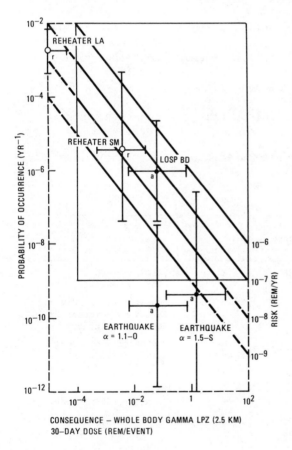

Fig. 8.   Summary risk plot from
          Volume VIII of AIPA
          Status Report

as a result of the detailed analysis.

4. Safety R&D recommendations. To derive the recommendations

for safety research and development, the results of the analysis need

to be examined in an organized fashion. This was done originally with

risk being the criterion. Table 5 shows the steps taken in a sorting

process to find which accident sequences were most important. The

broad spectrum of events in the column on the left was considered [6],

resulting in selection of the 17 initiating events indicated in the

next column. Analyses of these 17 events revealed that the three

events in the next column had the highest risks. The analyses for

those three initiating events yielded specific branches (shown in

Table 5. Method for safety R&D recommendation using initiating
          events and event trees

Broad spectrum of initiating events	Preliminary event trees [a]	Detailed event trees [b]	Relatively high risk branches	Area of safety R&D interest
Consider all sources of radioactivity	Seventeen initiating events obtained from Table 1	LOSP	BD	Fission product plateout activity
				Uncertainty on PCRV and containment integrity in core heatup transient
Consider about 50 initiating events				Probability of cooling system failure
		Reheater	SM	Operator action uncertainty
			LA	Reheater tube failures
		Earthquake	S	Uncertainty in earthquake occurrence probabilities
			0	Equipment survival probabilities

(a) 270 sequences in the event trees.
(b) 115 sequences in the event trees.

the table) for which a specific accident sequence resulted in the highest risk.

The high risk branches then need to be examined to determine what in the analysis of those branches contributed the most to the risk uncertainties. In Table 6, which shows the uncertainties for the whole body dose and frequency for branch BD, we can see that a major contributor to the uncertainty is the probability that the core auxiliary cooling system is not available in time. This probability therefore raises itself as a candidate for further work because the contribution of this uncertainty to the total uncertainty in the risk is great.

Table 6. Major uncertainties in whole body gamma risk for LOSP-BD; uncertainty parameter is (ln uncertainty factor)^2

Frequency 1 x 10^{-6}/yr Uncertainty factor 23		Consequence 0.07 rem Uncertainty factor 10		Risk 7 x 10^{-8} rem/yr Uncertainty factor 50	
Loss of offsite power	0.5	Containment leak rate	2.6	Total	15
Turbine remains on line	1.2	Weather	2.6		
Start CACS	8.0				

In Table 7 the thyroid risk and its uncertainties are displayed based on the interim revision reported in Volume VIII of the AIPA report [14]. Of course, the thyroid dose is not as important as the whole body dose on an equal rem basis, and the figure does not make any correction for the relative biological importance. However, it will be noticed that the uncertainty on the thyroid dose is relatively large. Therefore, the potential thyroid dose is high enough that it is

Table 7.   Major uncertainties in inhalation thyroid risk for LOSP-BD; uncertainty parameter is (ln uncertainty factor)^2

Frequency 1 x 10^{-6}/yr Uncertainty factor 23		Consequence 0.06 rem Uncertainty factor 93		Risk 6 x 10^{-8} rem/yr Uncertainty factor 250	
Loss of offsite power	0.5	PCRV plateout	10.1	Total	30
Turbine remains on line	1.2	Containment plateout	5.3		
Start CACS	8.0	Containment leak rate	2.6		
		Weather	2.6		

of evident merit to consider it for research. In particular, the plateout within the PCRV constitutes the greatest uncertainty and therefore raises itself as a research candidate.

The PCRV and containment exhibit outstanding integrity. However, the fact that there is insufficient knowledge regarding the long-term effects of the decay heat on these structures constitutes the basis for a further research recommendation.

The right-hand column of Table 5 summarizes the items explained in the preceding discussion, along with similar results from other branches. The uncertainties provide the principal focus for the types of research and development most appropriate for investigation of each high risk branch. The research areas of highest priority are listed first; in fact, ERDA-funded research is under way in the first three areas, with the greatest attention and the experimental planning being devoted to the first of these areas.

5. Public risk assessment. Much of the information discussed above can also be used to attempt an estimate of public risk of the

HTGR. Complementary cumulative distribution functions have been

derived for doses to a number of organs. Over a meaningful range of

frequencies (say, to 10^{-7} per year), no accident sequence considered

in the AIPA study is predicted to cause early or delayed fatalities

or illnesses at representative U.S. sites and only a small number of

delayed effects at urban sites (see Table 8). At a lower frequency of

10^{-9} per year, where the significance of results becomes somewhat

Table 8. Consequences of HTGR accidents having probability of about
1 in 10 million reactor years[a]

	Representative U.S. site		Urban site[b]	
Early fatalities	<1	(<1)[c]	<1	(<1)[c]
Early illnesses	<1	(<1)	<1	(<1)
Property damage, $ million	~1	(~0)	~50	(~0)
Relocation area, sq miles	N/A		N/A	
Decontamination area, sq miles	0.05		0.05	
Latent cancer fatalities	<1	(<1)	<5	(<5)
Thyroid nodules	<1	(<1)	<10	(<20)
Genetic effects	<1	(<1)	<1	(<1)

(a) Some results may be modified at the end of FY-77 based on results
of additional work performed since publication of the Status
Report. Although an attempt was made to quantify the uncertainties associated with the predictions in the study, the absolute
values listed here depend on a number of assumptions that could
not be qualified accurately. Therefore, these results should be
viewed as indicative of the levels of risk due to HTGR accidents
as opposed to absolute risk levels.

(b) Site not licensable under existing U.S. LWR guidelines. Reported
consequences are for a meteorologically unfavorable site, population distribution extreme.

(c) Values in parentheses represent consequence change without post-
accident population evacuation.

questionable for a variety of reasons, it is predicted that a small number of latent cancer deaths will occur during the approximate 30-year period following an accident at a representative site.

One must, of course, exercise caution in using these results since the study is continuing. Hence, these results should be viewed as indicative of the levels of risk due to HTGR accidents as opposed to absolute risk levels. However, there are basic reasons why the risks are so low. It is not only that the probabilities of the events of interest are reasonably low, but it is also that the doses appear to be low.

The low doses are attributed not to the engineered safety features within the HTGR but rather to the inherent safety features of the HTGR. The first of these is the high heat capacity and the high temperature capability of the graphite core, which does not melt but allows the heat from fission product decay to be stored for many hours before the heat causes the release of the fission products into the pressure vessel. This allows decay of the shorter lived fission products in the core. Second, the pressure vessel is not damaged severely and therefore provides significant (though not fully assessed) retention of the fission products other than noble gases. Finally, a further major inherent safety feature is that the coolant, being gaseous helium, is not able to deliver a sufficiently high content of energy and resulting pressure to the containment to threaten it with higher leak rates. Therefore, the prediction that the doses are relatively

low is believed to be valid even though additional work is needed to

refine the analyses. Before leaving this discussion, it should be

pointed out that the discussion of inherent features is oversimplified

since a detailed discussion is considered to be outside the scope of

this paper.

6. Conclusions. The methods of probabilistic risk assessment

give valuable insights into many aspects of nuclear safety. Justifi-

cations for safety research become clarified. New levels of objectiv-

ity become possible in risk assessments. The impact on safety of

design changes can be better quantified and, in some cases, plant cost

reductions may be found that cause no significant reduction in overall

safety.

REFERENCES

[1] F. R. FARMER, Siting criteria --- a new approach, in Proceedings
        of IAEA Symposium on Containment and Siting of Nuclear
        Power Plants, April 3-7, 1967, (paper SM-89/34), pp. 303-329.

[2] R. J. MULVIHILL et al., A Probabilistic Methodology for the
        Safety Analysis of Nuclear Power Reactors, USAEC Report SAN-
        570-2, Planning Research Corporation, February 28, 1966.

[3] V. JOKSIMOVIC, Fault analysis of nuclear reactors, Ph.D Thesis,
        London University, 1970.

[4] Committee on Public Engineering Policy, National Academy of Engi-
        neering, Perspectives on Benefit-Risk Decision-Making,
        Report of a Colloquium, April 26-27, 1971, National Academy
        of Engineering, Washington, D.C., 1972. References to p.1
        and to paper by Chauncey Starr, Benefit-cost studies in
        sociotechnical systems, pp. 17-42.

[5] N. C. RASMUSSEN, Statement of N.C. Rasmussen, Director, Reactor
    Safety Study, before the Joint Committee on Atomic Energy
    Hearings on Nuclear Reactor Safety, September 1973.

[6] HTGR Accident Initiation and Progression Analysis Status Report,
    ERDA Report GA-A13617, Volumes I-VIII, General Atomic Com-
    pany, 1975-77.

[7] The Reactor Safety Study - An Assessment of Accident Risks in
    U.S. Commercial Nuclear Power Plants, USAEC Report WASH-
    1400, October 1975.

[8] A. W. BARSELL, V. JOKSIMOVIC, and F. A. SILADY, General Atomic
    Company, An assessment of HTGR accident consequences, sub-
    mitted for publication in Nuclear Safety, 1977.

[9] K. N. FLEMING, A reliability model for common mode failure in
    redundant safety systems, Proceedings of the Sixth Annual
    Pittsburgh Conference on Modeling and Simulation, April 23-
    25, 1975. General Atomic Company, January 1977, p. 3-4.

[10] Diesel Generator Operating Experience at Nuclear Power Plants,
     USAEC Report OOE-ES-002, June 1974.

[11] O. B. MONTEITH, Preliminary Reliability Prediction for the FFTF
     Conceptual Building Electrical Power Systems, BNL 813,
     Battelle-Northwest Laboratory, August 1968.

[12] The Reactor Safety Study - An Assessment of Accident Risks in
     U.S. Commercial Nuclear Power Plants, USAEC Report WASH-
     1400, Appendix III, October 1975.

[13] AEC abnormal occurrence reporting file, including approximately
     105 reactor years for eight plants, Office of Operation
     Evaluation, Directorate of Regulatory Operations, United
     States Atomic Energy Commission, March 1, 1969, through
     March 1, 1974.

[14] HTGR Accident Initiation and Progression Analysis, Responses to
     Comments on AIPA Status Report, ERDA Report GA-A13617, Vol-
     ume VIII, General Atomic Company, January 1977, p. 3-73.

PRODUCTIVITY ASSESSMENT OF NUCLEAR GENERATION SYSTEMS

M. E. LAPIDES*

Abstract. Selection of programs to improve the productivity of
generating units has been a concern of the Electric Power Research
Institute (EPRI) since its inception. The process in use emphasizes
direct communications between the Institute, its sponsoring utilities,
and equipment suppliers, supplemented by analytical methods for program selection. The analyses, based on modeling of operating experience, illustrate the potential value of formal reliability analyses
for individual generating units and for selecting productivity improvement R&D programs. This paper summarizes the status and future
of these analytical assessments.

1. Introduction. Modeling nuclear generation unit performance

mandates evaluation of a maintained system and is as much concerned

with the times and fiscal consequences of maintenance, replacement,

and repair as with the probability of failure per se. The applicable

cost/benefit relations ultimately depend on factors such as generation

system mix, demand forecast, fuel prices, and anticipated regulatory

climate. The operating experience data available are usually incom-

plete in the statistical sense, and must be critically disaggregated

to detect changes in engineering practice or external factors in-

fluencing maintenance times. Although these considerations appear

quite complex, there are reasonable prospects that existing formal

reliability analysis methods can be adapted to plant practice bene-

ficially. This paper examines the data, methodology, and applications

---

*Electric Power Research Institute, Palo Alto, CA 94303.

that have been evaluated since 1974.

2. Productivity Analysis of Nuclear Generating Units, General.
The nuclear industry has placed heavy emphasis on the use of reliability-based disciplines in performing those technical risk assessments
which examine the relationships between sets of postulated in-plant
accident sequences and ex-plant consequences. Comparatively less
emphasis has been placed on the same formalism when the "risk" is
loss of generating capacity. Achievable generation capability, in
turn, is the basis for measuring the productivity of an investment in
that capacity.*

The productivity assessments may use risk assessment methodology,
but the focus shifts from safety-related subsystems to normally
operating subsystems; the output 'risk' functions must be examined as
a function of service life; they must be portrayed in ways that satisfy
the communication and management needs of all groups dealing with productivity (e.g. engineering, plant operations, production planning,
system planning, R&D). Some of these representative interests are
summarized in Table 1.

---

* In baseload units not subject to a high fraction of discretionary
power reductions (e.g. load-following), the unit capacity factor is a
reasonable measure of productivity.

Table 1

REPRESENTATIVE UTILITY AVAILABILITY AND PRODUCTIVITIY ASSESSMENT REQUIREMENTS

APPLICATION	REQUIRED DATA	UTILIZATION
A. PROCUREMENT		
1. Economic evaluation of generation alternatives	Achievable capacity factor averaged over unit service life (30-40 years)	Fixed equipment charge computation
2. Selection of unit size, forecast of reserve margin capability	As above, plus estimate of annual outage, power reduction characteristics	Input data to system planning models
3. Fuel purchase optimization	Achievable annualized capacity factor	Determination of time base for fuel optimization
4. Spares purchases	Projected utilization, impact of on-site spare availability on productivity	Spares purchase policy
5. Plant design specifications	Comparative productivity assessment	Life cycle cost optimization of design (notably for "fixed" elements such as containment vessel, isolation valves)
B. OPERATIONAL*		
1. Production utilization of unit	Same as A1, A2, but on 1-2 year planning horizon	Inputs to production planning
2. Retrofit and auxiliary equipment purchases	Same as A5	Remaining life cycle worth versus required downtime economic assessment
3. Maintenance actions during refueling outage	Anticipated maintenance requirements as derived from malfunction forecast	Input to refueling outage management plan
4. Maintenance personnel requirements	Statistical assessment of annual malfunctions and repair requirements	Estimate of radiation exposure levels during maintenance
5. Optimal plant rating	Influence of power level on capacity factor	Plant operation optimization
6. Refueling interval optimization	Same as B1, plus projection of "refueling" outage duration	Fuel specifications, production planning

*All supplement plant operating data

How much of the time is the nuclear unit available for service?
And at what fraction of its normal power rating can the unit operate
when it is available? These are the fundamental questions addressed.
A conventional reliability analysis which addresses only failure
frequency yields only one part of the answer. What must also be considered are items such as: How long does it take to diagnose and
repair a malfunction? How much maintenance time is scheduled to prevent malfunctions—or to replace limited lifetime elements such as
nuclear fuel? And finally, how does a varying environment, both
operating and regulatory, influence these attainments?

EPRI's initial examination of productivity analyses proved to be
quite disappointing. Available literature was quite sparse; applicable
data were even sparser; historical behavior patterns could be utilized
only for gross identifications. Hence, the EPRI utilization program
turned out to be a developmental effort and is being continued on this
basis.[1]

3. R&D 'Pointing' Functions. The performance of any class of
generating equipment during a substantial portion of its lifetime is
better characterized by a distribution function than by an 'average'
value. A contemporary performance distribution is shown in Figure 1.
Given this, it is comparatively easy to solicit a broad spectrum of
opinions as to what needs to be done to improve the situation; a natural

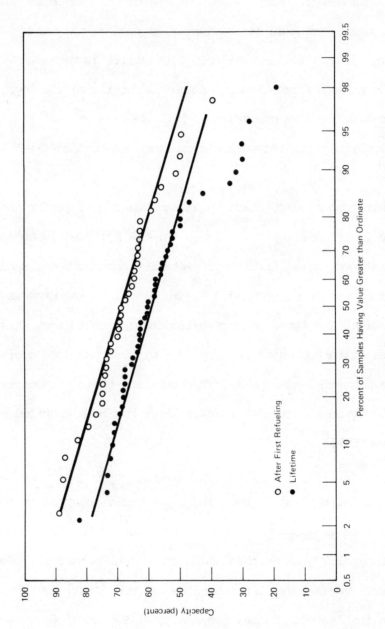

Figure 1 Comparison of cumulative lifetime capacity factor and capacity factor after first refueling as of January 1, 1977

bias is introduced by the interviewer's position on the distribution
curve when interviewed. More valid opinions can be collected on what
needs to be done - and what can be justified economically than can be
fitted to any visible monies available. Hence, it is necessary to
discriminate operating experience information until a) the logical
elements for quantitative comparison can be defined and b) the
priorities and prospects for possible changes can be identified.

Discrimination studies start by disaggregating of performance
data (Figure 1) by unit type and size. This is followed in sequence,
by further refined 'cuts' until representative component and subsystem
characteristics can be determined. Figure 2 depicts the process in
flow-sheet form. The three primary malfunction outputs from this
analysis are: a) a component failure rate with an assigned confidence
level; b) a mean-time-to-repair (MTTR) term for these components; and
c) the frequency-intensity distribution which has been averaged to
establish the MTTR.

A substantial portion of the outage presently observed in many
component classes is accounted for by a restricted number of major
incidents (nominally greater than 500 hours outage duration). Some
of these large events have illuminated needed modifications that have
already been instituted. Hence, the analysis (Figure 2) focuses on
accounting for the status of problem resolution, and on any resulting

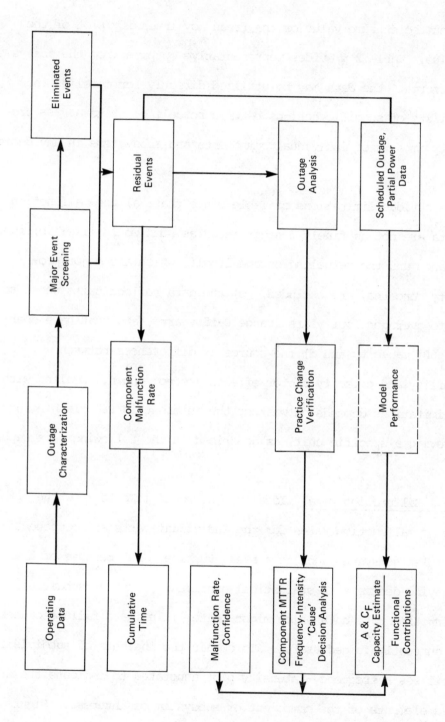

Figure 2 Data analysis

effect on the extreme value of the frequency-intensity plot of the

MTTR value. Table 2 provides representative outputs from the analysis.

In this format, the data may be utilized directly for modeling the

availability of specific designs between refueling, or to assess pro-

ductivity impacts of individual components and subsystems in any design.

The obvious limitations of Table 2 are that: a) the malfunction

rate data are not as finely disaggregated as some would like; b) fuel

performance and the refueling outage itself, which are major pro-

ductivity concerns, are excluded; c) the data reflect units whose ser-

vice life averages four years (range 0-15 years), and that have oper-

ated at 60% capacity and d) the source of differences between

availability and capacity factor effects are not shown. Dealing with

these limitations depends on whether the interest is R&D planning or

in improving a specific unit, as described in the following paragraphs.

3a. Malfunction Data. Ideally, one would like to know the

failure or malfunction rates for the individual parts and components

of a complex subsystem such as a steam turbine. But because of the

accompanying decrease in statistical confidence, further subdivision

of malfunction rate data which might better illuminate failure causes

is not currently judged useful. To obtain the insights of subdivision,

EPRI utilizes a frequency-intensity plot annotated to indicate the

affected elements of the component or subsystem of interest. Figure 3

Table 2

MALFUNCTION RATE AND MEAN-TIME (MTR)
DATA FOR NUCLEAR PLANT AVAILABILITY ESTIMATION

COMPONENT OR SUBSYSTEM	λ (INCIDENT/YEAR)	MTTR (HOURS)	λC_{90} (INCIDENT/YEAR)	MTTR PLOT 10%	MTTR PLOT 90%	NOTES
A) Derived From This Study						
Turbine	2.01	42	3.6	see appendix		
Generator	.19	130	.27	1120	14	1
Condenser	.73	22	.90	82	6	2
Steam Generator	.62	90	.75	fig. 16		3
Reactor Vessel and Core (assembly)	.79	15	.92	100	1.5	
Pumps						
Condensate and Booster (BWR)	.008	17	.021	-	-	
(PWR)	.020	21	.041	-	-	
Recirculation (BWR)	.32	58	.43	360	8	
Reactor Coolant (PWR)	.16	190	.22	580	60	
Condenser Coolant	.004	6	.014	-	-	
Make-up/Charging (BWR)	NF	NF	-	-	-	
(PWR)	.007	50	.037	-	-	
Heater/Drain	.0027	(90)	.01	-	-	
Main Feedwater (BWR)	.087	51	.16	170	7	
(PWR)	.224	15	.33	36	5	
Component Closed Cooling Water	.004	-	.02	-	-	
Service Water (BWR)	.008	(30)	.04	-	-	
(PWR)	NF	-	-	-	-	
All Pumps (BWR)	.048	-	-	-	-	
(PWR)	.056	-	-	-	-	
Valves						
Stop/Isolation	.003	85	.005	-	-	
Safety/Relief	.03	62	.062	-	-	
Main Steam Isolation	.057	50	.07	180	14	
Bypass	.006	42	.008	280	7	
Control/Regulation	.033	37	.037	220	7	
Check	.012	53	.016	-	-	
Governor	.011	34	.015	100	4	
Discharge	.005	17	.007	-	-	
Intercept	.008	24	.016	-	-	
All Valves	.012	30	-	-	-	
B) Derived From EEI Data Sources						
Plant Piping	.00113	20	-	-	-	4
Feedwater Heaters	.11	54	-	-	-	5
Switchgear and Electrical	.140	18	-	-	-	
Operator Error	1.2	7	-	-	-	

Table 2 (continued)

I.   NOTES - GENERAL

   a) These data are for events causing full power outage - or mandating
       full isolation of the component in the case of a multi-loop system.

   b) λ = Observed Malfunction Rate/Active Operating Year
       MTTR   = Mean-Time-To-Repair (Hours)
       λ_{90} = Malfunction Rate at 90% Confidence Level

   c) Note that data for supporting systems (e.g. radwaste) and safety
       equipments are not included.

   d) Conversion of availability to inherent capacity factor data requires
       empiric assessment of scheduled outage, refueling and factors which
       decrease output factor. See Tables 2 & 5, Figure 9 for estimating
       data.

   e) 'MTTR Plot' are approximate values for log normal distribution.

   f) ( ) in MTTR column indicates average value, too few points for
       distribution.

II. NOTES FOR SUBSYSTEM

   1) Condenser repairs usually occur under partial power conditions.
       Data shown are full forced outage only.

   2) Steam generator data expected to show considerable variation.
       Data heavily weighted by problem resolution.

   3) Reactor assembly data exlude major internal modifications
       such as B&W specimen tube, G.E.-LRPM vibration which are
       believed to have been resolved. Also, see Table 6 for reactor
       assembly further breakdown.

   4) 1975-1976 nuclear unit data are $\lambda = 1.2$, MTTR = 19. Higher
       malfunction rate may be result of definition, but trend appears
       correct due to added number of electrical subsystems and safety
       regulations.

   5) 1975-1976 nuclear unit data are $\lambda = 1.3$, MTTR = 8
       1976 $\lambda \sim$ 50% 1975

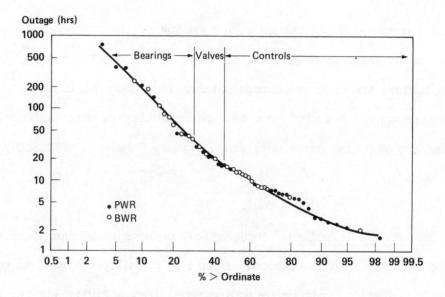

Figure 3   1975 Steam turbine outages

is a representative example. Priority R&D interests are defined from
these data by rating malfunctions by consequential impact (nominally
loss of generating capacity). Annual comparisons provided a method
of updating and measuring status changes.

A log-normal distribution with some 80% of the consequential
outage being associated with 20% or fewer of the events usually
characterizes the observed outage pattern for component classes. The
low-frequency, high impact events, which usually require technically
intensive resolution campaigns, are a primary domain of EPRI R&D
interests.[2]

3b. Fuel & Refueling. Nuclear fuel defects rarely manifest themselves directly in unit outages. A unit is normally designed to permit safe operation, without exceeding regulatory activity limits, while
containing one percent defective fuel. If this target is met, unim-
peded unit operation until the next refueling outage can usually be
achieved. The impact of fuel defects is to reduce capacity factors
(i.e. by derating of unit to maintain activity limits, or restricted
rates of power-level-change to minimize cladding failures), and to extend inspection activities during refueling (e.g. 'sipping', disassembly of fuel bundles, operations in high-radioactivity environment).
Gauging these impacts cannot readily be accomplished from easily available data sources, but a representative picture can be developed from

historical reviews and the results of in-depth EPRI programs.

Through January 1976, defective fuel was responsible for,
approximately, a 5% capacity factor loss in domestic LWRs. This loss
was largely a result of generic fuel abnormalities, such as faulty
welds, hydriding, densification, fretting and rod bowing all of which
have been the subject of corrective actions by equipment suppliers. A
recognized current issue is the relation between fuel failure rate and
power transients which appears to center around interactions between
fuel pellet and zirconium clad during start-up or maneuvering. The
overall fuel picture is suggested by Figure 4, which illustrates:
a) historical experience and the pellet-clad problem in the form of a
conventional failure rate presentation; b) the present impact of fuel
defects; and c) the "fuel conditioning" efforts in use to ameliorate
failures. In these terms, a rapid rate of power change is a 'severe'
service condition yielding an increased failure rate. Severity can be
reduced by restricting the power change rate--at the expense of a reduction in capacity factor due to extended time requirements.
Alternately, fuel with improved transient capability may be desirable.
The conduct of these trade-offs by development of fuel performance
modeling methods and conduct of verification testing is currently a
major EPRI activity.[3] The inhibition in the process is the comparatively low exposure to-date of statistically relevant quantities of
modified fuel, the uncertainty impact of immature statistics being

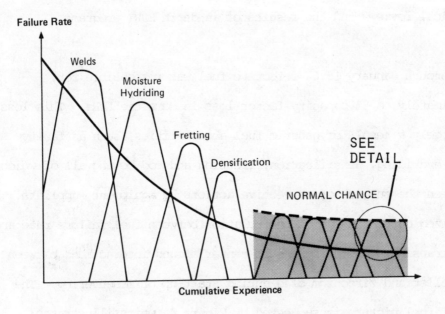

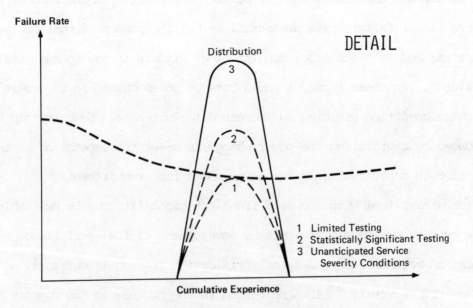

Figure 4   Nuclear fuel defect experience

illustrated in Figure 4.

Refueling currently represents the major source of nuclear unit
outage, but such nomenclature tends to obscure more than it reveals,
because a large number of unit service and maintenance activities,
only one of which is actual refueling, are conducted during this
period. Such activities include major equipment overhaul (e.g. turbine), surveillance and inspections of safety-related subsystems
and equipment modifications. The essential data necessary to select
R&D activities in this area are contained in the annual operating reports for each unit and in the outage management plan that utilities
develop for each refueling. These reports have limited statistical
significance at this time. The data that do exist suggest a 1700-hour
average refueling period, which has gradually been reduced to 1200-
1650 hours for PWR and BWR units respectively. Current R&D efforts
relate primarily to unit design and performance modeling studies aimed
at achieving 'high availability' units. These contractual programs
consider factors such as impact of containment design, turbine overhaul procedures and plant layout on refueling duration.[4]

3c. Environmental Severity. Table 2 contains comments on
'environmental severity' that may influence either the malfunction
rate or the MTTR values as further service experience data are
accumulated. The concern reflected is either that some failure modes

('wear-out') have not yet been revealed by operational experience--
or that other factors, notably radiation environment, will increase
maintenance time. Possibly the most substantial consideration is
stress-corrosion kinetics associated with primary and secondary cooling
water control chemistry. Subsystems that might be affected include
piping, steam generators, pressure vessels and, in some cases, the
steam turbine. Because the consequential costs of these possible
malfunctions are so large, existing R&D programs are frequently justified on the basis of potential rather than observed impacts. The R&D
emphasis is on detailed evaluation of specific designs, and on in-depth
failure analyses of service observations that may be signalling more
significant incidents.

The effect of increase in radiation environment as operating time
accrues is presently somehwat conjectural. Data on radiation levels
encountered during incidents are not readily accessible, nor is
activity build-up uniform throughout plant subsystems. Activity build-
up studies indicate two limiting situations: one linear with energy
generation, the other becoming asymptotic after approximately four
full-power-years of operation.[5] The MTTR data noted (Table 2) are
probably adequate for the asymptotic case and optimistic for the linear
model.

3d. Output Factor. There is a substantial difference between the

availability and capacity factor reportings for nuclear units. Only a
comparatively small part of the difference has been associated historically with load-following--or equivalent discretionary outages
which do not reflect on the units' inherent reliability (this behavior
is in contrast to that of fossil-fired units, for which load-following
has been substantial). It is generally difficult to identify all
causes of partial power reductions from readily accessible data sources.
An EPRI program was initiated to provide perspective on the subject.
A sample output from this program during 1975 is shown as Table 3. As
noted, the reporting units had an average output factor (ratio of
capacity factor to availability factor) or 94.3%, in contrast to an
historic average of 84% for all plants. About one-third of the
partial power reductions are associated with equipment malfunctions.
This includes, as examples, condenser repairs conducted under partial
power conditions by isolation, or repairs to off-line control systems.
A somewhat smaller fraction is associated with nuclear fuel defects
including radioactivity release limitations and 'time to come to full
power' (i.e. restrictions on rate of power change to prevent fuel
failures). Regulatory and discretionary outages will vary, dependent
on: a) number of units approaching refueling' b) system considerations (e.g. percent of total capacity that is nuclear), and c)
Nuclear Regulatory Commission evaluations.

   4. Plant Design & Operations. Plant productivity analyses such

Table 3 (continued)

PARTIAL POWER REDUCTIONS: Detail

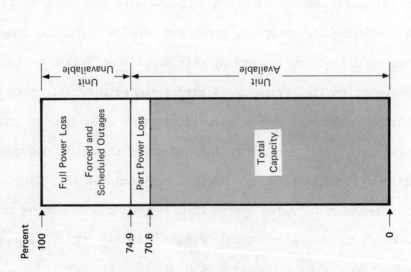

Table 3

SELECTED PARTIAL POWER OUTAGE EXPERIENCE

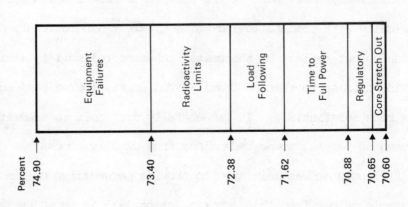

as illustrated in Table 1 draw on the same data and methodologies used for R&D identifications, but are modified to reflect specific unit interests rather than those of generic concern to the industry. Two of the more important differentiators are: a) the detailed design and operating information that the unit operator has available, notably for assessment of high frequency-low impact events and for maintenance actions that do not involve power curtailment, and b) the manner of dealing with the low confidence inherent in the low frequency-high impact events.

Returning to Table 1, it can be observed that virtually all of the procurement-oriented applications can be performed on at least a rudmentary basis using previously illustrated data and basic "single-chain" reliability models for specific units. This application has been illustrated in earlier reports.[1,6] With further acquisition of experience data, notably for refueling and other scheduled outages, considerable refinement is feasible. However, existing techniques already provide a good vehicle for communication and comparisons, these being the primary values of a formal modeling exercise.

4a. Design Analysis. In some of the procurement applications, notably acquisition of major spares, the formal utility evaluation process will focus on those low frequency-high impact events where it is not practically possible to acquire statistically valid data for

decision purposes. In such cases, a useful utility option is a
decision-or risk-analysis formatting of frequency intensity data, such
as shown in Figure 3. For example, for steam turbines, a present
estimate is that 2 failures/year will occur with about a 5% chance
that the outage will equal or exceed 1000 hours. Assuming a consequential impact (power replacement cost) of 0.5 million dollars per
1000 MWe-Day, the annual risk is equal to or greater than:

$$2 \ \text{x} \ .05 \ \text{x} \ 1000 \ \text{x} \ \frac{\$.5 \ \text{x} \ 10^6}{24 \ \text{hrs.}} = \$2.08 \ \text{million}$$

An analysis of potential turbine failure modes can suggest
several options that might reduce the outage duration. These options
could take the form of improved early diagnostic capabilities or
spares. The cost and benefit of these options can be compared to
determine preferred utility methods.

From a line of reasoning similar to that just illustrated for
turbines, the utility may also find it beneficial to utilize comparative statistical confidence treatment to assess competitive equipment
offerings such as fuel.[7]

4b. Data Bases. For operational applications (Table 1), two
major utility requirements appear to be: a) accelerated information
feedback on major forced outages, and b) development of malfunction

prediction techniques to aid planning and conduct of refueling outages. This is perhaps a deceptively simple statement—and one that appears to fly in the face of the much more highly-publicized claim of need for more extensive and comprehensive data bases. However, the perception appears to be correct based on consequential cost and probable benefit considerations.

It has been noted previously that the larger fraction of forced outage incidents do not result in a substantial contribution to the total forced outage duration (e.g. 80% of the events yield only 20% of the total outage). This establishes a benchmark on need for more elaborate reporting of these events. It is reasonable to assume that 80% of the events are those for which maintainability responses are reasonably well in hand. A more refined incident reporting system might do little in the way of reducing their productivity impact. Conversely, a higher priority interest is to provide a means of accelerating information dissemination on major outages and their precursors. EPRI is presently investigating advanced information system technology in lieu of fixed format computer data bases for this purpose.

An underlying consideration is flexibility. Because of purchasing, operating and system requirement variations, utilities have different incentives and constraints imposed on problem resolution. The utility wants to know as rapidly as possible what aggregate ex-

perience others may have available on the subject. A primary interest
is the remedies attempted and the success of those actions. This is
not the classic R&D situation (where many degrees of freedom may
exist for exploration). More commonly, the valid utility objective
may be finding an interim fix that permits deferral of a major
maintenance action until a scheduled refueling. From this perspective,
the information system being sought resembles a near-real-time communication capability emphasizing full-text retrieval of applicable
data.

4c. Refueling. Possibly the most significant area for utility
productivity improvement is in the refueling outage. A distinction
exists between units designed at a time when spacious layouts for
maintenance could not be justified (i.e. those introduced when it was
not yet clear that the nuclear unit would be the most economic generator in the system, and those of more modern vintage). The former
may be inherently limited on space available for parallel maintenance
functions. Ideally, the utility would like to perform virtually all
plant maintenance functions on normally-operating components during
refueling. To achieve this objective, outage planning may be initiated some 6-8 months prior to planned shutdown, the plan being based
on operating inputs with revisions as incidents are encountered during
the planning period. The largest source of extension of refuelings
is usually an unplanned action that is recognized after firm manpower

and equipment plans have been established. Some of these events, such as fuel-defect-related operations (whose magnitude cannot easily be judged prior to disassembly) may be mitigated by instrumentation techniques rather than modeling methods. However, it also seems clear that the development of optimal maintenance strategy techniques could prove highly beneficial. Such approaches have only been developed in a very limited way in the literature. A possible key input is the actual time between failures for a specific component. Such data exist in utility records, but are not otherwise readily accessible.

5. Industry Performance Maturity. Many people associated with the electrical generation industry have a justifiable interest in knowing what performance is expected without intimate exposition of why this is the expectation. Detailed, reliability-based analyses are not the answer here, but the use of the basic principles to interpret and extrapolate today's information shows considerable promise.

The development of electrical generating capacity is similar to classical high-reliability practice, but has at least two substantial differences. One is the practice inability to perform full-scale, system verification tests without using the steam-raising capability of the delivered product; the second is the comparatively high number of units committed before statistically significant service experience can be acquired. As a result, sampling of industry performance during specified time intervals will almost always include a heavy bias from units which are 'maturing'. Predicting the future from these data is frequently difficult since an equilibrium state is rarely established. So there is an interest in seeing if the generating industry has characteristic maturity patterns.

EPRI has attempted to model maturity patterns in terms of maintenance practice development. In effect, an operating utility: a) has an existing baseline of experience and know-how which it can apply to the planned operation of a new acquisition, and b) expects to learn about needed modifications during initial operations. This plan is perturbed by unscheduled outages which, if minor, will reduce performance, but not substantively modify maintenance planning. If a major scheduled outage is encountered, the operator may choose to modify his annual operating plan to best accommodate it; the duration of this outage may be highly dependent on whether it has occurred before, and what industry experience can be brought to bear on its rapid resolution.

Existing data on industry buying patterns and the frequency and duration of outages discussed in earlier sections can be utilized to model this process and to indicate how the accumulation of service experience damps the perturbing effect of unscheduled outages. For fossil-fired units, fairly simple models yield surprisingly good re-

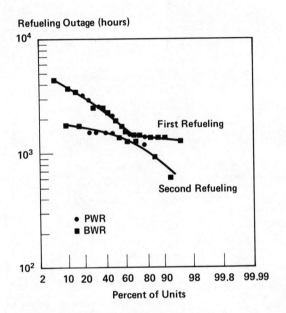

Figure 5 Domestic LWR "refueling" outages (June 1974–December 1976)

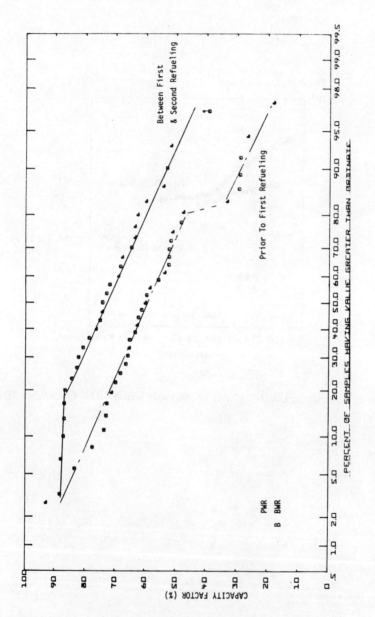

Figure 6 Capacity factor of LWR units during early operations (refueling outages eliminated)

plication of historic data and explain the wide variations in performance encountered in studies that look simply at annual industry
figures. Nuclear units, which cannot be examined in precisely the
same way, nevertheless appear to follow similar patterns as shown in
Figure 5-6. (In nuclear unit operation, the prospects for shifting
refueling to mitigate the impact of a major unscheduled outage are
minimal). Therefore, independent studies of operational and refueling
maturity are required. Combining these observations for an overall
projection is feasible, but verification awaits the accumulation of
further operating data for the distinctly different designs which
exist.

6. Conclusions. The performance of nuclear generating units
throughout their lifetime is the result of an interaction of design,
operational and management decisions that may be without parallel in
technical endeavors in their complexity. Classical reliability-based
analyses cannot be utilized to predict performance behavior accurately.
However, the application of basic principles, with particular emphasis
on maintenance efforts, can be used effectively to guide efforts
needed to interpret and improve the productivity of these units.

REFERENCES

(]) M. LAPIDES and E. ZEBROSKI, Use of Nuclear Plant Operating
          Experience to Guide Productivity Improvement Programs, EPRI
          Special Report SR-26R, November 1975.

(2)   E. ZEBROSKI, EPRI Programs in Nuclear Systems and Materials Which
      Bear on Component Design, American Power Conference,
      Chicago, IL, April 1976.

(3)   H. OCKEN, et al, EPRI LWR Fuel Performance Program, EPRI Special
      Report NP-370-SR, January 1977.

(4)   EPRI Research Project 894, Limiting Factor Analysis of High
      Availability Nuclear Plants, initiated April 1977.

(5)   R. W. SHAW and D. L. UHL, EPRI Programs in Radiation Control at
      Nuclear Power Plants, International Water Conference,
      Pittsburgh, PA, October 1976.

(6)   M. LAPIDES, Nuclear Unit Productivity Analysis, EPRI Special
      Report SR-46, August 1976.

(7)   L. JOHNSON et al, Developing Methods for Establishing Improved
      Plant Reliability, American Power Conference, Chicago, IL,
      April 1976.

RELIABILITY MODELLING OF ELECTRICAL AUXILIARY SYSTEMS IN NUCLEAR POWER STATIONS

R. BILLINTON*, R. N. ALLAN** AND M. F. DEOLIVEIRA***

Abstract. The electrical auxiliary system in a nuclear power station is a complex configuration involving transformers, load buses protection and isolation devices, normally closed and normally open paths, stand-by and parallel redundancy and a multitude of individual loads. The availability of the various loads can have a considerable influence on the generating unit and station output. This paper illustrates the development and application of a quantitative technique for evaluation of the adequacy of a typical nuclear station electrical auxiliary system.

1. Introduction. Total power system economics dictate that the availability of a nuclear power station be as high as possible. The impact of generating unit unavailability can be readily assessed in terms of additional installed capacity requirements to meet a given load level at a specified risk[1] and the expected energy replacement costs associated with satisfying the load by a more expensive energy source. The installed capacity aspect can be easily observed using the conventional loss of load expectation technique[1,2]. It has been estimated[3] that an increase of one percent in the Equivalent Forced Outage Rate of an Ontario Hydro 3,400 MW nuclear generating station coming in service in 1986 has a penalty whose net percent worth in 1986 dollars is about 100 M$ including both system reserve and energy costs. A one percent increase in scheduled outage rate for the same situation has a penalty of about 75 M$ in capacity and replacement costs. The energy replacement cost penalty can be easily seen using the Ontario Hydro system in which the fueling cost of the CANDU reactor is approximately $1/MWhr and

* Electrical Engineering Dept., University of Saskatchewan. Saskatoon, Canada. ** Department of Electrical Engineering and Electronics. UMIST Manchester, England. *** Department of Electrical Engineering, University of Porto, Portugal. This research was supported financially by the Central Electricity Generating Board, England.

coal cost in a typical thermal station is $12/MWhr. The loss of a 500
MW nuclear unit therefore results in an increased energy cost of
$924,000/week. The benefit to the system in terms of capacity and
energy costs can be evaluated for a unit and subsequently allocated to
the various subsystems which together form the unit. Most subsystems
can be modelled using a two state representation (available or unavailable) and the analysis while detailed is often relatively straightforward. This is not usually the case, however, with nuclear station
electrical auxiliary systems which provide the normal electrical supplies
required to ensure safe operation and shut down. The latter aspect is
usually of primary concern due to the potential risk to operating staff,
society at large and the environment. In order to achieve high availabilities and low failure frequencies, these systems are very complex
and incorporate many aspects that make reliability evaluation difficult.
It is, however, vital that the adequacy and potential risk is evaluated
as precisely as possible so alternative designs may be compared with
each other and with specified reliability targets. The approach utilized
should not make unrealistic assumptions and approximations and should
permit many possible systems and operating policies to be compared.
Sensitivity studies should also be performed to identify weak areas in
the system and to assess the impact of inaccurate component reliability
data.

The paper describes the techniques that the authors have developed
to perform realistic reliability modelling of all the components in an
electrical auxiliary system and shows how these models can be used to
evaluate the adequacy of a typical system in a nuclear power station.

The word reliability in this paper is associated with an overall assessment of subsystem and unit adequacy portrayed by the three steady state indices of failure rate, average outage duration and average annual outage duration. These indices provide a measure of unit and station adequacy which is absolutely necessary when considering the implications of alternate designs and should not be confused with the time dependent probability values associated with safety assessment. The average failure rate associated with a particular subsystem and with the unit can however provide useful input to the safety assessment procedure.

2. System Failure Modes. The bulk of the literature which deals with reliability evaluation of electrical systems considers each system component to be represented by 2 states; one in which the component operates normally and one in which it has failed and is being repaired. This model does not recognize the multitude of component failure modes and restoration procedures and does not relate the failure modes to their impact on the system. It is, however, impractical to attempt to identify every conceivable failure and restoration mode in a large system as the required data collection scheme would be extremely complex and the computational effort required for model solution and evaluation may be prohibitive. The need therefore for very efficient computational techniques is readily apparent when considering systems which may con-tain several hundred components.

Combination or Pooling of States. In an electrical distribution network, each component can have several failure modes, i.e., normally closed circuit breakers may fail or inadvertently open, cables may suffer a phase/phase, phase/ground or open circuit fault due to a host of physical phenomena.

It does not follow however, that all component failure modes must be identified separately since, in practice, many failure modes are mutually exclusive but cause identical system effects. In such cases, the failure modes causing identical system effects can be pooled to create a single failure mode for which the expected failure rate is the sum of the rates of occurrence of the constituent failures.

Two and Three State Models. In actual practice it may not be possible to pool all the failure modes of a given component because of the different system effects which may occur. Such components must therefore be represented by several failure modes although each of these modes may be created by the pooling of individual failure events that have the same system effect. The number of component states to be considered computationally is then equal to the number of pooled component failure modes plus one; the latter corresponding to the up or operating state.

In the case of a normally closed circuit breaker, three pooled modes can be identified:

(i)   false breaker operation (opening)

(ii)   all short circuits, i.e. bushing failures, internal breakdown

(iii) failure to operate, this is defined as a stuck breaker.

Failure modes (i) and (ii) can be identified with most electrical components, i.e. open circuit faults and short circuit faults respectively. These faults can usually be pooled as mode (i) does not usually cause other circuit breakers to operate whereas mode (ii) will initiate action in the entire protection zone around the failed component and remove other healthy components and branches from service. These two failure modes have been designated[6] as passive and active respectively. Failure mode (iii) is associated only with protective devices such as

breakers and manifests itself when a need for operation arises, i.e.
when an active failure occurs in a component that it is meant to protect. In some cases the failure rate of some modes may be zero, i.e.
busbars usually have a passive failure rate of zero.

The cycle of events associated with modes (i) and (ii) is that,
following a passive failure, the component must be repaired before
service can be restored whereas, following an active failure, although
the failed component must itself still be repaired, the tripped breakers
can be reclosed once the failed component has been isolated. This
philosophy leads to a component state-space diagram in which, following
an active failure and a switching operation, the failed component resides
in a state identical to that directly following a passive failure.
These system states can be identified in the system shown in Figure 1
in which the breaker feeding load point L2 actively fails.

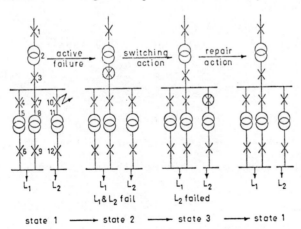

Figure 1.   Example Configuration

The general model for an electrical component should therefore
consist of the three states, up, switching and repair. Most literature
neglects the second state though it has been recognized and included in

some recent papers[6-10]. Neglecting this state is justified in
certain cases when only the unavailability of the system is required
since the average residence time in this state is usually very small.
When the failure rate is also required as in the assessment of nuclear
station auxiliary systems, then the inclusion of this state is vital.
A component which fails only in a passive mode can be represented by
the usual 2-state model.

Reduction of System States. The previous discussion suggests that
a 3-state model of each component is necessary to realistically assess
the electrical systems in nuclear stations. This can lead to an extremely large number of system states. The authors have minimized the
number of system states and the computational effort without introducing
any significant error by utilizing the following assumptions:
a) The probability of occurrence of two overlapping switching states
   is negligible.
b) The probability of occurrence of two overlapping stuck breakers is
   negligible.
c) The load point failure events can be identified by the load point
   minimal cut sets associated with the minimal paths from the load
   point to the system sources. The authors have published[11] an
   efficient algorithm for detecting minimal cut sets.

3. Simulation of Passive and Active Failures. Direct application
of the minimal cut set technique is not possible except in the case of
passive failures because the technique assumes a 2-state component
model and therefore cannot identify the switching effects caused by
active failures.

In order to illustrate how the technique can be applied to both

2-state and 3-state models, consider the network shown in Figure 1.

In the following analysis P, A and S after a component number represents a passive event, active event and stuck breaker respectively. Visually the following failure events of load point L_2 can be identified.

a) 1P, 2P, 3P, 10P, 11P, 12P are failure events of the load point; they are the minimal cut sets associated with the minimal path 1-2-3-10-11-12.

b) 1A, 2A, 3A, 10A, 11A, 12A are also failure events of the load point; they are the same minimal cut sets as in (a).

c) 4A, 7A are failure events because they cause breaker 3 to trip and all paths (in the example only one path) are broken.

d) 5A and 4S, 6A and 4S, 8A and 7S, 9A and 7S again cause the same effect as in (c).

    The load point failure events are therefore: 1(P+A), 2(P+A), 3(P+A), 10(P+A), 11(P+A), 12(P+A), 4A, 7A, 5A+4S, 8A+7S, 6A+4S, 9A+7S.

This simple example shows that in a practical reliability analysis, the passive failures lose their identity because they are always combined with active failures. This leads to the consideration of the component total failure rate (passive plus active) and the component active failure rate.

The example also illustrates that the minimal cut set technique can be used to detect, in one operation, the load point failure events due to all passive failures and those active failures having the same system effect as the corresponding passive failures. This direct simulation method can be used to deduce load point failure events due to the total failure rate of components up to any order. The authors have generally limited such simulations up to third order, i.e. load point failure events due to three overlapping forced outages.

The simulation of active failures and active failures overlapping a stuck breaker which have an impact on the load point due only to their switching effects requires additional techniques. This involves simulating an active failure on each system component one at a time, simulating the opening of the relevant primary protection breakers in the case of an active failure or of the back-up protection breaker in the case of an active failure overlapping a stuck breaker condition and detecting whether any of the minimal paths leading to the load point are broken. Following this simulation, active failures overlapping total failures can then be detected by deducing the minimal cut sets up to any desired order of the remaining, unbroken minimal paths. In the case of forced outages (total or active) overlapping a maintenance out-age, the component considered on maintenance is removed computationally and the simulation of total failure events and the simulation of active failure events and stuck breaker conditions carried out as previously described.

4. Restoration Modes. The previous section described the ways in which the components and the electrical auxiliary systems may fail and the modelling approach used to consider these situations. This method-ology includes restoration events involving both repair (or replacement) and isolation (switching) of the failed elements and systems. This leads to the concept of three distinct types of failure/restoration modes:

a) A failure/restoration mode in which components are either out for repair or for maintenance. Service can only be restored by replacing at least one of the components out.

b)   A failure/restoration mode for which one component is actively

   failed and other components are out either for repair or for

   maintenance. Service can be restored by isolating the actively

   failed component and re-energizing the rest of the system.

c)   A failure/restoration mode similar to (b) but one for which the

   outage of the actively failed component overlaps a stuck breaker.

   In the electrical auxiliary systems associated with nuclear

stations, additional restoration modes are possible in order to

achieve high availabilities of the load points and of the units.   These

additional restoration modes are easily identified in the simplified

auxiliary system shown in Figure 2.   In the event of a load point

failure due to overlapping forced outages or forced outages overlapping

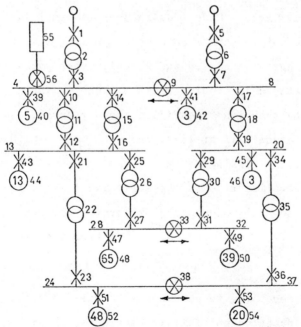

Figure 2.   Typical Electrical Auxiliary Network

a maintenance outage (failure/restoration mode (a)), it may be possible
to restore supply to the load point by closing one of the normally
open breakers or by starting emergency stand-by equipment such as diesel
generators or gas turbines. This type of restoration mode must be
modelled in order to make the assessment realistic and leads to a fourth
type of failure/restoration mode.

  d) A failure/restoration mode similar to (a) but one for which service
      can be restored by closing a normally open breaker or starting a
      stand-by generator.

  5. Modelling Normally Open Paths. Normally open paths are
generally[6,8,12,13] assumed not to fail during their required operating
time. This may not be acceptable when the normally open paths are
required to operate for relatively long periods or when their failure
rates are relatively large. Modelling these paths is relatively simple
if they are assumed not to fail when closed. During the tracing of the
minimal paths, all normally closed and normally open paths are
identified. Following a load point failure event in the normally closed
paths, the modelling technique identifies whether any of the normally
open paths remain intact. If at least one such path remains intact, the
restoration mode is the closing of this normally open path (mode (d)
above) and the restoration time is the switching time of the normally
open path. If all normally open paths are destroyed, restoration is by
repair action (mode (a) above) and the restoration time involves repair
times only.

The following modelling technique can be used when the assumption
that normally open paths do not fail is considered invalid. This
approach includes failures of the normally open path and also the

probability that the normally open breaker does not close i.e. a stuck
condition.

Define the two events, A and B as:

A - a failure event that can be eliminated by closing any of the
normally open paths p_1, p_2, ... p_n.

B - an event defined as, at least one of the normally open paths
p_1, p_2, ... p_n does not fail before restoring to service one
component of the failure event A.

Using conditional probability:

$$P(A) = P(A/B) P(B) + ((A/\overline{B}) P(\overline{B})$$

where $P(A/B)$ is the probability of event A assuming that at least one of
the normally open paths is always available, i.e. the indices given by
mode (d) and $P(A/\overline{B})$ is the probability of event A assuming that none of
them is available to restore service, i.e. the indices given by mode
(a). For each of the events A/B and A/$\overline{B}$, the contribution to the
average failure rate λ, average outage duration r and average annual
outage time U due to forced outages and due to forced outages overlapping a maintenance outage (λ'', r'', U'' respectively) are evaluated.
These indices are:

event A/B   - λ_1, r_1, U_1 and $\lambda_1'', r_1'', U_1''$

event A/$\overline{B}$ - λ_2, r_2, U_2 and $\lambda_2'', r_2'', U_2''$

where $\lambda_1 = \lambda_2 =$ failure rate of event A (similarly $\lambda_1'' = \lambda_2''$)

To evaluate $P(B)$ and $P(\overline{B})$, the same set of indices are evaluated
for the event represented by $\overline{B}$ by deducing the minimal cut sets of the
paths p_1, p_2, ... p_n. From these cuts those that are also normal mini-

mal cut sets of the load point are eliminated. If load point indices
up to third order are being evaluated, then these minimal cut sets
should preferably be evaluated up to second order if event A is of first
order, up to first order if event A is of second order and not evaluated
if event A is of third order. The authors have found, however, that the
error is generally negligible if the minimal cut sets of the paths p_1,
p_2, $\ldots$ p_n are evaluated only up to first order irrespective whether
event A is of first or second order. This approximation permits considerable reduction in computational effort.

From the remaining cut sets associated with paths p_1, p_2, $\ldots$ p_n
the reliability indices of the "element" equivalent to the normally open
paths are evaluated. Let these be λ_e, r_e, U_e and λ_e'', r_e'', U_e''. The
normally open breakers which are first order cut sets of the paths p_1,
p_2, $\ldots$ p_n are considered "in series" in order to evaluate the stuck
probability (P_S) of this "equivalent element".

Since the average annual outage time of event A, when expressed in
years/year, represents $P(A)$, the following equations are obtained using
conditional probability:

$$U_A = U_1 [1 - (U_e + P_S)] \; + U_2 (U_e + P_S)$$

$$U_A'' = U_1'' [1 - U_e''] \; + U_2'' \, U_e''$$

It can also be reasoned that:

$$\lambda_A = \lambda_2 (1 + \lambda_e r_2)$$

$$\lambda_A'' = \lambda_2'' (1 + \lambda_e'' r_2'')$$

$$r_A = \frac{U_A}{\lambda_A} \quad \text{and} \quad r_A'' = \frac{U_A''}{\lambda_A''}$$

This technique is applied to every minimal cut set of each load point that can be eliminated by closing a normally open path.

6. Modelling Stand-by Generators. Stand-by generating plant can be modelled in an identical manner to that of normally open paths, i.e. during the tracing of minimal paths a stand-by generator is identified as a normally open path. When a mode (d) load point failure event occurs in the normally closed paths, this type of normally open path is considered with all other normally open paths to evaluate their impact on the load point reliability indices. The only difference is that the relevant component reliability parameters of the stand-by facility must be modified as follows:

a) Stand-by facility

defined failure rate = running failure rate of the stand-by facility

i.e. number of failures/year of running

defined repair time = average repair time of the stand-by facility following a running failure

b) normally open breaker

switching time = average time between a load point failure and the instant when the load point is fed from the stand-by facility, i.e. the time includes the average start-up time of the facility.

stuck probability (P_S) = the combined probability of a stuck breaker condition (P_b) and a running-up failure of the stand-by facility (P_g), i.e. $P_S = P_b + P_g - P_b P_g$.

7. Modelling Branches Off the Busbars. The branches off the bus-

bars cannot be ignored as they can have a significant impact on the busbar indices, particularly the busbar failure rate. This is due to the
effect of active failures and active failures overlapping a stuck
breaker condition in the branch components being reflected back to the
breakers on the input side of the busbars. The number of branches off
a busbar in a typical nuclear power station could exceed 60 with each
branch containing at least three components, i.e. a breaker, a cable
and a motor. Several hundred extra components would then be introduced
to adequately represent these branches. The solution time and storage
requirements are a function of the number of components. Individual
inclusion of all these additional components will seriously degrade the
computational efficiency.

This difficulty can be overcome by representing the branches off
each busbar by an "equivalent branch". The active failure rate of each
component in this equivalent branch is obtained by summing the active
failure rates of the similar components existing in the real branches.
The switching time and stuck breaker probability are the same as any of
the similar real components. Using this technique the busbar reliability
indices of each busbar are calculated precisely and the increase in
computer time and storage requirements is minimal.

8. Component Data and Load Point Indices. The following component
reliability data is required in order to evaluate the reliability
indices (average failure rate, average outage duration and average
annual outage time) of each failure event and of each load point:

a) Total failure rate - average total number of component failures
    per year that require the component to be removed from service for
    repair due to any of its failure modes.

b) Active failure rate - average number of component failures per year
    that cause breakers to open and therefore tripping of other healthy
    components.

c) Average repair time - average time taken to repair all kinds of
    component failure modes.

d) Switching time of actively failed components - average time between
    the occurrence of an active failure and the instant when the failed
    component is isolated and all possible healthy components are
    restored to service.

e) Switching time of normally open breakers - average time between the
    occurrence of a load point failure event and the instant when the
    supply to the load point is restored by closing the normally open
    breaker.

f) Maintenance outage rate - average number of occasions per year that
    a component is taken out of service for preventive or scheduled
    maintenance.

g) Average maintenance time - average duration of all preventive
    maintenance outages.

h) Stuck probability - probability of a breaker or a switch failing to
    open or close when called upon to operate.

The reliability indices of each failure event and of each load
point can be evaluated using the above component data and the equations
previously published[6,14,15].

The load point indices are valuable to both designer and operator.
When calculated for the important busbars in the system, they provide
a quantitative appraisal of the average number of occasions and duration
of time that each subsystem will not be able to perform its intended
function. This can be vital information in the safe operation of a

nuclear station.   If the failure results in excessive perturbation of

station operation then the outage duration will depend upon the time to

restore stable reactor operation which may exceed the actual failed

component repair time.   An alternative method of assessment is to

evaluate relevant busbar indices neglecting the impact of any stand-by

generators.   The indices are now, not the true unreliability of the

busbar, but instead define the time and frequency required of the

stand-by systems in order to provide an adequate reliability level.

9.   Unit Reliability Indices.   The busbar indices provide a

quantitative assessment of the reliability associated with each load

point and indicate the number of occasions that safety problems may

arise.   They do not, however, provide an assessment of the overall impact

on the unavailability of the unit due to load point failures.   The

unavailability of a nuclear generating station is an important system

parameter to which failures in the electrical auxiliary system contri-

bute.   The overall impact of the auxiliary system on the unavailability

of the unit can be evaluated as follows.

In a nuclear power station the load points have a common objective;

this being the output of the unit.   The consequence of losing a busbar

or busbars can therefore be expressed in terms of the output power

reduction of the unit.   Each failure event in the auxiliary can however

cause the failure of more than one busbar and therefore the failure

events associated with the unit cannot be obtained by simply summing

the independent failure events of all the busbars.   An algorithm pub-

lished previously[10] permits the unit independent failure events to be

deduced from a knowledge of the busbar independent failure events.

This algorithm[10] requires the impact of the loss of a busbar or

combination of busbars on the output of the unit to be specified, i.e.
how much the output of the unit is reduced when a load point busbar
fails. Consequently, each busbar, when lost, is assumed to have one of
the following impacts on the output power of the unit:

a)   no impact on the unit (= 0% reduction)

b)   reduction of the output power by a value between 0% and 100%, i.e.
     it causes a derated state (= x% reduction)

c)   complete shutdown of the unit (= 100% reduction)

The reduction of power due to the loss of any combination of two
busbars is given by the power reduction states shown in Table 1 and the
reduction of power due to the loss of three or more of the x% - type
busbar assumed to be 100%.

Table 1.   Reduction states due to loss of 2 busbars

busbar i $\longrightarrow$	0%	x%	100%
busbar j 0%	0%	x%	100%
y%	y%	z%	100%
100%	100%	100%	100%

where $0\% \leq z \leq 100\%$

Using this algorithm[10], the unit independent failure events can
be established and the reduction of output power and reliability indices
associated with each failure event deduced.   The reliability indices
(expected encounter rate λ_s, average duration r_s and annual residence
time U_s) of each derated state can then be evaluated by summing the
relevant indices of all unit failure events causing that derated state,
i.e.

$$\lambda_{s(c)} = \sum_{i=1}^{m} \lambda_i, \quad U_{s(c)} = \sum_{i=1}^{m} U_i, \quad r_{s(c)} = \frac{U_{s(c)}}{\lambda_{s(c)}}$$

where m is the number of unit independent failure events causing derated
state c.

This set of indices is not suitable for comparing different
systems with different sets of derated states. One comparative technique is to evaluate the average annual cost associated with unit availability[4]. This cost can be related to the expected annual energy not
supplied by the station. Since this energy not supplied must be provided by a station lower down the merit order, a financial penalty is
directly related to the unreliability of the first station. The energy
not supplied by the station due to the electrical system not being able
to perform its intended function can therefore be considered as a
consistent indicator associated with alternate electrical auxiliary
systems. The expected annual energy not supplied R_t can be obtained
from the derated state indices using:

$$R_t = (\sum_{i=1}^{n} R_i \, A_i) \times C \text{ MWhr/year}$$

where n is the number of derated states, R_i is the reduction of power,
expressed as a p.u. value of the unit capacity, associated with derated
state i, A_i is the annual residence time (hr/yr) or availability of
state i and C is the capacity of the unit (MW).

10. Reliability Analysis of Typical Case Studies. The techniques
described in this paper have been implemented in a computer program and
used in the analysis of many systems, one of which is shown in Figure 2.
The assumed component reliability data is shown in Table 2 and the
assumed consequences of busbar outages are provided in Table 3. In
practice it may be desirable to study many variations to the basic
system together with various sensitivity analyses.

Table 2. Consequences of busbar outages

number of failed busbars	busbar number	reduction of output power percent
1	4, 13, 24, 28	100
1	8, 37	30
1	20	20
1	32	0
2	8 + 20	40
2	8 + 37	50
2	20 + 37	45
3	any	100

Table 3. Reliability data (maintenance not considered)

component	failure rate (f/yr) total	active	repair time (hr)	switching time (hr)	stuck probability
n/c breakers	0.020	0.015	20	1	0.001
n/o breakers	0.020	0.015	20	1	0.013
transformers	0.020	0.015	800	1	-----
busbars	0.005	0.005	10	1	-----
stand-by generators	12.00	-----	78	1	-----
breakers 56 (a)	0.020	0.015	20	1	0.113
motors	0.050	0.040	--	1	-----

(a) includes probability of start-up failure of standby generator

The following six studies have been performed to provide an indication of the range of possible investigation:

(i)   a base case study of the system shown in Figure 2 considering up to third order failure events and assuming that normally open paths may fail

(ii)   as the base case but assuming that normally open paths do not fail

(iii) as the base case but omitting the stand-by generator from the system

(iv)   as the base case but omitting the branch containing components 14, 15 and 16

(v)   as the base case but including a second identical branch between busbars 8 and 20

(vi)   as the base case but omitting breaker 38.

Unit Availability. As discussed previously, the most suitable
reliability index for assessing the total impact of auxiliary system
failures on the availability of the unit is to evaluate the expected
energy not supplied. These indices for case studies (i) - (vi) are
shown in Table 4 assuming a 660 MW unit together with the details of
encounter rate, average duration and annual residence time for the base
case. It can clearly be seen that a considerable difference exists
between the values of expected energy not supplied and that some significant features are discernable.

Table 4. Expected energy not supplied

Case Study	Expected Encounter Rate/Yr.	Average Duration Hr.	Annual Residence Time Hr/Yr.	Derated State %	Capacity Out MW	MWhr/yr Not Supplied
(i)	0.06500	254.36	16.533	20	132	2182
	0.36598	1.7668	0.64661	30	198	128
	0.00500	8.8633	0.04432	40	264	12
	0.14036	2.6835	0.37666	45	297	112
	2.35640	1.4257	3.3595	100	660	2217
(i)					TOTALS:	4651
(ii)						4038
(iii)						4833
(iv)						15710
(v)						2354
(vi)						21523

It is evident that the failure of normally open paths can have a
significant impact (studies (i) and (ii)) and, if greater precision is
required, such failures should be included in the assessment. It is
also evident that the network topology, as expected, has a most marked
effect on the value of energy not supplied; a 10:1 ratio existing for
the systems studies in this paper. This particular index is therefore a
very pertinent measure for assessing different designs and permits

station designers to make an objective engineering judgement in relating
the capital cost and unreliability cost of various alternative proposals.

Busbar Indices. As an example of the indices that can be obtained
for individual busbars, the reliability indices for busbar 13 are
shown in Table 5. These include a detailed set of indices for base
case (i) and the overall indices for cases (ii) to (iv). Clearly case
studies (v) and (vi) will have no impact on the indices of busbar 13.

Table 5. Reliability Indices of Busbar 13

Case Study	Failure or Restoration Process	Failure Rate/Yr	Average Duration Hr.	Annual Outage Time Hr/Yr
(i)	repair action	0.01023	12.924	0.13222
	n/o path	0.06006	1.0576	0.06352
	active failure	0.36000	1.0000	0.36000
	active plus stuck breaker	0.0084	1.0000	0.00084
(i)	TOTALS	0.43113	1.2910	0.55658
(ii)		0.43107	1.2830	0.55306
(iii)		0.43123	1.8636	0.80364
(iv)		0.46087	37.525	17.294

In the case of this busbar it is seen that failures of normally
open paths have only a very small effect on the indices (cases (i) and
(ii)) though this is not a general conclusion for all busbars. It can
also be seen that the stand-by generator improves the availability of
the busbar (cases (i) and (ii)) by about 44%; this improvement being due
entirely to a reduction in the average outage duration since the busbar
failure rate increases very slightly. Finally, as expected, removal of
the branch containing components 14, 15 and 16 increases dramatically
the unavailability of the busbar (cases (i) and (iv)) due to an
increased number of first order events, all of which involve a repair
action since no normally open paths can be used.

Conclusions. This paper has presented some novel techniques for realistic evaluation of electrical auxiliary systems in power stations. The reliability indices of failure rate, average duration and expected annual outage time can be used to compare the adequacy of alternate configurations in terms of their impact on the unit output. The recognition of active failures is an important requirement in the reliability assessment of electrical distribution networks. The minimal cut set technique developed by the authors is very efficient and can be used to analyze single unit systems and also multi-unit plant configurations.

REFERENCES

[1] R. BILLINTON, Power System Reliability Evaluation, Gordon and Breach Science Publishers, New York, N.Y., 1970.

[2] R. BILLINTON, Bibliography On The Application Of Probability Methods In Power System Reliability Evaluation, IEEE Trans., PAS-91, No. 2, pp. 649-660.

[3] J. KRASNODEBSKI AND R. BILLINTON, Reliability and Maintainability In Nuclear Power Generation - Viewpoint of a Utility, Society of Reliability Engineers Symposium, 1976, Ottawa, Ontario.

[4] R. BILLINTON, D. O. KOVAL, D. R. CROTEAU, R. S. WEAVER AND V. PRASAD, Application of Reliability Concepts In The Selection Of Transformers For Large Generating Units And Stations, IEEE Winter Power Meeting, January 1977, New York, N.Y.

[5] R. BILLINTON AND S. Y. LEE, Availability Analysis Of A Heat Transport Pump Configuration Using Markov Models, IEEE Winter Power Meeting, January 1976, New York, N.Y.

[6] M. S. GROVER AND R. BILLINTON, A Computerized Approach to Substation And Switching Station Reliability Evaluation, IEEE Trans., PAS-93, 1974, pp. 1488-1497.

[7] J. ENDRENYI, Three State Models in Power System Reliability Evaluations, IEEE Trans., PAS-89, 1970, pp. 1909-1916.

[8] M. B. GUERTIN AND Y. LAMARRE, Reliability Analysis of Substations With Automatic Modelling of Switching Operations, IEEE Trans., PAS-94, 1975, pp. 1599-1605.

[9] R. N. ALLAN, R. BILLINTON AND M. F. DEOLIVEIRA, Reliability Evaluation of Electrical Systems With Switching Actions, Proc. IEE,

123, 1976, pp. 325-330.

[10] R. N. ALLAN, R. BILLINTON AND M. F. DEOLIVEIRA, Reliability
      Evaluation of the Auxiliary Electrical Systems of Power
      Stations, IEEE Winter Power Meeting, 1977, paper F77 008-6.

[11] R. N. ALLAN, R. BILLINTON AND M. F. DEOLIVERIRA, An Efficient
      Algorithm For Deducing the Minimal Cuts and Reliability
      Indices of a General Network Configuration, IEEE Trans.,
      R-25, 1976, pp. 226-233.

[12] J. ENDRENYI, P. C. MAENHAUT AND L. E. PAYNE, Reliability Evaluation
      Of Transmission Systems With Switching After Faults -
      Approximations and a Computer Program, IEEE Trans., PAS-92,
      1973, pp. 1863-1875.

[13] R. N. ALLAN, R. BILLINTON AND M. F. DEOLIVEIRA, Reliability
      Assessment of Power System Networks, PSCC, 1975, Cambridge,
      England, Paper No. 1.2/4.

[14] R. BILLINTON AND M. S. GROVER, Reliability Assessment of Trans-
      mission and Distribution Schemes, IEEE Trans., PAS-94, 1975,
      pp. 724-732.

[15] M. S. GROVER AND R. BILLINTON, Quantitative Evaluation of
      Permanent Outages in Distribution Systems, IEEE Trans.,
      PAS-94, 1975, pp. 733-741.

DETERMINATION OF INITIATING EVENTS AND SEQUENCES OF REACTOR ACCIDENTS BY A BARRIER ANALYSIS

A. CARNINO AND J. DUBAU[*]

Abstract. From a barrier analysis (fuel cladding, primary circuit, containment), a systematic search for the potential initiating events and the accident sequences is conducted.

The method is divided into three parts :
- The barrier analysis - For each barrier, an event tree is constructed having for top event the barrier loss.
- The search for initiating events of accidents (an accident has occurred when all the barriers are broken or failed).
- Accidents event trees - From an initiating event, the accident sequence is derived.

To illustrate the method, the failure of all the electrical power supplies in a 900 MW PWR is considered.

Possible extensions of the method to other risks are suggested.

1. Introduction. In order to analyse the safety of a nuclear reactor, a method has been developed for many years and used in France in the safety deterministic analysis and in the presentation of safety reports. It is based upon a refined analysis of the "barriers defences" which form a physical obstacle between the radioactive materials and the environment. In the nuclear island of a nuclear power station, the three barriers are :

- First   : the fuel cladding.

---

[*]Ingénieurs au Commissariat à l'Energie Atomique - France -
Institut de Protection et de Sûreté Nucléaire
Département de Sûreté Nucléaire.

- Second   : the primary circuit including : reactor vessel, steam
             generators, pressurizer, reactor coolant pumps, isolation
             valves, pipes, etc...
- Third   : the containment.

The safety of the plant depends upon maintaining the integrity of
each of these barriers in every type of functioning of the reactor. This
is accomplished in three steps : anticipating and removing potential
problems by proper design and quality control, by a continuous survey
and control of the barriers, and by intervention actions, in case the
limit values are exceeded.

A full probabilistic approach to the accidents of a nuclear reactor has shown the need for an exhaustive and systematic search for the
potential initiating events [1]. This search may be based on a list of
events such as the ANS classification of accidents [2]. In this paper,
it is proposed a different approach which is closely related to the
safety analysis by the defence of the barriers mentioned above. Each
barrier is characterised by a set of parameters : mean value of functioning, design basis values and upper limit values which, if exceeded,
would cause the failure of the barrier.

The accident, by definition, occurs when all the three barriers
are lost. From the analysis of each barrier we may study the behaviour
of the barrier after the occurrence of a potential initiating event of

an accident. It is therefore possible to draw an event tree to
show the different (external and internal) barrier aggressions and their logical combinations which will lead to values of the
barrier parameters outside their limits with resultant failure of the
barrier.

   2. "Loss of barrier" analysis.   A systematic search was conducted
to identify all the possible initiating events which could result in
the loss of a barrier. A list of all possible elementary initiating e-
vents might be so large and so intricate that it would not be possible
practically to demonstrate its completeness $[3]$, $[5]$. In the field of
nuclear reactor accidents, there is a lack of statistics and therefore
the only way to determine potential initiating events is based on "im-
agining" what could happen. Thus it is impossible to show that such a
list is exhaustive and represents reality. The search reported herein
is based on a systematic review of the aggressions that could occur
against the barriers.

When determining these possible barrier aggressions, the level of
detail must be addressed. This analysis, illustrated in Figs. 1 to 10
was limited to identifying classes of intiating events which are well
enough defined to enable one to recognize the family of elementary
events associated with each class. When examining the list of abnormal
occurrences and the operating experience, it is obvious that most of

the elementary events and their coincidences are more or less random.
It is therefore useless to try to obtain a high level of details and
refinement in determining the initiating events. What is most important
is that all the initiating events listed represent families of elementary occurrences. These families can be associated with accident sequences and these sequences can encompass all the associated elementary
initiating events. The only difficulty is to allocate a corresponding
probability to a family of events. The probability of a family however
is dominated by the major contributors and thus one can ignore all the
constituent elementary events of very low probability. Using families
of events, it is therefore possible to draw an event tree in the same
manner as a fault tree to show the different aggressions (external or
internal) and their logical combination.

The reactor can be in different states : functioning, refueling cold
shutdown, hot-shutdown ... For each of these states, one has to draw
different generalized event trees : the barriers might be different
with regard to consequences and thus the loss of each one leads to a
different tree. But this paper presents only the case when the reactor
is under normal operation at full power.

The trees have two levels : at the first level are the physical
causes of failures of the barrier ; at the second level are the classes
of events which produce the physical cause. From the former generalized

event trees it is easy to check that one can obtain all the accidents
already taken into account in a safety report. The major advantage of
this method is its parallel with the usual safety analysis by barriers
which protect the environment against nuclear radioactivity and, therefore, it is logical to describe an accident in terms of the barriers
[4].

The construction of the generalized event trees needs a good definition of what constitutes the failure or the loss of the barriers.
An initial feeling was that a definition could be made using the upper
limit values of the physical and environment parameters of the barriers.
But after several attempts to implement this idea, it was found difficult to accomplish. As an example, for the third barrier (containment),
the design limit value is well known, but the upper limit value which
when exceeded defines the failure of the containment is not known
with accuracy. In such a case, it was found prudent to be pessimistic
and use the design value. In general the upper limit values of physical
and environmental parameters of the barriers, when these values were
known, were used. All these definitions of failure are important when
considering the probability of initiating events or of accident sequences.

3. Accident chronology : timing and sequences.   After determining the possible initiating events of accidents by the systematic

review of the barriers, one has to deal with the problem of the chronology of the accident. In this chronology two parts are important : the
timing and the paths or sequences. The timing is necessary in order to
know how and to what values the physical and environmental characteristic parameters of the barriers are driven. This implies use of thermal-
neutronics-dynamics-codes. The sequences are also necessary for determining the actions or interventions that can take place to avoid an
accident, and the "failures" of such acting systems and their consequences. From the initiating event, another type of systems event tree
can be drawn similar to those in WASH 1400 [3] .

For each barrier, there are systems for surveying and monitoring
the barrier. There are also systems designed for intervention where
needed for barrier protection. After an initiating event has occurred, the
loss of each barrier was considered, one after the other in sequential
order.

For each of the barriers, the different barrier sensors which
show that the initiating event is occurring are evaluated. Probabilities
of functioning and non-functioning are allocated to these sensors. Possible interventions are postulated and the delay in which they must have
been acting are calculated. If the barrier is effective, the accident
path is terminated. If the intervention has failed, the consequences
to the other barriers are determined. The other barriers are evaluated

in a similar manner.

It is then possible to draw event trees showing the timing and delays for getting the loss of the barriers and for demonstrating which
are accident sequences or paths. The probabilities of failure of the
sensors and of safeguard systems allow us to calculate the probability
of any of these paths leading to an accident. As an example of this
approach, a tree corresponding to the initiating event "Loss of electrical supply on a PWR" is developed (Figures 11,12,13).

4. An illustration : loss of external electrical supply on a PWR.
It is assumed that the initiating event occurs when the reactor is operating at full power in normal conditions. During the first phase of the
accident, considered here to be approximatively the first minute, it is
impossible by manual action to control the reactor and the safeguard
systems.

In the event of lack of external electrical sources, the reactor
coolant pumps, the feedwater pumps and the condensate pumps will stop
slowly due to the inertia in their fly-wheels. The chemical and volume
control system will stop injecting water into the reactor coolant pumps
seals and the seals will start leaking. In a few seconds after the initiating event, the signals, "lack of external electrical power", "low
primary flowrate", "low secondary flowrate" etc.. reach the control-rod
circuit breakers which command the scram.

There is a finite probability that these signals do not reach the circuit breaker (common-mode failure) or that some control rods do not fall
for some very improbable mechanical reason. During the following transient, less than ten seconds later, the pressurizer control and safety
valves will be activated ; the failures of these valves might produce a
LOCA. At the same moment the diesel generators reach their running-
speed. They may now be connected to the safeguards systems one after
the other, unless they fail (failure of one diesel or common-mode failure of two diesels). If any of the diesels start, the emergency core
cooling system (ECCS), the chemical and volume control system (CVCS)
and the containment spray system will become operational in approximatively 30 seconds after the first signal detection. Just before that
moment, 20 seconds after the initiating event, the control rods electrical generators will run down under their own inertia. Therefore, if
the scram signal has not already been effective, there will be a scram
without signal due to lack of power to the control rod units.
(Figure 11).

The only important branches in the accident sequences are those
corresponding to "No diesels", because for them the ECCS, CVCS and
containment spray system will not be operational. However the auxiliary
feedwater system might still be available for it has a steam turbine
driven pump. The main consequence of the loss of external and internal
power is therefore the failure of the CVCS to inject water into the

reactor coolant pumps seals. The second barrier will lose its integrity

through these seal ($\approx$ 20m^3/h) (Figure 12). In the case where none of

the external or internal electrical sources are recovered before an hour,

the water levels in the pressurizer and in the steam generators drop

and the auxiliary feedwater system cannot cool down the reactor. The

second barrier is completely lost and subsequently the first barrier

will be lost.

The accident sequences are not yet finished because the third

barrier is still untouched. It might still be possible to recover some

electrical power and in that case to use the containment spray system.

Otherwise the temperature will increase quickly in the containment until

the third barrier is lost (Figure 13).

5. Classification of accident sequences.   Using the above method,

one may determine all the significant accident sequences. It is possible

to order the systems with regard to their number of interventions. From

the number of failed barriers and the sequence in which they are lost,

one can evaluate the consequences and then give a classification by

order of importance. In this way, "incidents" or "abnormal occurrences"

where at least one of the barriers remains intact may also be classi-

fied.

A tree showing all the possible combinations of the failed-unfailed

states of the three barriers in the order "first to third", has $2^3 = 8$

possible combinations and is illustated below.

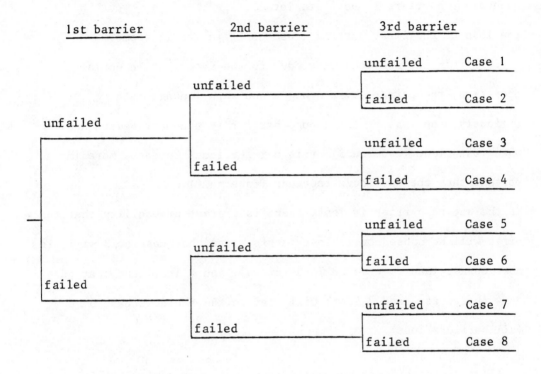

From all these 8 cases, case no. 1 is the normal one ; 6 of the
others correspond to incidents and the 8th to an accident. When setting these cases in order of importance as safety related, we take into

account that :

- the loss of the third barrier only (case no. 2) has no Safety consequences as barriers 1 and 2 are intact.

- the loss of the first barrier (case no. 5) in itself seems more
  serious than the previous case, due to the fact that in normal
  operation, the primary system has a non zero leakage rate.

- Obviously, the loss of the second barrier is the most severe and
  when lost alone (case no. 3), this barrier leads to cases more important than the two above together (case no. 6).

- If the second barrier is lost, there is a great probability that the
  first will be subsequently lost : this means that case no. 3 will, in
  most cases, lead to case no. 7 (barriers 1 and 2 lost) and that case
  no. 4 (barriers 2 and 3 lost) will lead to the accident case no. 8
  (all barriers lost).

All these considerations would then give a possible classification
with regard to safety consequences of the incidents and accidents :

2 . 5 . 6 . 3 . 7 . 4 . 8 .

$\longrightarrow$

increasing severity

6. Conclusions. In the present paper we have illustrated a
method based on the barriers to determine initiating events and their
possible chronology to an accident. The method has not here been

applied to all the operating states of a nuclear reactor. It would

be implemented in the same manner for each state, however. It can also

be extended to any type of nuclear plant.

The method proceeds in the same manner as our safety analysis :

- determination of initiating events aggressing the barriers.

- protection (design, survey, monitoring of the barriers or non-pro-

    tection in the sequences leading to potential accidents).

- interventions or actions by safety systems.

- other preventive actions of the failures of these systems which

    lead to accidents.

It is also a very useful method to show the timing in the chronol-

ogy as it indicates what modes of failure of the barriers and systems

have to be calculated. As shown in the example, in the case of an acci-

dent which takes a long time to evolve, the method takes into account

the possibility of recovering some protective functions. This might be

very important as it is then possible, when needed, to use stochastic

process models in these sequences instead of trees.

The method can be a help for quickly ordering the incidents and

safety related accidents. In the same manner, the method can also be

applied to the risk for nuclear power plant workers, for whom the third

barrier could be considered as unfailed as, in most cases, their risk

of exposure to radioactivity would most likely result from the loss of

one and/or two of the first barriers. The method would then be implemented for the combination of the loss of these two barriers only.

After the determination of initiating events and of accidents
sequences, the accident probability can be calculated by knowing the
probability of the initiating event and the failure probabilities of
protection, safeguards and vital systems. Such probabilities can be
calculated either by fault trees, using a computer code like PATREC
[6,7] or by Markov and/or stochastic processes.

The main difficulty in the method is to determine the level of
detail to which the elementary initiating events are to be considered.
The review of all initiating events might lead to an enormous number
of events if no grouping is done at some level. Another difficulty is
the calculation of the evolution of all characteristics - physical and
environmental - of the barriers. One should try at this stage to perform complete probabilistic assessments of the final values obtained
by method like "best estimates" and "sensitivity analysis". From such
analysis of initiating events and accident sequences one can go back
to the design of the plant in order to optimise the systems with respect to a given probability number and to their safety importance [8].

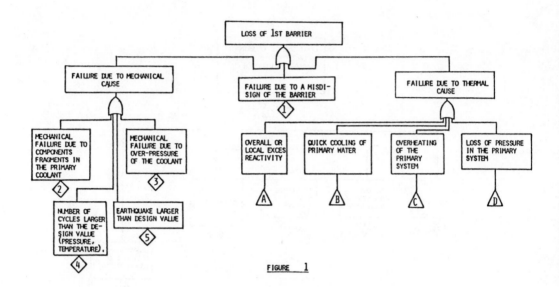

FIGURE   1

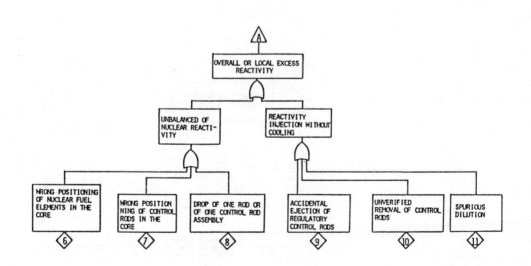

FIGURE   2

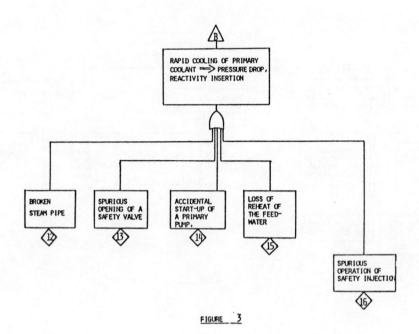

FIGURE 3

FIGURE 3

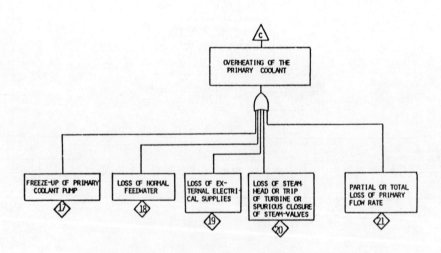

FIGURE 4

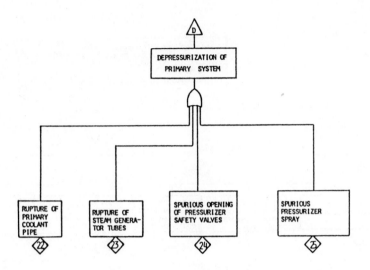

FIGURE   5

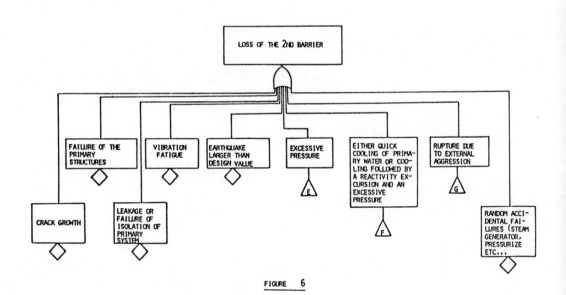

FIGURE   6

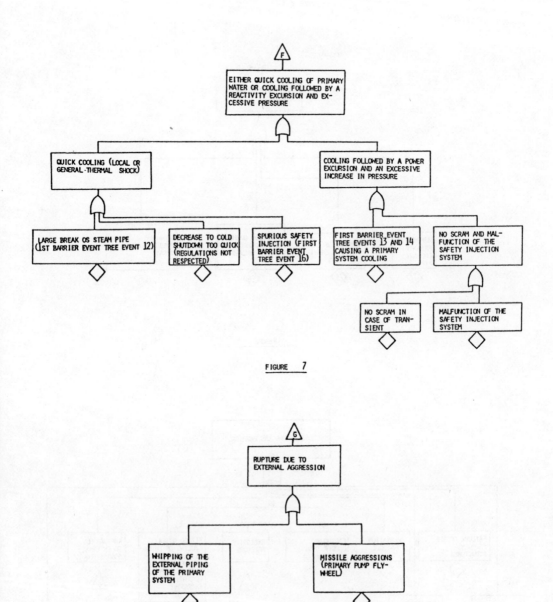

FIGURE   7

FIGURE   8

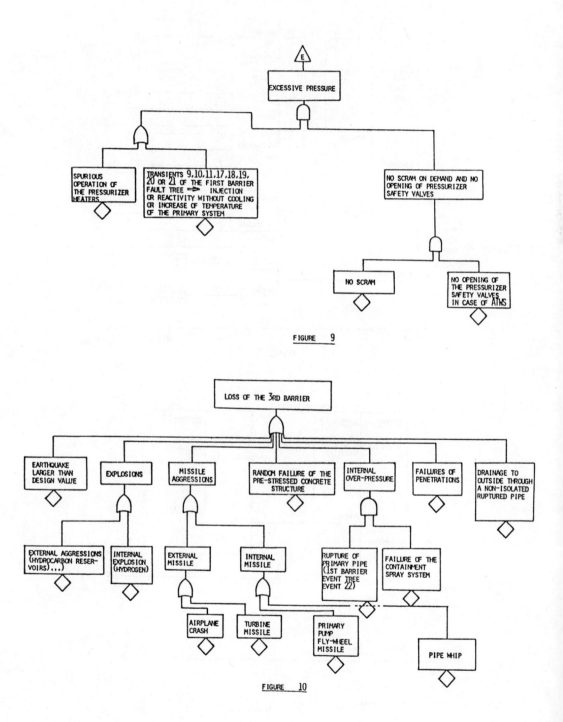

FIGURE   9

FIGURE   10

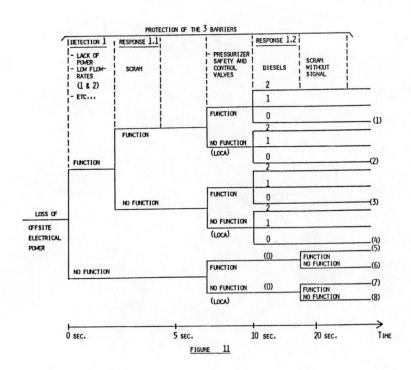

FIGURE   11

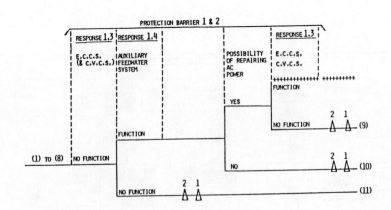

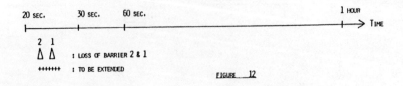

FIGURE   12

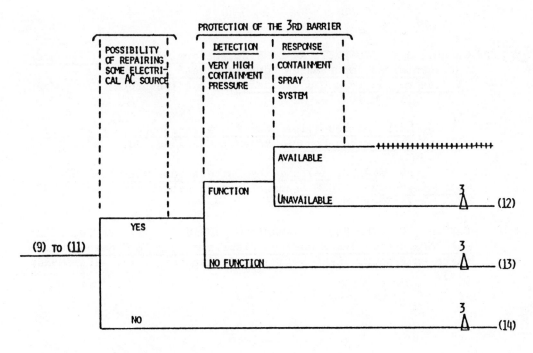

FIGURE 13

REFERENCES

[1] P. TANGUY, The impact of WASH 1400 on Reactor Safety Evaluation, ANS/ENS International Conference Washington, Nov. 1976, DSN-125-E, CEA-FRANCE.

[2] American National Standard, ANSI No 18-2-1973, (Revised in 1975).

[3] Reactor Safety Study, An Assessment of Accident Risks in US Commercial Nuclear Power Plants, WASH 1400, USAEC, 1974.

[4] J. BOURGEOIS, Aspect Technique de la Sûreté des Installations Nucléaires, Annales des Mines, Janvier 1974.

[5] B. GACHOT, J.P. SIGNORET, A. CARNINO, Détermination des Séquences Accidentelles Importantes pour la Sûreté d'une Centrale Nucléaire du type Eau Ordinaire, IAEA Innsbrück Conference, 1975, AIEA-SM 195/10.

[6] B.V. KOEN, Méthodes Nouvelles pour l'Evaluation de la Fiabilité, Reconnaissance des Formes, Rapport CEA-4368, Juin 1972.

[7] B.V. KOEN, A. CARNINO, Reliability Calculations with List Processing Technique, IEEE Transactions on Reliability, April 1974, Vol. R-23 No 1, p.43.

[8] A. CARNINO, C. COUDERT, B. GACHOT, G. JUBAULT, J.F. GREPPO, Bilan et Perspective des Etudes de Fiabilité dans le Domaine Nucléaire en France, IAEA Salzbourg Conference, 1977, AIEA-CN-36/243.

Topic 3
COMMON CAUSE FAILURE ANALYSIS

DISCUSSION BY THE EDITORS

Topic 3
Common Cause Failure Analysis

"Common cause failure (CCF)" is being rapidly accepted by reliability and safety engineers as the appropriate descriptive phrase for failure of multiple components or systems due to a single secondary event. The phrase "common mode failure (CMF)" is just as rapidly being accepted as describing CCF of redundant components or systems. The distinction has arisen due to the fact that different components, which can fail due to a single secondary cause, may have no failure mode in common.

Common mode failure analysis (CMFA) of a complex system is a more manageable problem both qualitatively (determining potential CMFs) and quantitatively (assessing the probabilistic contribution of CMF to system malfunction) than is its counterpart, CCFA. Qualitatively, those sets of components subject to CMF usually comprise a relatively small subset of those subject to CCF and hence are often much easier to determine. A number of methods exist for the qualitative portion of both CMFA and CCFA. One approach that has received recent attention by a number of investigators is to examine fault tree minimal cut sets to determine those containing component failure modes all of which could occur due to a single secondary event. Several computer codes have been developed to implement this technique. Quantitatively, more data exists concerning CMF than CCF. Moreover, the k-of-n type logic encountered in dealing with redundancy offers a convenient mathematical base from which to construct models for treating CMF quantitatively.

To avoid costly design modifications, these types of analyses should be performed as early as possible in the plant design stage and should be made an integral part of the entire design process. Component diversity can be designed into the system to reduce the potential for CMF. Another commonly employed defense against CMF is periodic testing of components during plant lifetime. However, the first paper in this section, by E. P. Epler, presents examples of CMF which illustrate that diversity and periodic testing may or may not have succeeded in preventing the failure. Situations are described in which operator perversity played a crucial role in the failure. The author suggests that operator perversity may be a limiting factor in defense against CMF since factors are involved which probably cannot be anticipated by designers.

The second paper of this section, by D. P. Wagner, C. L. Cate and J. B. Fussell, presents a methodology for circumventing some of the most serious difficulties encountered when qualitatively analyzing fault trees for CCF potential. That is, when large trees are involved, the number of minimal cut sets is usually enormous. Computer codes for obtaining minimal cut sets are usually limited to locating those of small order because of the long running time required to obtain all the minimal cut sets. Thus a CCFA requiring minimal cut sets as input is usually incomplete since minimal cut sets of higher order have been discarded. The approach presented in the second paper locates minimal cut sets of any order which could fail due to CCF. Moreover, this can be done without examining all the minimal cut sets. The ability to perform CCFA on large fault trees has important application in risk analysis. Event tree sequences can be transformed by AND gates into equivalent fault trees. The resulting fault trees can then be analyzed for CCF to determine single secondary events that could eliminate combinations of the event tree branches.

The third paper, by W. E. Vesely, develops statistical estimation techniques for CCFA. The multivariate exponential Marshall-Olkin model [1] is specialized to produce an efficient estimation technique for dealing with the sparse data usually available for quantitative CCFA. Implicit in the use of the Marshall-Olkin model are the assumptions that each failure cause has an exponential distribution for its first time of occurrence and all possible failure causes are assumed to be competing; i.e. the observed component failures are determined by the failure cause which first occurs. The Marshall-Olkin model is specialized by assuming that the component population is homogeneous in the sense that it consists of components which are similar and are subject to similar failure causes. Two cases are considered within the homogeneous model: failure rates for common causes which are (1) constant, and (2) binominal. The model for the latter case is illustrated via an example unavailability problem for a boiling water reactor scram system.

The fourth paper in this section, by B. B. Chu and D. P. Gaver, presents two types of stochastic models for systems susceptible to both random failures and CMF. The models can be used to assess reliability and availability of redundant, repairable systems. Results are also presented to illustrate the effects of CMF on system mean-time-to-failure and system survival probabilities.

REFERENCE

[1] A. W. Marshall and I. Olkin, "A Multivariate Exponential Distribution," JASA, 62 (1967), pp 30-44.

DIVERSITY AND PERIODIC TESTING IN DEFENSE AGAINST COMMON MODE FAILURE

E. P. EPLER

Abstract. The probability of failure to scram has been estimated
to be as high as 10^{-4} per reactor year; WASH 1270 has proposed that
this be improved to 10^{-7} per reactor year. To accomplish this both
the application of diverse shutdown systems and frequent periodic
testing have been proposed. An examination of a number of systemic
failures which have occurred in both development reactors and current
LWR's, leads to the conclusion that operator perversity continues to
be a dominant mechanism leading to failure. Conditions are thereby
established which cannot be anticipated by the designer, with the result that limited improvement would be gained through the application
of diverse systems.

1. Introduction. During the early years of the reactor develop-

ment program twenty [1] or more events occurred in which a critical

core assembly was made subcritical by means of a violent disassembly

into a less reactive configuration. In two of these incidents a mis-

hap during manual assembly resulted in loss of life. Procedures were

then adopted requiring that such operations be performed by remote

manipulation with the operator safely shielded. Still, even with

carefully engineered reactor control systems, failures continued.

Overall, the rate of prompt-critical excursions with violent core dis-

assembly occurred at a rate greater than one per year. For the first

1000 reactor years this was substantially greater than 10^{-2} per year.

Much has been learned as a result of these early failures. Reac-

tor shutdown systems have now reached a high state of development.

WASH 1270 [2],"Anticipated Transcients Without Scram," has estimated the probability of failure to scram to be 10^{-4} per year, which would constitute a vast improvement, a factor of 100, over the early record.

Failure to scram has never occurred in any Light Water Reactor (LWR) in commercial operation in the U.S., nor would be expected to occur. Even if the rate were as high as 10^{-4}, which is widely believed to be pessimistic, no more than two failures would be expected to occur in 20,000 reactor years, by which time LWR's would be expected to be phased out in favor of the breeder. The absence of failures has encouraged the belief that the existing failure probability is much lower than 10^{-4}; however, WASH 1270 has proposed an objective of 10^{-7} per year through the application of diversity, i.e., a secondary shutdown system using different hardware or a different principle of operation. In the belief that the estimate is grossly pessimistic, the industry has concentrated its attention on the Common Cause Failure wherein the failure of identical components would be detected by frequent periodic testing.

The improvement from 10^{-2} to the estimated 10^{-4} shutdown system failure probability is no small achievement; to improve by another factor of 1000 would be an almost unbelievable feat. Inasmuch as 10^{-4} can barely be demonstrated before LWR's are phased out, the objective of 10^{-7}, if achieved, would be without benefit of any additional failure experience. This being so, it is important that existing experience, from whatever source, be fully exploited.

In an attempt to estimate how much improvement can be reasonably expected, a number of system failures will be examined in an effort to determine how much has been learned from the early failures. Experience with test and research reactors has almost invariably been rejected as being not applicable to current LWR's. This may be justifiable insofar as the failure experience would be used to establish a rate for LWR failure to scram; however the mechanisms underlying those failures must be carefully considered to insure that they do not reappear. A number of system failures, which have occurred in research and development reactors as well as in commercial power reactors, will therefore be examined. Because of the limited number of shutdown system failures, the sample has been augmented by the inclusion of systems having characteristics essentially similar to those of the shutdown system.

2. Failure at the Input or Failure of Sensing Devices. An early paper on Common Mode Failure (CMF) [3] described three protection system failures which occurred on Oak Ridge National Laboratory (ORNL) reactors. Inasmuch as that paper contributed to the present Anticipated Transients Without Scram (ATWS) concerns, it would be worthwhile to assess the likelihood of applying a remedy which would prevent repetition of those failures.

1. At the Clinton X-Pile, the world's first operating reactor, a different means was selected for accomplishing each reactor shutdown in order to test and exercise the reactor shutdown system. In this

instance the operator chose to use the neutron thermopiles.   To insure that this feature alone accomplished the shutdown, he bypassed
all other diverse protection, and withdrew control rods to establish a
positive period. The response of the thermopiles is characteristically
quite slow so that, by the time scram occurred, the power had risen by
a factor of three. As a result, several fuel slugs were ruptured.

Operator perversity will always be with us and we have seen many
examples of it. Although this particular error will not likely be repeated, we can conclude that periodic testing would not be a remedy,
and in general, should diversity stand in the way of accomplishing a
desired result, it would be defeated as was done in this case. Neither
periodic testing nor diversity can be assured of success against operator perversity. It is noteworthy that this event was caused by an
inappropriate test procedure.

2. At the ORNL Tower Shielding Facility, the bare reactor is
hoisted to an elevation of 200 ft. where it is suspended by a cable.
Should the cable become slack and be fouled in the gears of the hoist
drum, it could be severed and drop the reactor, a truly maximum credible accident. To guard against a slack cable, a bar was mounted
parallel to the axis of the drum, which would be activated by any slack
loop in the cable. Redundant switches monitored the actuation of this
bar and would serve to stop the hoist motor.

The reactor was unshielded and the hoist house therefore was un-
inhabitable. When trouble developed, it was discovered that the cable

had indeed become slack and in flailing about with the rotation of the

hoist drum, it had dislodged the bar which was intended to monitor this

condition, and had knocked it clear across the room. By a miracle,

the cable, although completely severed, had fouled to such an extent

that the reactor was not dropped.

3. A system for reactor protection, developed at ORNL, had been

applied to a dozen or more different reactors. Although the system

had been proven in use, the availability of solid state components led

to the development of a second generation reactor protection system,

which would be used on all subsequent ORNL reactors. The new system

was tested for a year or more on existing operating reactors before

being applied to the High Flux Isotope Reactor (HFIR), where it per-

formed flawlessly. An on-line testing system made it possible to per-

form tests in a few minutes so that the entire system was tested on a

daily basis.

After the new reactor protection system had been in operation for

about two years, a similar system was installed on a fast burst reac-

tor. On the very first burst, the overload destroyed the field effect

transitors at the input to the flux amplifiers. The HFIR had in fact

been unprotected for two years against a short period power excursion,

although the system had been expressly designed to provide that pro-

tection.

In seeking a remedy it proved to be difficult to find components

which would meet requirements and at the same time be immune to the overload which would be caused by a fast transient. The problem was compounded by the rapid obsolescence of transistors which made it necessary to solve the same problem repeatedly. Here again is an example of the design error wherein the protection is destroyed by the excursion. Again, armed with the knowledge of the mechanism underlying the failure it would be possible to devise a diverse system, but not without some difficulty. As an alternate to neutron flux, prompt gamma flux would be capable of the required fast response; however it would be necessary to revert to the obsolescent vacuum tube amplifiers to provide diverse protection to insure against the overload problem. Without this prior knowledge it would not be likely that diversity would succeed. With prior knowledge, the design error would not have been committed.

This warning should be heeded. Improved components and techniques will undoubtedly be incorporated to update older reactors.

In the preceding three examples, periodic testing would be completely ineffective and diversity, which in fact did exist in one instance, would be of questionable value. In the example to follow the conclusion would be much the same.

4. The protection system used on the Heat Transfer Reactor Experiment 3 (HTRE3) [4] at the National Reactor Testing Station had been used on two earlier reactors and would be considered to be well qualified. In Nov. 1958 the reactor was shut down by a reactivity loss of more than 2% Δ k/k due to the redistribution of fuel resulting from

melting. The event occurred as the reactor power was being escalated,

each day to a higher level, in an approach to rated conditions. On

this day the reactor power as indicated by instruments had reached ap-

proximately the desired level, then unaccountably decreased, when in

fact the power was rapidly increasing. Several factors contributed to

the failure: (a) a resistance-capacitance noise filter had at some

time been inserted in each of the ionization chamber power supply cir-

cuits, which limited the current to a value less than that required to

produce a scram, (b) the power supply voltage was adjustable and was

set at a low value and (c) the same ionization chambers and amplifiers

were being used for both control and protection, which was not an ac-

ceptable practice. A failure would assure that the controller would

cause rod withdrawal and at the same time defeat the ability to protect.

The practice of using the same devices for both protection and

control is today allowed by IEEE Standard 279-71, providing "the re-

maining redundant protection channels shall be capable of providing the

protective action, even when degraded by a second random failure." The

HTRE system exceeded this requirement for protection against random

failure alone, and provided protection against both random and common

mode failure in the form of a diverse temperature measurement. On this

occasion however, the reactor was operating in an unusual mode, at re-

duced power and reduced coolant flow. The thermocouples were located

at the coolant exit, which would be the normal location of the highest

temperature; however with no coolant flow the hot spot would be at the

center of the core. As a result the diverse temperature protection
was ineffective.

The preceding four failures which occurred in early reactors would
not have been prevented by periodic testing inasmuch as no component
had failed before the event. In two instances, failure occurred in
spite of existing diversity and it is doubtful, lacking prior knowledge, whether diversity would have been effective in the remaining
two. In each case the failure occurred because of conditions not anticipated or beyond the control of the designer.

It is noteworthy that the four preceding failures occurred at the
input to the system, relating to the instrumentation or sensing devices; the following examples would fall into the category of the
system of logic which stands between the instrumentation and the actu-
ators.

3. Logic System Failures.

5. At the Kahl reactor, an American built Boiling Water Reactor
(BWR) in Germany, all relays became immobilized because of a manufacturing error wherein a chromate primer was insufficiently cured. The
failure was discovered by periodic test and did not result in failure
to scram on demand. This is the only example of CMF of components discoverable by testing which has come to light, and it would appear that
either hardware or functional diversity would also be an effective
remedy. The ease with which the CMF of components is discovered by

testing or even casual surveillance readily accounts for the fact
that component failure either random or common mode, has never been a
major contributor to shutdown system failure.

6.   The 860 Mw(e) Hanford N Reactor [5], for several years the
world's largest in electric power generation, failed to scram on demand and although component failure was a contributing factor, the
major factor was design error. The reactor has 87 rods, each with a
selector switch which establishes the primary function of the rod. The
switch positions are (1) Manual, (2) Power Setback, (3) Safety, (4)
Withdrawal, and (5) Off. The positions WITHDRAWAL and OFF permit a rod
which is in the withdrawn position for maintenance purposes to be immobilized and its insertion into the core thereby prevented. The im-
mobilization is accomplished by switching the rod from the scram bus to
a holding bus. The switching is aided by diodes which serve to isolate
the holding bus from the scram bus. The four diodes for each rod are
mounted as a package on an octal socket, and as has been demonstrated,
could all be caused to fail by momentarily creating, with a screw-
driver, an electrical short in a rod safety circuit. The failure of
the package of diodes would properly be considered as the failure of a
single component rather than the CMF of four. These diodes were not
included in the periodic tests, in fact, following the incident, additional diodes were found to be in a failed condition. The failure to
scram occurred when the OFF position had been selected for a rod having
a failed set of diodes which then caused all other rods to be connected

to the HOLD bus and thereby immobilized. The major factor would not be the component failure, but a design error wherein the single failure criterion was violated 87 times. Each rod contained a component which by its failure could cause the failure of all rods.

In this example, periodic testing, as performed, was again ineffective; however, diversity in the form of poison balls was effective in shutting down the reactor.

7. At Zion I [6] a safety injection system failed on demand although it responded to periodic tests. It was found that the wiring shown on one drawing did not agree with three earlier drawings or with the actual rack wiring. This single drawing was returned to the vendor where it was revised and reissued as an as-built drawing. It turned out that this drawing had been correct but had been changed to agree with the incorrect drawings. The system, as checked, did agree with the incorrect as-built drawings. Periodic tests were not performed from input to output in a single test, but were performed by parts as is customary for logic systems. Thus the tests failed to disclose that the parts were incorrectly connected. This is an unusually complex system of logic, although essentially identical to shutdown system logic. It is significant, however, that it baffled the vendor, the architect engineer, and the operators. This same error was later found in another reactor system at another plant.

Systems employing coincidence are almost invariably tested by parts as was done in this case. Inasmuch as a signal inserted at the

input would be unable to pass through the coincidence logic network,
it is necessary to insert a second signal at that point to test the
remainder of the system. The Zion event illustrates how this practice
can assure that no component has failed, but not necessarily assure
that the components are correctly connected. A state of the art system
[7] which avoids this problem has elsewhere been developed and proven
in use. This system can be tested on line in a single operation which
exercises the sensor, the logic, and the actuation device.

In this systemic failure no component had failed and the condition
was not discovered by periodic tests. However, diversity either in
hardware or in function would almost certainly have succeeded.

8. At Oyster Creek [8] in 1972, after four years of operation, a
design error was disclosed by the separate failures of two systems.
In one case this destroyed the capability to isolate secondary containment when this capability was repaired. In the second case the
racking out of a breaker for one safety injection pump disabled not
only the pump removed from service, but its redundant counterpart as
well. In the case of the safety injection system, the error resulted
from an arrangement intended to prevent both pumps from running at the
same time; to accomplish this an auxiliary contact in the movable portion of the circuit breaker was used. As a result of the failure,
reactor shutdown could not be effected by either of the redundant
trains either from the control room or by means of the local switch.
This, it must be observed, is a design error and in addition, a

violation of channel independence, which somehow escaped detection
during review and four years of operation and testing. The safety injection system is the currently required diverse shutdown system. The
amount of effort expended to assure the reliability of this system is
not impressive.

The first group of four failures related to the input or the instrument portion of the system, and came about because of conditions
not anticipated or not controllable by the designer. For that reason
a diverse system could not be counted on to succeed. The second group
of failures occurred at the logic level, where diversity would likely
be effective in all cases. The two systemic failures to follow would
challenge any effort to anticipate and prevent on the part of the designer.

4.   Operator Perversity.

9.   The EBRI [I] had the distinction of being the first nuclear
plant to deliver electrical energy to a public utility network. In
conducting a test, in 1955, to observe the kinetic behaviour of the
reactor, a power excursion was terminated by actuating the wrong switch.
The resulting delay in shutting down caused the core to be melted.
This is another example of operator perversity which could not be mitigated by periodic testing nor anticipated by the designer so as to be
prevented by the application of diversity. It would be expected, in
current reactors, that it would be unlikely that a similar operation

would be conducted, depending solely on operator response to terminate the test. The following however is one example of current procedures.

In May 1972 the water level in Quad Cities Unit 2 [9] was to be lowered for a final visual inspection before putting the vessel head in place. The use of jumpers was authorized to prevent automatic startup of emergency systems. The jumpers however were installed on Unit I in error. As a result Unit 2 experienced startup of its emergency systems and Unit I became unprotected.

The improper jumper placement was made after the required two independent checks. As a measure of the difficulty in finding a way to prevent repetition of this error, the Safety Committee could do no more than to recommend a third independent check.

Redundancy has been used to minimize the effect of component failure and in reactor shutdown systems, this has been quite effective as component failure has not contributed to systemic failure. The use of redundancy however introduced a new and troublesome problem in that the wrong channel was often serviced in error. We now have a new and greater problem; instead of the wrong channel, the wrong generating unit is serviced and not one, but two units are in trouble.

10. At the Zion [10] station on April 20, 1976 the turbine lube oil cooling system was turned off, by mistake, on the operating unit instead of the shutdown unit as intended. A typewriter ribbon was being changed at the time so that the mistake was not annunciated, and as a result the main generating unit bearings were lost.

Operator perversity which was an important factor in the preceding examples of reactor protection system failure but is also an important contributor to the failure of more conventional systems such as the turbine bearing lubricating systems. Loss of these bearings is a disaster to a central power station and great pains are taken in prevention, including the use of diversity.

At HB Robinson 2 [11], a Pressurized Water Reactor (PWR), the 50 HP d.c. emergency lube oil system was placed on test, the test to be terminated in two hours to avoid depleting the charge of the station battery. The test was not terminated and after four hours the battery charge was depleted to the point that the d.c. lube oil pump became inoperative and for lack of d.c. the circuit breakers could not be operated to restart the normal a.c. powered system. Also for lack of d.c. at the scram breakers, a fail-safe scram occurred, thereby shutting down the generating unit. As the speed of rotation slowed, the shaft-driven oil pump was unable to deliver sufficient oil and the main unit bearings were lost. In this instance of operator perversity, three "independent" and diverse systems failed. It is paradoxical that it was the performance of a periodic test that initiated the series of events.

Also at the 950 Mw(e) Bull Run fossil plant at Oak Ridge, the main unit bearings were lost. A serious fire threatened the integrity of the building structure and was thought to be caused by a jet of flame from a broken oil line. The fire occurred at shift change time when

plenty of help was available. It turned out that the fire was caused,
not from an oil line, but by oil flowing from a sump over a leaking air
duct. No possible explanation for the loss of the turbine bearings has
been found other than the supposition that, with two shifts on hand,
someone went around closing valves in an effort to stop the oil leak.

It may well be that so long as bearings must be lubricated by
means of systems of pumps and valves which are accessible to people,
occasional failures must be accepted.

Operator perversity may take a variety of forms ranging from simple forgetfulness to the willful wrong-headed act as in the SLI incident where an operator, with considerable effort, manually withdrew a
control rod and, it is believed, by intent, destroyed the reactor.
Willful wrong-headedness can however be destructive without intent.

At the EDF St. Laurent des Eaux [1], a 500 Mw(e) natural uranium
gas cooled reactor, fuel is charged on-line by a computer controlled
charging machine. On Oct. 17, 1969 the computer repeatedly refused to
carry out the operator's instructions, and in frustration he turned off
the computer and carried out the operation using manual control. Instead of charging a fuel element, he charged a flow restrictor, an
error the computer had refused to commit. As a result, substantial
fuel melting occurred.

At another place, which shall remain nameless, on orders from the
front office and over the objections of reactor control engineers, a
temporary line was run from the reactor building to supply instrument

air to an adjacent building. The air pressure was raised from 65 to
100 lb/sq in, to accomodate the parasitic load, which the compressor
was perfectly capable of delivering, but the air drier was not. As a
result all pneumatic instruments in the reactor systems became loaded
with finely divided alumina.

A pattern is clearly emerging; operator perversity will always be
with us and will not diminish as systems become more complex. It can
be believed that this will ultimately establish a limit on attainable
system failure probability.

5. Perversity the Limiting Factor. The Reactor Safety Study has
found the probability of core melt to be 1 in 20,000 reactor years,
which is about the same as the estimated failure to scram. By coincidence, 20,000 reactor years of operation will see the end of LWR's.
If failure experience exceeds this predicted rate, we can be sure that
remedies will be demanded. If however the failure probability is significantly better than predicted, the accomplishment can never be
demonstrated. This makes 1 in 20,000 a number we may have to live
with.

The failure rate of four-engine commercial aircraft is well known
and nothing more can be done to alter it. The failure of the essential
component, the jet engine, is about 10^{-5} per flight, and the independent
failure of three, which would force the plane down, would be 10^{-15}.
The successive independent failure of three engines should never bring

a plane down, even on an overseas crossing. Failures come, not from
components, but are caused by a navigation error wherein a plane is
regularly flown into a mountain, at a rate of about 10^{-5} per flight.
Here 10^{-5} is the magic number. Although a jet plane is flown into a
mountain once per month and invariably results in the death of all on
board, nothing more can be done to the hardware to alter the situation,
so we live with it. With 1000 flights per plane per year, this failure
rate would be no better than 10^{-2} per year, about the same as the original, wholly unacceptable, rate of reactor failure. In reactor systems, unlike the jet plane, we will not have the benefit of failure
experience either to guide us in making improvements, or to convince
us that further improvements are not to be realized.

We can however draw some conclusions regarding failure mechanisms
which would likely become dominant. The four failures attributed to
the logic system would likely be remedied by the application of diversity. The Zion I wiring error would not likely occur in diverse systems. The Oyster Creek logic error should not be permitted to recur.
A diverse system indeed was successful in shutting down the Hanford N,
and at Kahl, either diversity or periodic testing would succeed in
minimizing failures to the logic system.

Contrary to widespread apprehensions, no failure to scram has ever
occurred because of sticking rods or rod drive malfunction, and the
Kahl event is the only important failure attributable to component
failure. We can conclude that in these areas as in the logic system,

matters are well in hand or can readily be remedied by the application

of either hardware or functional diversity. This however is not the

case for failures at the input to the shutdown system, which were

brought about by conditions unanticipated by the designer and which

would be much more likely to affect both diverse systems.

Periodic testing has adequately prevented component failure from

becoming a contributor to systemic failure. In several instances sys-

temic failure would have been prevented by performing the right test.

However, had the right test been performed just once, the fundamental

defect would have been disclosed and corrected, and the systemic fail-

ure would have been averted. It is noteworthy that in several instances

the failure was initiated by improper testing.

Diversity in instrumentation has always been employed, yet in

spite of this, failures have occurred. The four failures of early re-

actors, illustrating failure at the input or instrument level, were of

two kinds. Those at the Tower and HFIR were attributable to design

error and we can be confident that, as protection systems continue to

evolve, designers will be able to minimize errors in successive de-

signs. The HTRE failure however resulted from uncoordinated changes to

equipment during operation and the X-Pile failure, from operator per-

versity. These factors create conditions that cannot be anticipated by

the designer.

The several instances of error causing main unit bearing failure,

indicate that operator perversity has always been with us, and now in addition to redundant trains to be serviced in error, we have multiple, identical generating units. It is conceivable that a crew could pull switches and proceed to dismantle the protection systems on the wrong reactor.

Whereas system designs are created in closely coordinated groups, and centralized in four major vendors, the operating organizations are many and widely dispersed. This creates a problem which has been troublesome in early reactors, but has not yet fully made its appearance in commercial operations. This is illustrated by the HTRE failure.

It is likely that little more can be accomplished through the application of diversity and like the navigation error which causes the commercial jet plane to be flown into a mountain, and the human factor which has repeatedly been the cause of ruined turbine bearings, operator perversity may prove to be the limiting factor for reactor protection systems.

In the ten examples cited, several failures occurred in spite of existing diversity and in other instances diversity would not be assured of success. It might be believed, should a shutdown system fail for reasons not anticipated by the designer, that a diverse system would succeed in 9 of 10 cases. However to improve the failure probability from 10^{-4} to 10^{-7}, success must be achieved in 999 cases in 1000. This is not believable in the light of one third century of experience with reactor shutdown systems.

It is also noteworthy that of the ten Common Mode Failures only one, the Kahl event, was a Common Cause Failure discoverable by periodic testing.

REFERENCES

[1] W.R. STRATTON, A Review of Criticality Accidents, Progress in Nuclear Energy, Series IV, Vol. 3, Technology Engineering and Safety, Pergamon Press, Inc., New York, 1960.

[2] Regulatory Staff USAEC, Anticipated Transients Without Scram, WASH 1270, Sept. 1973.

[3] E.P. EPLER, Common Mode Failure Considerations in the Design of Systems for Protection and Control, Nuclear Safety 10 (1), Jan.-Feb., 1969.

[4] E.P. EPLER, HTRE3 Excursion, Nuclear Safety 1 (2), 57-59, Dec. 1959.

[5] GEORGE R. GALLAGHER, Failure of the Hanford N Primary Scram System, Nuclear Safety 12 (6), 608-614, Nov.-Dec., 1971.

[6] Reactor Operating Experience. Wiring Error in Safety Injection System.

[7] W.D. BROWN, S.H. HANAUER and R.E. WINTENBERG, A Coincidence Electromagnet for Reactor Safety, Trans. ANS. 7 (1), 135, June, 1964.

[8] Jersey Central Power and Light Co., Letter to Division of Reactor Licensing, April 20, 1972.

[9] Commonwealth Edison Co., Letter to Directorate of Licensing and Regulation, May 4, 1972.

[10] S.M. ZIVI, Personal Communication to E.P. Epler.

[11] Carolina Power and Light Co., Letter to Division of Reactor Licensing, April 7, 1971.

[12] Electricite de France, St. Laurent Reactor, Fuel Meltdown Incident, Oct. 17, 1969.

COMMON CAUSE FAILURE ANALYSIS METHODOLOGY
FOR COMPLEX SYSTEMS*

D. P. WAGNER, C. L. CATE AND J. B. FUSSELL

Abstract. Common cause failure analysis, also called common mode failure analysis, is an integral part of a complex system reliability analysis. This paper extends existing methods of computer aided common cause failure analysis by allowing analysis of the complex systems often encountered in practice. The methods presented here aid in identifying potential common cause failures and also address quantitative common cause failure analysis.

1.0. INTRODUCTION.

Common cause failure analysis, also called common mode failure analysis, is an integral part of a complete system safety and reliability analysis. A common cause failure is any occurrence or condition that results in multiple component failures. Epler[1] reported that "the common mode failure may be dominant by as much as a factor of 10^5. . . ." Taylor[2] reported on the frequency of common cause failures in the U.S. power reactor industry. "Of 379 component failures or groups of failures arising from independent causes, 78 involved common mode failure of two or more components." The reported dominance and frequency of common cause failure events illustrates the need for an effective approach to common cause failure analysis.

---

*This work was supported by Westinghouse Electric Corporation under contract number 54-7WM-227400-S with The University of Tennessee.

A structured qualitative approach to common cause failure analysis has been developed in concert at the Idaho National Engineering Laboratory (INEL) and the University of Tennessee.[3,4] This existing methodology scopes the analysis to consider only those common cause failures that can result in system failure. The approach is designed for computer aided analysis. The extension of this existing methodology to highly complex systems and the introduction of quantitative considerations are the subjects of this paper.

The extensions to the qualitative methodology presented are necessary because the existing methodology for computer aided common cause failure analysis requires that all the system hardware minimal cut sets (see Section 2.1) be determined. While developing a logic model that reflects all appropriate hardware malfunctions is generally feasible, determining all minimal cut sets (including the highest orders) from such a logic model for a complex system is generally not possible because of computer or other limitations. The methodology presented here does not require all such minimal cut sets be determined although the results are the same as if these cut sets were determined.

The method for including the results of the qualitative common cause failure analysis in the quantitative system safety and reliability analysis for complex systems is also discussed. The results of the qualitative common cause failure analysis supply the analyst with a portion of the required information for the quantitative analysis. The information is added as new one event cut sets to the list of hardware minimal cut sets to be used in the quantitative analysis. These new one event cut sets will represent the dependencies within the system brought about by the common cause candidates. Computer programs exist for use in quantitative system safety and reliability analyses.[5,6]

2.0. CONCEPTS OF COMMON CAUSE FAILURE ANALYSIS.

2.1. Minimal Cut Sets.

The fault tree analysis[7] technique is used as the basis for a structured approach to common cause failure analysis. Fault tree analysis is a technique of system reliability methodology. A failure logic model, called a fault tree, is developed for the system failure of interest, called the TOP event. The TOP event is usually a highly undesirable system failure situation. In fault tree analysis, the modes of system failure are known as minimal cut sets.[8] A minimal cut set is a smallest group of basic component failures, called basic events, that are collectively sufficient to cause the TOP event to occur. The occurrence of each basic event in the minimal cut set is necessary if the occurrence of the TOP event is a result of the failure of the minimal cut set in question. Computer programs for determining minimal cut sets from existing logic models are available.[5,9,10]

2.2. Common Cause Definitions.

Strictly, common cause failure is any occurrence or condition that

results in multiple component failures. For this paper the term
secondary failure will be used to identify the categories of component
malfunction pertinent to common cause failure analysis. A significant
common cause event[11] is a cause of secondary failure that is common
to all basic events in one or more hardware minimal cut sets. If a
minimal cut set has a significant common cause event and, if all the
components represented by the basic events in that minimal cut set
share a "common physical location," that minimal cut set is called a
common cause candidate. Components share a common physical location if
no barriers that are capable of insulating the components from the
failure cause are present. Components may share a common physical
location irrespective of the physical distance separating them.
Classification of causes of secondary failure and details on determining common locations are discussed in Sections 2.2.1 and 2.2.2.

In some cases a significant common cause event may not be
required for identifying a common cause candidate. The common cause
candidate may be identified on the basis of a "special condition." A
special condition is a condition that closely links all the basic
events in the minimal cut set. The special condition gives rise to
increased failure probability of a minimal cut set in such a way that
barriers are not applicable. An example of a special condition is all
the components implied by the basic events in the minimal cut set being
produced by the same manufacturer. The minimal cut set can then be
identified as a common cause candidate without concern for physical

location. Also, the special conditions identify common cause

candidates resulting from dependencies among the components of the

minimal cut set, without requiring a significant common cause event and

common physical location. Suggested special conditions are listed in

Table 1.

2.2.1. Classification of Causes of Secondary Failure. A large

number of secondary failure causes can be found that would cause com-

ponent failures. Therefore the analyst is directed toward the generic

cause of component failure rather than the specific event that results

in the component failure. For example, the events "water hammer" and

"pipe whip" can be represented by the cause "impact." These causes

are termed generic causes. To further aid the analyst, three broad

categories of generic causes have been suggested:[12] mechanical-

thermal, electrical-radiation, and chemical-miscellaneous. Table 2[3]

shows potential failure causes of a mechanical-thermal nature.

Suggested lists of failure causes for all categories can be found in

Reference 3.

2.2.2. Details on Defining Common Physical Location. For common

cause failure analysis, the geographic area containing the system

should be divided to indicate the barriers against specific generic

causes. The domain of the generic cause is the geographic area

divided to indicate these barriers. Usually there are many natural

barriers in a building, for example, walls and cabinets. However, a

barrier for one generic cause may not be a barrier for another generic
cause. For example, a liquid spill may be confined to one room, while
a fire may affect every room in the building.

Consider the floor plan shown in Figure 1. This floor plan must
represent the finest resolution of areas described by all the generic
cause domains. All the barriers indicated in Figure 1 may not be
applicable for a specific failure cause. For example, the walls
surrounding room 101 and cabinet A in room 103 are thermally insulated.
While these would represent barriers against thermal causes, they may
not represent barriers against vibration. The domains for thermal
events are:

DOMAIN 1 - 101

DOMAIN 2 - 103A

DOMAIN 3 - 102, 103, 103B, 104, 199

while the domain for vibration events is:

DOMAIN 1 - 101, 102, 103, 103A, 103B, 104, 199

Details on constructing domains are found in Reference 3.

2.2.3. Quantitative Considerations for Common Cause Failure

Analysis. Most generic causes of secondary failure can be assigned a
time dependent occurrence rate in each domain. Often these occurrence
rates are identical to hardware failures (usually hardware external to
the system being analyzed) that result in the generic failure cause.
The occurrence rate for the generic cause temperature could correspond
to the rate of cooling system failure. Unlike component reliability

characteristics, the common cause candidates reliability characteristics are not completely described by the time dependent occurrence rate of the failure cause. Each occurrence of a cause of secondary failure will not cause system failure. Common cause candidates are ranked by their sensitivity to each cause of secondary failure.[3,4,11] The sensitivity ranking is used to determine the fraction of the occurrences of a cause of secondary failure that actually cause system failure. The failure rate of the common cause candidate is the occurrence rate of the cause of secondary failure weighted by the fraction of system failures which result from the cause of secondary failure. The common cause candidate sensitivity rank is determined by the sensitivity rank of the least sensitive component since all components in a minimal cut set must fail for system failure.

A time dependent repair rate can also be determined for the new cut set that results from the common cause event. The repair of the cause of secondary failure must be considered in addition to the repair of the components implied by the common cause candidate.

Some special conditions cannot be represented by an occurrence rate. Special conditions such as manufacturer and similar parts are examples. These links among all the events in a hardware minimal cut set increase the failure probability of that minimal cut set. They are usually treated as having a constant probability of contributing to system failure. The reputation of the manufacturer or the similar part in common influences the probabilistic effects of the common cause candidate on the system reliability characteristics.

2.3. Current Approach to Computer Aided Common Cause Failure Analysis.

A current approach to computer aided common cause failure analysis consists of five basic steps:

1. Determine the list of hardware minimal cut sets for the system being analyzed.

2. Obtain the qualitative failure characteristics for each basic event in the minimal cut sets.

3. Search the complete list of hardware minimal cut sets for common cause candidates using a computer.

4. Include the probabilistic effects of the common cause candidates in the quantitative system analysis.

5. Form conclusions and recommendations based on the results of the qualitative and quantitative analyses.

This approach makes maximum use of computer aided analysis. Computer programs exist for determining the minimal cut sets from a fault tree.[5,9] Two programs have been developed for common cause failure analysis using minimal cut sets as input; COMCAN,[3] developed at INEL, and BACFIRE,[4] developed at The University of Tennessee. These programs also interface with the existing computer programs used in the quantitative system analysis.[5,6]

For complex systems, determining the list of minimal cut sets becomes a difficult and often an impossible task. Computer time and storage capacity become prohibitive. To overcome this difficulty a new procedure for common cause failure analysis is defined.

3.0. A PROCEDURE FOR AUTOMATED COMMON CAUSE FAILURE ANALYSIS OF COMPLEX SYSTEMS.

3.1. Concepts of the New Procedure.

A problem in computer aided common cause failure analysis of complex systems is that all minimal cut sets must be determined for the TOP event. This problem is overcome by dissecting the fault tree and determining minimal cut sets for individual branches and then synthesizing common cause failure analysis results for the TOP event. The minimal cut sets found for the individual branches of the fault tree are termed intermediate minimal cut sets. Common cause failure analysis of the intermediate minimal cut sets yields intermediate common cause candidates. A dummy event must then be defined to describe all the intermediate common cause candidates for a particular branch of the fault tree. The dummy event contains all the information necessary for analysis of the next level of the fault tree. At the completion of the analysis, the dummy events are expanded and common cause candidates for the TOP event are constructed. The new procedure is a step-by-step analysis, advancing from the bottom to the top of the fault tree through its individual branches. This procedure is equivalent to an analysis using all the minimal cut sets for the TOP event.

An example problem is defined to illustrate the procedure. This example problem is used extensively in Sections 3.2 through 3.7, which are a detailed discussion of the procedure. Figure 2 is the fault

tree for the example problem. The qualitative failure characteristics

for the basic events in Figure 2 are listed in Table 3. The sensi-

tivity rank of the component to each qualitative failure characteristic

is given below the susceptibility. Figure 3 shows an example floor

plan. The domains of the generic failure causes are listed in Table 4.

3.2. Dissecting the Fault Tree.

There are two guidelines in deciding the point of dissection in

the fault tree:

1. The number of Boolean Indicated Cut Sets (BICS)[5] of the

    dissected branch must be small enough to allow determination

    of the intermediate minimal cut sets.

2. The intermediate minimal cut sets preferably should contain

    two or more basic events.

The number of BICS in a fault tree is easily found.[8] By keeping

the number of BICS low, only minimum computer requirements are

necessary. The maximum length of the BICS can be determined in a

similar manner.

By definition, a common cause failure involves the development of

more than one component malfunction. By obtaining two-event or larger

intermediate minimal cut sets and analyzing them for common cause

failure, all intermediate minimal cut sets which are not candidates for

common cause failure may be eliminated. This further reduces the

amount of information which must be carried to the next step of the analysis. The dissected fault tree for the example problem is shown in Figure 4.

3.3. Intermediate Analyses.

Once the fault tree has been properly dissected, the intermediate minimal cut sets may be found by conventional means. Treating the dummy events as ordinary basic events, standard computer programs will correctly determine the intermediate minimal cut sets for all segments of the dissected fault tree. Table 5 lists the intermediate minimal cut sets for the three segments of the dissected fault tree of the example problem.

The intermediate common cause failure analysis must proceed in a stepwise fashion from the lower levels of the dissected fault tree to the top level of the dissected model. The lowest segments of the dissected fault tree contain no dummy events. The existing approach to common cause failure analysis[11] may be applied directly to the lowest segments of the fault tree to obtain the intermediate common cause candidates. The intermediate common cause candidates contain all the information necessary to the next level of the analysis and are represented by the dummy events.

In the example problem segments C and L are the lowest segments of the dissected fault tree. Segments C and L contain no dummy events and may be analyzed directly. The intermediate common cause candidates found for segments C and L are listed in Table 6.

3.4. The Dummy Event.

Wherever a branch of the logic model is dissected, a dummy event must be added to the remaining portion of the original fault tree. The dummy event is added to the fault tree at the point of dissection. The dummy event represents the dissected branch of the fault tree through-out the remainder of the analysis. For the example problem, the necessary dummy events have been identified as C and L, and are shown in the dissected fault tree in Figure 4.

All of the special conditions identified in the intermediate common cause candidates are assigned to the dummy event. The dummy event is also assigned all of the generic cause susceptibilities of the intermediate common cause candidates.

Extreme care must be exercised in defining the physical location of the dummy event. The location assigned to the dummy event must accurately represent each intermediate common cause candidate in the proper area of its identified generic cause domain. Three methods have been defined for determining the physical location of the dummy event. From the information in Table 6 dummy events C and L can be defined as they appear in Table 7.

3.5. Constructing Common Cause Candidates for the Top Event*

Once the top segment of the dissected fault tree has been analyzed for intermediate common cause candidates, the dummy events in the

*Actually a multi-tiered analysis is possible (rather than the two-tiered analysis described) by cascading the procedures given here.

results must be expanded to obtain the common cause candidates for the
TOP event. Expansion of the dummy events produces common cause candidates for the TOP event in terms of the minimal cut sets for the
original fault tree. This maintains the original resolution of the
analysis.

Constructing common cause candidates for the TOP event consists of
three steps:

1. Expanding the dummy event to the intermediate common cause
    candidates which it represents.

2. Determining potential common cause candidates for the TOP
    event.

3. Obtaining the common cause candidates for the TOP event.

Consider the intermediate common cause candidates found for the
TOP event in the example problem, listed in Table 8. Dummy events C
and L must be expanded to represent the intermediate common cause
candidates found in their segments of the logic model. The first
intermediate common cause candidate for the TOP event, (1, C), contains
the dummy event C. Referring to Table 6, C can be expanded, resulting
in

$$
1, C \rightarrow
\begin{array}{l}
1, 1, 4 \\
1, 3, 4, 12 \\
1, 1, 3, 12
\end{array}
$$

This must be done for each occurrence of the dummy event in the intermediate common cause candidate list.

The expanded intermediate common cause candidates, known as

potential common cause candidates, are analyzed to eliminate redundancies and supersets. Any repeated events within a potential common
cause candidate may be eliminated. For example, the (1, 1, 4) expansion
becomes (1, 4) and (1, 1, 3, 12) becomes (1, 3, 12). Also, any potential common cause candidate which is a superset of another potential
common cause candidate is eliminated. The cut set (1, 3, 4, 12) is a
superset of (1, 3, 12) and is discarded. All common qualitative failure characteristics of (1, 3, 4, 12) are contained in the subset (1, 3,
12) since the characteristics must be shared by each basic event in the
common cause candidate. This establishes the potential common cause
candidates for the TOP event in terms of the minimal cut sets for the
original fault tree.

Next, the qualitative failure characteristics of the basic events
within the potential common cause candidates are compared. Consider
the remaining cut sets of the (1, C) expansion. The qualitative failure characteristics of (1, 4) and (1, 3, 12) are compared within the
candidate for common characteristics. Intermediate common cause candidate (1, C) is identified for generic cause susceptibility S and
special condition 01. The cut set (1, 4) is identified as potential
common cause candidate because the basic events share generic cause
susceptibility S. The physical locations of the basic events must be
compared to determine if they are in a common domain for the potential
common generic cause. If no common domain exists, the potential common
cause candidate is discarded. (1, 4) is in the same domain for generic

cause S and is retained as a common cause candidate for the TOP event.
The (1, 3, 12) expansion of C shares special condition O1 and is also
retained.

Once the construction process is complete, the complete list of
common cause candidates for the TOP event is obtained. The complete
list of common cause candidates for the TOP event of the example
problem is found in Table 9.

Figure 5 is a flowchart identifying the steps in the procedure.

3.6. Quantification of Common Cause Failure.

The information needed to determine the input to a quantitative
evaluation of common cause failures is the qualitative failure
characteristics of the common cause candidates. New one event cut sets
are generated for each generic cause/domain pair, and for each special
condition.

3.6.1. Quantification of Generic Causes. For the example problem,
Impact was identified as a susceptibility for two cut sets, 1 and 2
(Table 9). The generic cause, Impact, is in the same domain for both
common cause candidates. One new one event cut set is added to the
list of hardware minimal cut sets for this potential cause of failure.
The failure rate for this event will be the occurrence rate for Impact
in domain 2 IMP weighted by the probability of failure of a common
cause candidate, given the potential cause of failure. The probability
of failure is determined using the sensitivity ranking of the

susceptibility to the common cause candidate. When there is more than

one common cause candidate, the probability of system failure, given

the potential cause of failure, is determined from the maximum common

cause candidate's sensitivity ranking. For common cause candidates 1

and 2, the maximum sensitivity ranking for Impact is 4. The failure

rate for the one event cut set for Impact in domain 2 IMP is 10^{-6}/hr.,

the occurrence rate times 1/100, the probability of system failure,

given the potential failure cause. The repair rate determined for

system failure due to Impact in domain 2 IMP is determined to be

10^{-1}/hr. For system repair, a minimum of one basic event in each

failed minimal cut set must be repaired.

For the other potential causes of system failure, there are no

common domains. Each potential cause of failure is added to the list

of cut sets for the system. The failure rate is determined from the

occurrence rate of the potential cause of failure and the common cause

candidate's sensitivity ranking. See Table 10 for the new cut sets and

their quantitative failure data.

3.6.2. Quantification of Special Conditions. Operator number 1,

O1, has been identified as a special condition for cut set 6. Each

occurrence of O1 in a common cause candidate can be represented by a

new one event cut set with the probability of causing the TOP event.

An alternative to this is to represent the probabilistic effects of O1

on the system by only one new cut set whose probability is the sum of

the probabilities for each occurrence of O1 in a common cause candidate.

4.0. SUMMARY AND CONCLUSIONS.

The step-by-step procedure provides a method of assessing common cause failure both qualitatively and quantitatively in complex systems. The procedure satisfies the objectives of the analysis by:

1. Reducing the amount of information handled in the analysis.

2. Maintaining the original resolution of the system fault tree.

3. Eliminating the need for a complete list of minimal cut sets for the TOP event.

4. Illustrating the input for quantitative analysis.

5. Making full use of current automated procedures.

One disadvantage is that the qualitative procedure is not fully automated at the present time. Another prominent difficulty is the lack of failure data, both qualitative and quantitative, that exists for common cause failures. Even with full automation and available data, the procedure could "bog down" in complex systems that are very prone to common cause failure. However, the procedure offers a highly organized approach for assessing potential common cause failures in complex systems.

TABLE 1. SPECIAL CONDITIONS RESULTING IN DEPENDENCE BETWEEN COMPONENTS

Symbol	Special Condition	Example Situations That Can Result in System Failure when All Basic Events in a Minimal Cut Set Share the Special Condition
E	Energy source	Common drive shaft, same power supply
C	Calibration	Misprinted calibration instructions
F	Manufacturer	Repeated fabrication error, such as neglect to properly coat relay contacts
I	Installation contractor	Same subcontractor or crew
M	Maintenance	Incorrect procedure, inadequately trained personnel
O	Operator or operation	Operator disabled or overstressed, faulty operating procedures
T	Test procedure	Faulty test procedures which may affect all components normally tested together.

TABLE 2. GENERIC CAUSES OF A MECHANICAL OR THERMAL NATURE

Symbol	Generic Cause	Example Sources
I	Impact	Pipe whip, water hammer, missiles, earthquake, structural failure
V	Vibration	Machinery in motion, earthquake
P	Pressure	Explosion, out-of-tolerance system changes (pump overspeed, flow blockage)
G	Grit	Airborne dust, metal fragments generated by moving parts with inadequate tolerances
S	Stress	Thermal stress at welds of dissimilar metals, thermal stresses and bending moments caused by high conductivity and density of liquid sodium
T	Temperature	Fire, lightning, welding equipment, cooling system faults, electrical short circuits

TABLE 3.   QUALITATIVE FAILURE CHARACTERISTICS FOR THE EXAMPLE PROBLEM

Basic Event	Special Condition	Generic Cause Susceptibilities	Physical Location
1	E2, O1, I1 8 6 2	I, S 4 6	101
2	E1, T1, F1 3 1 2	I, S 4 6	105
3	O1, I3 3 1	I, T 2 9	104
4	T2, I3 6 9	S, T 8 5	104
5	O2, I3 1 9	T, V 7 3	102
6	I2 9	I, V 3 3	102
7	O2, I2 4 8	I, V, S 4 3 4	105
8	T2, I3 1 9	I, V, T 5 3 6	105
9	E1, I3 3 9	S 3	199
10	O2, I2 4 8	I, V, S, T 7 8 4 2	106
11	T1, I3 6 9	S, T 8 5	103
12	O1, I1 6 2	T 1	104

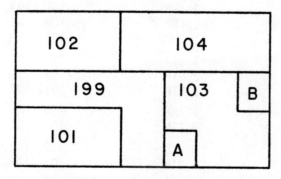

Figure 1.   Example Floor Plan

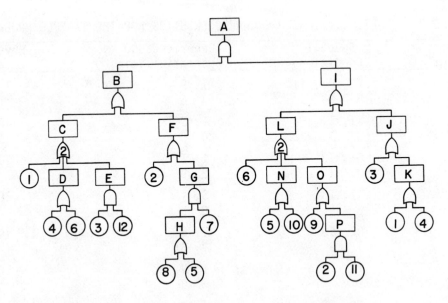

Figure 2.   Fault Tree for the Example Problem

TABLE 4.   GENERIC CAUSE DOMAINS FOR THE EXAMPLE PROBLEM

Generic Cause Susceptibility	Domain	Physical Location
I	1 IMP	102, 104
	2 IMP	101, 103, 105
	3 IMP	106
	4 IMP	199
S	1 STR	103, 105, 106
	2 STR	199
	3 STR	101, 102, 104
T	1 TEM	106
	2 TEM	101, 102, 103, 105, 199
V	1 VIB	102, 104, 106
	2 VIB	101, 103, 105, 199

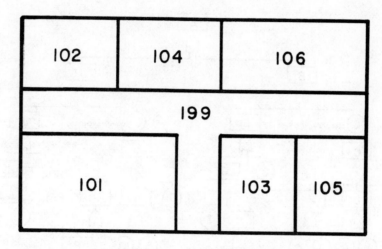

Figure 3. Floor Plan for the Example Problem

TABLE 5. INTERMEDIATE MINIMAL CUT SETS FOR THE EXAMPLE PROBLEM

Segment of Fault Tree	Intermediate Minimal Cut Sets	
A	1, C	1, 5, 7
	1, 2	1, 7, 8
	2, L	3, 5, 7
	2, 3	3, 7, 8
	2, 4	4, 5, 7
	3, C	4, 7, 8
	4, C	L, 5, 7
	C, L	L, 7, 8
C	1, 4	3, 4, 12
	1, 6	3, 6, 12
	1, 3, 12	
L	5, 6	9, 10
	5, 9	2, 5, 11
	6, 9	2, 6, 11
	6, 10	2, 10, 11

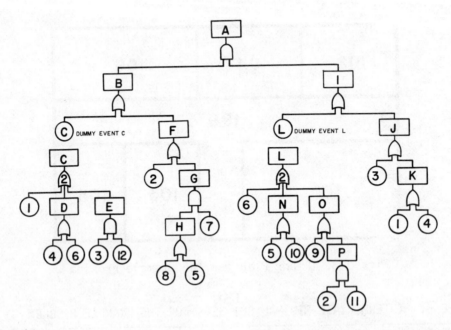

Figure 4.   Dissected Fault Tree for the Example Problem

TABLE 6.   INTERMEDIATE COMMON CAUSE CANDIDATES FOR FAULT TREE SEGMENTS
           C AND L IN THE EXAMPLE PROBLEM

Segment of Fault Tree	Intermediate Common Cause Candidate	Special Condition	Generic Cause Susceptibility	Physical Location	Domain
C	1, 4		S	101, 104	3 STR
	3, 4, 12		T	104	2 TEM
	1, 3, 12	01			
L	5, 6	I2	V	102	1 VIB
	6, 10		V	102, 106	1 VIB
	2, 10, 11		S	103, 105, 106	1 STR
	5, 9	I3			

TABLE 7.   DUMMY EVENT DESCRIPTIONS FOR THE EXAMPLE PROBLEM

Dummy Event	Special Conditions	Generic Cause Susceptibilities	Physical Location
C	01	S, T	104
L	I2, I3	V, S	106

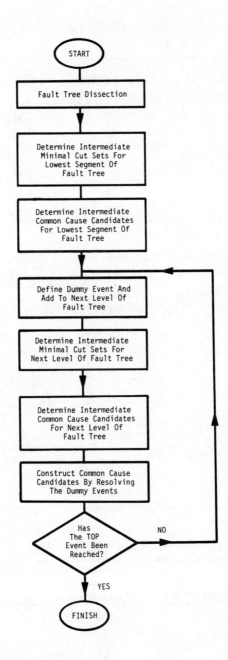

Figure 5. Flow Chart for Synthesis Procedure for Common Cause Failure Analysis

TABLE 8.   INTERMEDIATE COMMON CAUSE CANDIDATES FOR THE TOP EVENT OF THE
           EXAMPLE PROBLEM

Intermediate Common Cause Candidate	Special Condition	Generic Cause Susceptibility	Physical Location
1. 1, C	01	S	101, 102, 104
2. 2, L		S	103, 105, 106
3. 3, C	01	T	101, 102, 103, 104, 105, 199
4. 4, C		S	101, 102, 104
		T	101, 102, 103, 104, 105, 199
5. 1, 2	I1	I	101, 103, 105
6. 1, 7, 8		I	101, 103, 105

TABLE 9.   FINAL COMMON CAUSE CANDIDATES FOR THE TOP EVENT OF THE
           EXAMPLE PROBLEM

Common Cause Candidate	Special Condition	Generic Cause Susceptibility	Physical Location	Domain
1. 1, 2	I1 2	I 4	101, 105	2 IMP
2. 1, 7, 8		I 4	101, 105	2 IMP
3. 1, 4		S 6	101, 104	3 STR
4. 2, 10, 11		S 4	103, 105, 106	1 STR
5. 3, 4, 12		T 1	104	2 TEM
6. 1, 3, 12	01 3			

TABLE 10.   ONE EVENT CUT SETS FOR POTENTIAL COMMON CAUSE EVENTS

Cause of Secondary Failure	Domain	Failure Rate/hr.	Repair Rate/hr.
Impact	2 IMP	10^{-8}	10^{-1}
Stress	3 STR	2×10^{-9}	2×10^{-1}
Stress	1 STR	5×10^{-10}	2×10^{-1}
Temperature	2 TEM	5×10^{-8}	10^{-2}

REFERENCES

1. E. P. EPLER, Common Mode Failure Considerations in the Design of Systems for Protection and Control, Nuclear Safety, Vol. 10, No. 1 (1969), pp. 38-45.

2. J. R. TAYLOR, A Study of Failure Causes Based on U.S. Power Reactor Abnormal Occurrence Reports, Reliability of Nuclear Power Plants, IAEA-SM-195/16 (1975).

3. G. R. BURDICK, N. H. MARSHALL, and J. R. WILSON, COMCAN—A Computer Program for Common Cause Analysis, ANCR-1314 (1976).

4. C. L. CATE and J. B. FUSSELL, BACFIRE—A Computer Program for Common Cause Failure Analysis, The University of Tennessee, NERS-77-02 (1977).

5. W. E. VESELY and R. E. NARUM, PREP and KITT: Computer Codes for the Automatic Evaluation of a Fault Tree, IN-1349 (1970).

6. J. B. FUSSELL et al., SUPERPOCUS—A Computer Program for Calculating System Probabilistic Reliability and Safety Characteristics, The University of Tennessee, NERS-77-01 (1977).

7. D. F. HAASL, Advanced Concepts in Fault Tree Analysis, System Safety Symposium, June 8-9, 1965, Seattle: The Boeing Company.

8. J. B. FUSSELL, Fault Tree Analysis—Concepts and Techniques, Generic Techniques in Systems Reliability Assessment, Noordhoff, Leyden (1976), pp. 133-162.

9. J. B. FUSSELL, E. B. HENRY, and N. H. MARSHALL, MOCUS—A Computer Program to Obtain Minimal Sets from Fault Trees, ANCR-1156 (1974).

10. R. B. WORRELL, Set Equation Transformation Systems (SETS), SLA-73-0028A (1973).

11. J. B. FUSSELL, G. R. BURDICK, D. M. RASMUSON, J. R. WILSON, and J. C. ZIPPERER, A Collection of Methods for Reliability and Safety Engineering, ANCR-1273 (1976).

12. G. R. BURDICK and R. B. WORRELL, Qualitative Analysis in Reliability and Safety Studies, IEEE Transactions on Reliability, Vol. R-25, No. 3 (1976).

ESTIMATING COMMON CAUSE FAILURE PROBABILITIES IN RELIABILITY AND RISK ANALYSES: MARSHALL-OLKIN SPECIALIZATIONS

W. E. VESELY*

Abstract. The Marshall-Olkin approach to common cause failures is specialized to repairable situations where components have similar susceptibility to common cause mechanisms. Two cases are developed, the CFR case and the BFR case, which allow recorded failure data to be efficiently utilized in estimating common cause failure probabilities. The BFR case is applied to Boiling Water Reactor (BWR) scram failure data to estimate probabilities of multiple control rods failing.

1. Introduction. Common cause failures, or common mode failures as they are sometimes called, are dependent multiple failures originating from a common cause. Common cause failures can be important contributors to system unreliability and to accident risks. The Reactor Safety Study, (WASH-1400) [1], for example, states that the treatment of common cause failures is extremely important in assessments of the risk associated with nuclear power plant accidents.

In WASH-1400, common cause failure probabilities were evaluated using upper and lower probability bounds. Common cause failure probabilities can also be evaluated in a direct manner using the multivariate exponential model developed by Marshall and Olkin [2] and generalized by Lee and Thompson [3] and Block [4]. The Marshall-Olkin model is derivable from basic failure cause considerations and is applicable to reliability and risk evaluations.

* U.S. Nuclear Regulatory Commission, Washington, DC 20555.

In order to relate the Marshall-Olkin model to actual experience, reliability parameters in the model must be estimated from recorded failure data. Proschan and Sullo [5] discuss statistical estimation techniques based on observations of times to first failure. Two component redundancies are considered and maximum likelihood techniques are developed and compared. This paper develops statistical estimation techniques for those cases where common cause failures are repairable, which is the situation pertinent to nuclear power plant operations. As part of the development, the Marshall-Olkin model is specialized to allow efficient estimates to be made using sparse data.

2. Basics of the Marshall-Olkin Model. Consider an arbitrary group of m components which can fail from various failure causes and let the vector $\bar{x}$ be associated with a specific failure cause. The vector $\bar{x}$ contains ones for those components simultaneously failed by the cause and contains zeros for those components not affected by the cause. For example, for m = 3 components, the vector (1, 0, 1) describes a common cause simultaneously failing components 1 and 3 while not affecting component 2. The vector (1, 0, 0) describes an "independent" cause, affecting only component 1. The group of m components which is selected as the reference component set for the failure observations is termed here the "component population."

For the component population there will be a total of 2^m-1 possible failure causes, each of which is described by a unique vector $\bar{x}$. In the Marshall-Olkin model, each failure cause is assumed to have an exponential distribution for its time of occurrence;

$$f_{\bar{x}}(t) = \lambda_{\bar{x}} \exp(-\lambda_{\bar{x}}t), \tag{1}$$

where $\lambda_{\bar{x}}$ is the failure rate associated with cause $\bar{x}$. All the possible failure causes are assumed to be competing, and the observed component failures are determined by the failure cause which first occurs.

From the exponential and competing behaviors, multivariate exponential distributions are then obtained by Marshall and Olkin which describe the observed component failure times. For example, for two identical components the probability that neither component will fail in time t, $\bar{F}(t)$, is

$$\bar{F}(t) = \exp(-\lambda_1 t - \lambda_1 t - \lambda_2 t). \tag{2}$$

where λ_1 is the individual component failure rate (for $\bar{x} = (1, 0)$ or $(0, 1)$) and λ_2 is the common cause failure rate ($\bar{x} = (1, 1)$).

The basic Marshall-Olkin equations can be used to derive relations which are more pertinent to reliability and risk analyses. For example, with no repair performed on the two components, the probability $Q(t)$ that both components will fail in time t is to first order

$$Q(t) = (\lambda_1 t)^2 + \lambda_2 t. \tag{3}$$

When failures are repairable, the probability $\bar{A}$ that both components are down, i.e., the unavailability, is to first order

$$\bar{A} = (\lambda_1 \tau_1)^2 + \lambda_2 \tau_2 \tag{4}$$

where τ_1 is the average repair time for an individual component failure and τ_2 is the average repair time for the common cause failure.

The first terms on the right hand sides of Equations (3) and (4) are the standard independent failure contributions and the second terms are the common cause failure contributions. These more general equations replace the independent failure calculations used in reliability and risk analyses. Equations for two different components and for three or more components can likewise be developed from the general Marshall-Olkin formulations.

The Marshall-Olkin model thus lends itself to direct quantification of common cause failure probabilities as applied to reliability and risk analyses. In using the Marshall-Olkin model, all common cause failures are assumed to occur at the same time, which may be taken as an approximation for those situations where common

cause failures occur within short intervals of one another. To relate
the Marshall-Olkin model to actual experience, recorded failure data
must be used to estimate λ_1, λ_2, and $\lambda_{\bar{x}}$ in general. Since there are
often little data, the model needs to be specialized to allow efficient
estimation.

3. Estimation Techniques and Model Specializations. Assume from
hereon that whenever failures in the component population occur they
are repaired with essentially zero repair time and assume that the
failure causes can reoccur with the same exponential distributions[1].
From the exponential, reoccurring assumptions, the observed failure
times of the population therefore constitute a renewal process (Cox
[6]).

Let the random variable $N_{\bar{x}}$ be the number of times failure cause
$\bar{x}$ is observed to occur in a fixed time period T. Using the results
of [3] the following two theorems are straightforwardly obtained:

THEOREM 1: The number of occurrences $N_{\bar{x}}$ has a Poisson
distribution with parameter $\lambda_{\bar{x}} T$.

THEOREM 2: The numbers of occurrences $N_{\bar{x}_1}$, $N_{\bar{x}_2}$, ... $N_{\bar{x}_k}$,
$1 \leq k \leq 2^m - 1$ are independent Poissons with parameters
$\lambda_{\bar{x}_1} T$, $\lambda_{\bar{x}_2} T$, ... $\lambda_{\bar{x}_k} T$.

---

[1]The term, "zero repair time" means that the average repair time is
small compared to the minimum of the mean times of cause occurrences
$(1/\lambda_{\bar{x}})$. When the repair time is not small, actual operation time is
used as the observed time.

The above theorems allow standard Poisson statistical techniques (e.g., Johnson and Kotz [7]) to be used in estimating $\lambda_{\overline{x}}$. The estimation will generally give little information, however, since $N_{\overline{x}}$ will be observed to be zero for many of the vectors $\overline{x}$. Therefore, to obtain fewer parameters, we specialize the Marshall-Olkin model and assume that the component population m is selected such that the failure rates $\lambda_{\overline{x}}$ depend only upon the number of components failed, i.e., $\lambda_{\overline{x}} = \lambda_x$ where x is the total number of components simultaneously failed by the cause.

The assumption $\lambda_{\overline{x}} = \lambda_x$, x = 1...m, implies that the components in the population are similar and are subject to similar failure causes. This specialized model will therefore be called the homogeneous Marshall-Olkin model or more simply the homogeneous model. Identical components in the same environment, for example, constitute a homogeneous population for which the homogeneous model is applicable. Components whose failures are chiefly caused by the same maintenance error (e.g., miscalibration) also constitute an approximate homogeneous population. Many other examples could likewise be constructed. Common cause failures are most likely to occur in these homogeneous situations and hence the homogeneous model is a practicable model[2].

---

[2]It should be noted that the homogeneous model can be further generalized by assuming $\lambda_{\overline{x}} = \lambda_x$ for x $\geq x_0$. For example, for $x_0 = 2$, the homogeneous assumption only then applies to common cause failures affecting two or more components. All derived results will remain valid with the added restraint that numbers of failures be greater than or equal to x_0 for any failure rate used.

For the homogeneous model, let N_x be the number of times x

components simultaneously fail, where only the number of

simultaneous failures is noted at each observaton. Using the fact

that a sum of independent Poisson random variables is another

Poisson, the following theorem is obtained.

THEOREM 3:   In the homogeneous model, if N_x is the number of

occurrences of x components simultaneously failing then N_x

in time period T is Poisson with parameter $\binom{m}{x} \lambda_x T$.

The above theorem is useful since it allows λ_x to be estimated from

the grosser variable N_x using standard Poisson techniques. There

are still m parameters to be estimated (λ_x) and since m may be

large we will later examine simple, special cases of the homogeneous

model.

For reliability and risk analyses, common cause failure rates

are required for specific sets of components within the population.

For example, a particular safety system may contain two redundant

valves, where the two valves can be considered to belong to the

total population of valves of this type which exist in the plant.

To obtain the common cause failure contribution to the system

unavailability or the system unreliability, we require the common

cause failure rate for the two valves simultaneously failing, where

the other valves in the plant may either fail or not fail.

The following theorem allows the population m to be used to estimate the common cause failure rates for specific components:

THEOREM 4: For the homogeneous model, if N_{ij} denotes the number of occurrences in which i specific components simultaneously fail and j specific components do not fail, where the other components in the population may either fail or not fail, then N_{ij} in time period T is Poisson with parameter $\lambda_{ij}T$ where

$$\lambda_{ij} = \sum_{k=0}^{m-i-j} \binom{m-i-j}{k} \lambda_{i+k}, \quad 1 \leq i+j \leq m. \qquad (5)$$

The above theorem is obtained using the fact that a sum of independent Poissons is another Poisson. When i or j equals zero there is no restriction on the specific components failing (i=0) or the specific components not failing (j=0).

For k specific components, where i + j = k, the time of occurrence of a cause failing i of the components and not affecting the remaining j components is therefore described by an exponential distribution having the failure rate λ_{ij}. The failure rates λ_{ij}, i + j = k, can then be used for the population k in place of λ_x. For example in Equations (3) and (4), when the two components are part of a larger population then λ_{11} would replace λ_1 and λ_{20} would replace λ_2 and the equations would remain valid.

To obtain an estimator of λ_{ij}, denoted by $\hat{\lambda}_{ij}$, Theorem 3 can be used to obtain unbiased estimators of λ_x which are subsequently substituted in Equation (5):

$$\hat{\lambda}_{ij} = \sum_{k=0}^{m-i-j} \binom{m-i-j}{k} \binom{m}{i+k}^{-1} \frac{N_{i+k}}{T} \tag{6}$$

which can also be expressed as

$$\hat{\lambda}_{ij} = \frac{\displaystyle\sum_{k=i}^{m-j} \binom{k}{i} \binom{m-k}{j} N_k}{\binom{m}{i} \binom{m-i}{j} T} . \tag{7}$$

The estimator $\hat{\lambda}_{ij}$ is unbiased and consists of a weighted sum of independent Poisson variables. The variance of $\hat{\lambda}_{ij}$, V_{ij}, is

$$V_{ij} = \sum_{k=i}^{m-j} c_k \frac{\lambda_k}{T} \tag{8}$$

where

$$c_k = \frac{\left(\begin{array}{c} m-i-j \\ k-i \end{array}\right)^2}{\left(\begin{array}{c} m \\ k \end{array}\right)}$$

(9)

and where λ_k is the homogeneous failure rate for the overall population.

The distribution of $\hat{\lambda}_{ij}$ contains the unknown nuisance parameters λ_k making any additional statistical analysis difficult. Instead of pursuing conditional probability approaches to eliminate the nuisance parameters, we introduce two particular cases within the homogeneous model, the constant failure rate case and the binomial failure rate case, which allow simpler, but yet practicable, evaluations to be made.

4. The Constant Failure Rate Case. Within the homogeneous model the constant failure rate (CFR) case occurs when the common cause failure rates are independent of the failure numbers;

$$\lambda_x = \lambda, \ x \geq x_1,$$

(10)

where the equality is only assumed for numbers of failures greater than or equal to x_1. There is no restriction on the failure rates for $x < x_1$. For example, if $x_1 = 2$, the equality applies only to common cause failures in which two or more components simultaneously fail.

The equality $\lambda_x = \lambda$ will be applicable when common causes are equally likely to affect any specific component set (component combination). From Theorem 3, the failure rate equality $\lambda_x = \lambda$ implies that expected numbers of simultaneous failures will follow a combinatorial law since the expected value of N_x is proportional to $\binom{m}{x}$. The number of simultaneous failures most frequently observed will be approximately m/2 (the x at which $\binom{m}{x}$ is a maximum).

Failure causes for which the CFR case is applicable include random maintenance errors which could equally affect any specific components (such as due to miscalibration) and manufacturing defects which could be equally realized in any specific components (such as due to random quality control errors). Since the most frequent number of simultaneous failures is approximately m/2, for large m the constant failure rate model predicts that large numbers of simultaneous failures will likely occur if any do occur.

Theorem 3 produces the following corollary for the CFR case.

COROLLARY 1. For a CFR case, if $\tilde{N}_{x_1}$ is the number of occurrences of x_1 or more simultaneous failures, then $\tilde{N}_{x_1}$ in time period T is Poisson with parameter $K\lambda T$ where

$$K = \sum_{k=x_1}^{m} \binom{m}{k}. \tag{11}$$

The parameter K can be expressed in terms of the standard binomial probability $B(x_1, m, \frac{1}{2})$ of observing x_1 or more successes in m trials where each success has a probability of $\frac{1}{2}$ of occurring:

$$K = 2^m B(x_1, m, \frac{1}{2}).$$
(12)

K is therefore determinable from standard binomial tables [8].

The cumulative variable $\tilde{N}_{x_1}$ can thus be simply and efficiently used to draw inferences about λ using standard Poisson techniques. For example, an unbiased estimator of λ, $\hat{\lambda}$ is

$$\hat{\lambda} = \frac{\tilde{N}_{x_1}}{KT} \quad ,$$
(13)

and confidence bounds on λ are obtained from standard Poisson tables [9]. For the CFR case, Equation (5) for λ_{ij} simplifies to

$$\lambda_{ij} = \lambda \ 2^{m-i-j} \ , \ i \geq x_1 \ .$$
(14)

Since λ_{ij} is proportional to λ, inferences on λ_{ij} are directly obtained from inferences on λ.

The above estimators apply to a single population. More precise estimators may also be obtained by combining the experiences of n populations which have similar common cause behaviors and which are isolated from one another (i.e., common cause failures are negligible

across the populations). The quantity $K\lambda$ is the sum of all common
cause failure rates $(x \geq x_1)$ and represents the total common cause
frequency for the individual population. For n populations having
the same common cause frequencies $(K_i\lambda_i = K\lambda)$, $\tilde{N}_{x_1}$ summed over the
populations is Poisson with parameter $K\lambda(T_1+T_2+\ldots+T_n)$ where T_i is the
individual population time period. Hence the sum of $\tilde{N}_{x_1}$ can be used
to obtain the best, overall estimators of λ and λ_{ij}.

The CFR case thus allows simple evaluations to be performed. The
restriction of the CFR case is the assumption that $\lambda_x=\lambda$, which
involves engineering and failure cause considerations. If common
causes only affect isolated sets of components, the overall population
can be subdivided into n isolated populations in which the CFR
assumption is applied. With sufficient data, the most frequent
number of simultaneous failures $(m/2)$ can be used as an indicator of
the appropriate CFR population size. When the CFR case does not
seem applicable by any criterion, then the binomial failure rate
case can be considered which is another special case within the
homogeneous model.

5. The Binomial Failure Rate Case. For the binomial failure
rate (BFR) case, the equation for λ_x is obtained by factoring the
common cause failure rate into an overall occurrence rate and a
detailed effect probability. Let Λ be the sum of all the common

cause failure rates for $x \geq x_1$;

$$\Lambda = \sum_{x=x_1}^{m} \binom{m}{x} \lambda_x. \tag{15}$$

Assume that given a common cause failure occurrence, each component has a constant probability p of failing from the common cause. The failure rate λ_x is then given by

$$\lambda_x = \frac{\Lambda}{C} p^x (1-p)^{m-x}, \; x \geq x_1 \tag{16}$$

where the normalization constant C is the probability of all possible component failures;

$$C = \sum_{x=x_1}^{m} \binom{m}{x} p^x (1-p)^{m-x} \tag{17}$$

$$= B(x_1, m, p). \tag{18}$$

$B(x_1, m, p)$ is the binomial probability value defined earlier, now for a success probability of p.

In the BFR case, the failure rate λ_x as a function of x varies like the sequence probabilities for the binomial distribution. For $p = 0.5$, λ_x is independent of x and the BFR case reduces to the CFR case. For $p < 0.5$, λ_x decreases as x increases and for $p > 0.5$, λ_x increases as x increases. The BFR case is thus more general than the

CFR case, covering a wider variety of common cause failure behaviors.

The parameters Λ and p may be estimated in several ways. For the BFR case, Theorem 3 reduces to the following corollary:

COROLLARY 2. In the BFR case if N_x is the number of occurrences of x components simultaneously failing, $x_1 \leq x$, then N_x in time period T is Poisson with parameter $\Lambda_x T$ where

$$\Lambda_x = \frac{\Lambda}{C} \binom{m}{x} p^x (1-p)^{m-x} . \tag{19}$$

Maximum likelihood techniques (Bard [10]), may thus be used to estimate Λ and p from different observed values of N_x. The number of simultaneous failures most frequently observed will be approximately mp for C near 1.

For a second approach, the two parameters Λ and p may be separately estimated using the following two corollaries:

COROLLARY 3. In the BFR case if $\tilde{N}_{x_1}$ is the number of occurrences of x_1 or more simultaneous failures, then $\tilde{N}_{x_1}$ in time period T is Poisson with parameter ΛT.

COROLLARY 4. In the BFR case conditional on $\tilde{N}_{x_1}$, the variables N_x, $x_1 \leq x \leq m$, have a multinomial probability distribution with probability p_x, where

$$P_x = \frac{\binom{m}{x} p^x (1-p)^{m-x}}{C} \cdot \qquad (20)$$

Corollary 3 is obtained from Theorem 3, using Equation (15). Corollary 4 is obtained from standard Poisson-multinomial relationships [7]. According to Corollary 3, given that $\tilde{N}_{x_1} = n$, the numbers of simultaneous failures N_x are then multinomially distributed:

$$P(N_{x_1} = n_{x_1}, \ldots, N_m = n_m) = \frac{n!}{n_{x_1}! \ldots n_m!} p_{x_1}^{n_x} \ldots p_m^{n_m} \cdot \qquad (21)$$

Corollary 3 allows inferences to be made of Λ using standard Poisson techniques. An unbiased estimator of Λ, $\hat{\Lambda}$, is for example

$$\hat{\Lambda} = \frac{\tilde{N}_{x_1}}{T} \cdot \qquad (22)$$

If Λ is assumed to be the same for different populations then the sum of $\tilde{N}_{x_1}$ can be divided by the sum of the T's to obtain a more precise estimator of Λ.

Corollary 4 allows estimates and inferences to be made of p using multinomial associated techniques. The multinomial techniques include maximum likelihood techniques and least squares techniques where, for example, n_x/n can be used to estimate p_x. If different populations of equal size are assumed to have the same p then $\tilde{N}_{x_1}$, etc., summed over the populations can be used in Corollary 4.

When m is large and p is small, the Poisson approximation to the binomial may be used for p_x:

$$p_x \cong \frac{(mp)^x\ e^{-mp}}{x\ !\ C} \tag{23}$$

where the normalization constant C is now the sum of the respective Poisson probabilities. If logs of Equation (23) are taken for two different x values then

$$\ln n_{x_1} - \ln n_{x_2} = (x_1 - x_2)\ \ln mp + \ln x_2! - \ln x_1!, \tag{24}$$

where the estimates n_{x_1}/n and n_{x_2}/n are used for p_{x_1} and p_{x_2}, respectively. Linear least squares techniques may then be used to estimate p. The linear least squares estimate may also be used as a starting estimate in non-linear, iteration schemes.

With regard to λ_{ij}, for the BFR case Equation (5) becomes

$$\lambda_{ij} = \frac{\Lambda}{C}\ p^i (1-p)^j \tag{25}$$

and inferences on λ_{ij} can be obtained from influences on Λ and p. The inferences will be more involved, however they are manageable.

The BFR case is thus more involved than the CFR case, however it is adaptable to broader applications. Since the CFR case is a special case within the BFR model, when it is considered appropriate, hypothesis tests can be performed to determine the choice between the

CFR case and the BFR alternative. Engineering information and
engineering assessments should serve as the basis for any particular
approach which is used in a particular problem.

6. Reactor Scram Example. The reactor scram system is an
important system in nuclear risk considerations and multiple control
rods failing to insert due to common causes need to be evaluated in
the system analysis. Table 1 states the number of control rods and
operating times for Boiling Water Reactors (BWR's). The table is taken
from the EPRI report on BWR risk analyses [11]; 12 months has been
added to the operating times to approximately update the experience
through 1976. This 12 month addition does not account for the reactor
downtime; however, this effect is relatively small.

Table 2 is a compilation of control rod incidents which have
occurred in BWR's; the table is again taken from the EPRI report. The
reported incidents concern either slow or incomplete rod insertions;
in all these accidents, the control rods in question achieved partial
insertion. The reported mechanisms of failure include piston seal
leakages, possible filter plugging, and rod drive blockage. The control
rod failures to insert which resulted from the above mechanisms can
generally be classified as being mechanical rod failures which include
hydraulically related failures and drive mechanism failures. We will
use the data of Tables 1 and 2 to estimate the common cause probability
of multiple control rods failing to insert due to mechanical failure.

Table 1. Number of Control Rods and
Operating Times for Boiling Water Reactors

Reactor	Number of Control Rods	Operating Months Through Dec. 1976
Dresden 1	80	197
Big Rock Point	32	138
Humboldt Bay	32	160
Oyster Creek	137	83
Nine Mile Point	129	83
Dresden 2	177	41
Millstone 1	145	71
Monticello	121	64
Dresden 3	177	61
Quad-Cities 1	177	51
Quad-Cities 2	177	49
Vermont Yankee	89	48
Pilgrim 1	145	21
Duane Arnold	89	32
Peach Bottom 2	185	30
Cooper	137	30
Browns Ferry 1	185	20
Peach Bottom 3	185	25
Browns Ferry 2	185	13
Edwin L. Hatch 1	137	16
		1233

Table 2. BWR Control Rod Incidents

Reactor	Report Date	Discovery Mode	Malfunction
Dresden 2	9/70	test	2 rods inserted slowly, rods fully inserted and made inoperative
Dresden 2	11/70	test	2 rods inserted slowly, rods fully inserted and made inoperative
Dresden 2	12/70	unknown	1 rod inserted slowly, rod fully inserted and made inoperative
Millstone 1	3/8/73	test	2 rods inserted slowly, rods fully inserted and made inoperative
Millstone 1	3/19/73	test	2 rods inserted slowly, rods fully inserted and made inoperative
Oyster Creek 1	4/70	test	some rods stopped six inches from full insertion, all fully inserted manually
Monticello	6/72	unknown	1 rod stopped six inches from full insertion
Nine Mile Point	11/73	scram	11 rods stopped at position 02, operator inserted them manually
Nine Mile Point	12/73	scram	15 rods stopped at position 02, operator inserted them manually
Dresden 2	11/74	scram	93 rods stopped at position 02, 1 at position 04 and 2 at position 06, operator then inserted all 96 rods

Table 2. BWR Control Rod Incidents (Continued)

Reactor	Report Date	Discovery Mode	Malfunction
Dresden 3	3/75	test	3 rods exceeded 95% insertion time. Scram pilot valve failed on 1 rod. 1 rod left at position 48, 1 rod inserted and disarmed
Dresden 2	6/75	test	rods inserted slowly due to high regulated pressure in the scram valve air header
Duane Arnold	8/75	scram	2 rods stopped at 02 position

In the EPRI report, a conservative definition of failure is given as a control rod stopping at notch 04 or greater which is equivalent to a reactivity worth of 96% or less. (A fully inserted control rod would reach notch 00.) Based on this failure definition, the three (3) rods in Dresden 2 which stopped at notch 04 or greater are counted as failures. We thus have one common cause data point of 3 simultaneous rod failures ($N_3=1$).

For the data in Table 1, the component population m is the number of control rods in a given reactor. Dresden 2 has 177 control rods; if we assume the CFR model is applicable, then the most likely number of simultaneous rod failures is 89 (m/2). The fact that 3 simultaneous failures were actually observed indicates that the CFR model is not appropriate and that the more general BFR model should be used. More formally, a statistical hypothesis test could be performed and the CFR model would be rejected[3].

Based on the above considerations, we will use the BFR model and will take the cutoff failure number (x_1) to be 2 since we are interested in all common cause failures. In the BFR model, the two parameters that must be estimated are the total common cause frequency Λ and the probability p that a particular control rod will fail due to common causes. The estimate of Λ is obtained from Equation (22);

---

[3] In the CFR model when m=177, there is 95% probability that the number of simultaneous failures will be between 76 and 102 (i.e., the 95% tolerance region is approximately $177/2 \pm 2\sqrt{177/4}$).

$$\Lambda = \frac{1}{1233 \times 720} = 1.1 \times 10^{-6} hr^{-1} \tag{26}$$

where one month is assumed to contain 720 hours. All the BWR reactors are assumed to have the same common cause occurrence rate Λ and hence the total operating time (1233 months) is used in the estimate.

The estimate of p is obtained from Equation (20). Given that $\tilde{N}_2 = 1$ and m=177, the occurrence probability p_3 of 3 simultaneous failures, as observed in the Dresden 2 incident, is

$$p_3 = \frac{\binom{177}{3} p^3 (1-p)^{177-3}}{C} . \tag{27}$$

A value for p is obtained through maximum likelihood techniques by maximizing p_3. By using the iteration techniques for the conditional binomial in Johnson and Kotz [7] with an initial and naive estimate of p as 3/177=0.017, the final estimate of p is

$$p = 0.012 . \tag{28}$$

The corresponding estimate of C is

$$C = 0.63 . \tag{29}$$

By using the above estimates in Equation (25), the failure rates λ_{io} and unavailabilities q_{io} for various numbers of specific rods failing to insert can be calculated and are stated in Table 3. (Since only failures are considered, j=0 in Equation (25).) The unavailability q_{io} is obtained assuming monthly testing, i.e., $q_{io}=360\lambda_{io}$. The estimates in Table 3 are based on the Dresden 2 incident, however by presuming similar common cause probabilities, Table 3 is applicable to any BWR.

Table 3. Reliability Characteristics for
Specific Control Rod Failures

No. of Control Rods i	λ_{io} (per hour)	q_{io} (per demand)
2	2.6×10^{-10}	9.3×10^{-8}
3	3.1×10^{-12}	1.1×10^{-9}
4	3.7×10^{-14}	1.3×10^{-11}
5	4.5×10^{-16}	1.6×10^{-13}
6	5.4×10^{-18}	1.9×10^{-15}

The numbers in Table 3 are useful when specific control rods are of interest, for example, the three adjacent rods in the center of the core. In the Reactor Safety Study, the common cause probability (unavailability) for 3 specific rods failing from mechanical causes was predicted to be 1×10^{-9} (pp. 362, Appendix II, WASH-1400). The unavailability of 1×10^{-9} was obtained by using log normal bounding techniques and was used in the prediction of the BWR scram failure probability. In the Reactor Safety Study, scram failure was defined to consist of 3 adjacent rods failing and the value of 1×10^{-9} was multiplied by the number of adjacent rod combinations (3000) to obtain the overall rod contribution of 3×10^{-6}. The Reactor Safety Study result of 1×10^{-9} agrees very well with the 3 rod unavailability value given in Table 3.

By using Equation (19), and by summing over all the specific rod combinations, the failure rate and unavailability estimates for given numbers of multiple rod failures may be calculated. In Table 4, Λ_x is the failure rate for the simultaneous failure of any x rods, and q_x is the associated unavailability. The unavailability $q (\geq x)$ is the probability that any x or more rods are simultaneously failed. Table 4 is applicable to any BWR if the common cause probabilities are assumed to be similar.

Table 4. Reliability Characteristics For All
Combinations of Control Rod Failures[4]

No. of Rods x	Λ_x (per hour)	q_x (per demand)	$q(\geq x)$ (per demand)
2	4.8×10^{-7}	1.7×10^{-4}	4.0×10^{-4}
3	3.4×10^{-7}	1.2×10^{-4}	2.3×10^{-4}
4	1.8×10^{-7}	6.6×10^{-5}	1.1×10^{-4}
5	7.7×10^{-8}	2.8×10^{-5}	4.2×10^{-5}
6	2.7×10^{-8}	9.9×10^{-6}	1.4×10^{-5}
7	8.3×10^{-9}	3.0×10^{-6}	4.0×10^{-6}
8	2.2×10^{-9}	7.9×10^{-7}	1.0×10^{-6}
9	5.2×10^{-10}	1.9×10^{-7}	2.3×10^{-7}
10	1.1×10^{-10}	4.0×10^{-8}	4.0×10^{-8}

Table 4. Reliability Characteristics For All
Combinations of Control Rod Failures[4]

(Continued)

No. of Rods x	Λ_x (per hour)	q_x (per demand)	$q(\geq x)$ (per demand)
15	1.3×10^{-14}	4.8×10^{-12}	4.8×10^{-12}
20	3.1×10^{-19}	1.1×10^{-16}	1.1×10^{-16}

Table 4 is useful when the number of rods failing to insert is of
interest without regard to the specific rods. For example, in the
event of a transient in a BWR the failure of any 6 or more control rods
to insert will result in a slower insertion of reactivity and shutdown
problems may develop.

Confidence bounds associated with any of the previous results can
be obtained from confidence bounds on Λ and p. For example, based on
Poisson tables, the upper 95% confidence bound for Λ is approximately
a factor of 4.7 higher than the best estimate value (Equation (26)).
Based on the binomial, the upper 95% confidence bound on p is approximately a factor of 2.6 higher than the best estimate value of 0.012.
These factors can be used to obtain approximate upper confidence bounds
on any failure rate or unavailability which has been given.

----

[4] Values cited incorporate both non-adjacent and adjacent rod combinations.

7. Summary and Recommendations. The Marshall-Olkin formulations
have been simplified to allow their utilization in reliability and risk
analyses where sparse data often exist. Two special cases have been
particularly examined, the CFR case which has one failure parameter
and the BFR case which has two failure parameters. An example using the
BWR scram system illustrates the potential utility of the approaches
which have been developed.

The approaches which have been developed can be applied to a variety
of problems to obtain common cause probability estimates. Raw data
presently exists on common cause failure occurrences and the approaches
offer the potential of being able to incorporate some or all of this
detailed experience into reliability and risk evaluations. The limitations of the presented approaches, in additon to the basic Marshall-
Olkin assumptions, include 1) the assumption of equal failure rates for
the CFR case and 2) the assumption of the binomial behavior for the BFR
case.

Additional extensions of the approaches might be further explored
for broader utilizations. For example the failure rate λ_x might be
expressed as different parametric functions of x to describe common
cause failure behaviors not incorporated by the CFR or BFR cases.
Possibilities include negative binomial and discrete Weibull representa-
tions. The treatment of common cause failures is extremely important in
reliability and risk analyses and quantification approaches based on actual
data need to be further developed and need to be more extensively applied
to obtain more realistic predictions of common cause failure probabilities.

REFERENCES

[1] Reactor Safety Study - An Assessment of Accident Risks in U.S.
       Commercial Nuclear Power Plants, WASH-1400, (NUREG-75/014),
       October 1975.

[2] A. W. MARSHALL and I. OLKIN, "A Multivariate Exponential
       Distribution," JASA, 62(1967), pp. 30-44.

[3] L. LEE and W. A. THOMPSON, JR., "Results on Failure Time and
       Pattern for the Series System," Reliability and Biometry,
       SIAM (1974), pp. 291-302.

[4] H. W. BLOCK, "Continuous Multivariate Exponential Extensions,"
       Reliability and Fault Tree Analysis, SIAM (1975), pp. 285-306.

[5] F. PROSCHAN and P. SULLO, "Estimating the Parameters of a Bivariate
       Exponential Distribution in Several Sampling Situations,"
       Reliability and Biometry, SIAM (1974), pp. 423-440.

[6] D. R. COX, Renewal Theory, Methuen, London, 1962.

[7] N. L. JOHNSON and S. KOTZ, Discrete Distributions, Houghton
       Mifflin Company, Boston, 1969.

[8] Tables of the Cumulative Binomial Probability Distribution, Harvard
       University Press, 1955.

[9] E. S. PEARSON and H. O. HARTLEY, Biometrika Tables for Statisticians,
       Cambridge University Press, London, 1958.

[10] Y. BARD, Nonlinear Parameter Estimation, Academic Press, New York,
       1974.

[11] ATWS: A Reappraisal, Part II, Vol. II, BWR Risk Analysis, EPRI
       NP-265, August 1976.

STOCHASTIC MODELS FOR REPAIRABLE REDUNDANT SYSTEMS SUSCEPTIBLE TO COMMON MODE FAILURES

BOYER B. CHU AND DONALD P. GAVER*

Abstract. Redundant protective engineering systems are in principle susceptible to chance and common mode failures. The intent of this paper is to propose two Markov models which recognize both chance and common mode failures in quantifying system performance. These models have possible usefulness in common failure calculations in nuclear reliability and risk analyses. Model 1 incorporates chance failures and common mode failures of a catastrophic nature. Model 2 assumes that the common failure mode mechanisms increase the failure rates of the components. Solution methods for these two models are summarized, and some results are also given to illustrate the effects of common failure mode on the mean-time-to-system-failure and the system survival probabilities.

1. Introduction. Nuclear power electric generation stations are designed to achieve a reliable performance to assure the safety of the general public. Physical separation and redundant protective concepts are often incorporated in the design of various safety related engineering systems to prevent the occurrence of, and to mitigate the consequences of, hazardous accidents. However, engineering systems are susceptible to failures and plant operators are liable to commit human errors. System reliability could therefore be compromised.

Redundant protective engineering systems are in principle subjected to both chance and common mode types of failures. Chance

* Electric Power Research Institute, Palo Alto, California 94303.

failures are those failures that occur essentially independently to individual components, while common mode failures, as described in IEEE STD No. 279, result from a mechanism by which a single event can cause all or a large number of redundant components to be inoperable, perhaps essentially simultaneously. The intent of this paper is to present two models which recognize both chance and common mode failures in assessing the reliability and availability of redundant, repairable systems.

The models developed in this paper have evolved from the birth (repair) and death (failure) stochastic process model considering an additional common failure mode factor. Model 1 treats chance failures and also common mode failures of a catastrophic nature. Model 2 assumes that the common mode mechanisms could increase the failure rate of components over an exponentially distributed time. Solution methods for these two models are presented, and some results are also given to illustrate the effects of common failure mode on the mean-time-to-system-failure and the system survival probabilities.

2. Models for Characterizing Catastrophic Events -- Model Type I. Model I is intended to quantify the system availability similar to the engineering "Worst-Case" analysis. It assumes that the occurrence of a common failure mode initiator could disable the safety function of the entire system, regardless of the amount of system redundancy and the intensity of the common mode effect. Thus,

the results obtained from this model may be interpreted as the lower
bound of system performance.

2.1. Description and Derivation. A specific model that
embodies both common mode and chance failure causes may be described
as follows:

A system is made up of m (m $\geq$ 1) functionally identical
components or subsystems, each one of which has a chance
failure rate λ and rapair rate μ. In addition, there is a
common mode failure rate c and when the common mode failure
occurs the entire system is completely disabled. The
system may also fail from chance causes, and the system
operates successfully so long as k out of m (1$\leq$k<m) components operate, or until ℓ = m-k+1 units are down simultaneously.

Such a model has been discussed by Apostolakis [1] in the
context of nuclear power plant safety systems; a similar model was
analyzed by Gaver [2], and by Harris [3].

The above model can be calculated by using Markov processes.
Let D(t) be the state variable of the m-component redundant system,
D(t) = j implies that j identical components are down for repair at
time t. The state transition probabilities of the model for
D(t) = j, j=0,1,2,...,m, are given in Table 1.

Table 1. State transition probabilities for Model I

State Transitions   Probabilities

From $D(t)=j$ to $D(t+tdt)=$
$$\begin{cases} j-1 & \mu_j dt \\ j+1 & \lambda_j dt \\ j & 1-(\lambda_j+\mu_j+c)dt \\ m & cdt \end{cases}$$

Note that allowing the transition from state j to state m during the time dt is interpreted as the common mode failure effect. It also makes the model different from the conventional birth and death stochastic process. The differential equations for the state probabilties can be derived in a standard Markov manner. Denote $P_j(t)$ the state probability,

$$P_j(t) = P\{D(t) = j | D(0) = 0\} ; \qquad (2.1)$$

then

$$P_j(t+dt) = P_j(t)\left[1-(\lambda_j+\mu_j+c)dt\right] + P_{j-1}(t)\lambda_{j-1}dt \qquad (2.2)$$

$$+ P_{j+1}(t)\mu_{j+1} + 0(dt) \qquad j = 0,1...m-1$$

and

$$P_m(t+dt) = P_m(t)\left[1-\mu_m dt\right] + P_{m-1}(t)\lambda_{m-1}dt + \sum_{k=0}^{m-1} cP_k(t)dt+0(dt). \qquad (2.3)$$

The parameters λ_j and μ_j are defined by

$$\lambda_j = (m-j)\lambda$$

$$\mu_j = [\min(j,r)]\mu$$

and r specifies the number of repairmen, or repair teams.

In Equation (2.2), the first term represents the probability of no change of system state during (t, t+dt), the second represents a chance failure from j-1 to j and the last term represents a system improvement from j+1 to j via the repair mechanism. The last term in Equation (2.3) is the common mode failure effect. In the limit of dt approaching to zero, the set of difference equations reduce to the differential equations,

$$\dot{P}_j(t) = -(\lambda_j + \mu_j + c)P_j(t) + \lambda_{j-1}P_{j-1}(t) + \mu_{j+1}P_{j+1}(t) \qquad (2.4)$$

$$j = 0,1,\ldots,m-1 \quad,$$

$$\dot{P}_m(t) = \mu_m P_m(t) + \lambda_{m-1}P_{m-1}(t) + \sum_{k=0}^{m-1} cP_k(t) \qquad (2.5)$$

where $\dot{P}_m(t)$ denotes the derivative of P with respect to time. With appropriate initial conditions, equations (2.4) and (2.5) can be readily solved. For arbitrary k, the availability of a k-out-of-m

success system can be calculated by proper combinations of $\left\{P_j\right\}$.
Note that the equations (2.4) and (2.5) hold also when the failure
and repair rates are arbitrary constants.

   2.2. Survival Probability and Mean-Time-To-Failure. In the
presence of common mode failures, the system availability may not be
the only measure of interest in quantifying system performance. The
mean-time-to-system-failure or first passage time, and the system
survival probability without system failure before a specified time
could also be informative. The mean-time-to-failure (MTTF) or first
passage time is the expected time for the system to pass from a state
of no element down to a state of ℓ or more elements down for the
first time, where ℓ is specified by the operation logic.

   Let D(t) denote the number of components down at time t, and
now define the probabilities $\left\{P_j(t); \ t{\geq}0, \ j{=}0, \ 1, \ 2, \ \ldots\ldots, \ \ell{-}1\right\}$:

$$P_j(t) = P\left\{D(t) = j, \ 0{\leq}D(t'){\leq}\ell{-}1, \ \text{all} \ t'{\leq}t \big| D(0) = 0\right\} \qquad (2.6)$$

$$j = 0,1,2\ldots,\ell{-}1 \ .$$

Thus, $P_j(t)$ is the probability that a system of m units has j units
down at t, and that there have been no more than $\ell{-}1$ down simultaneously previously (i.e., at any time between zero and t). In a
similar manner as used for the derivation of equations (2.4) and
(2.5), it may be shown that $\left\{P_j(t)\right\}$ satisfies a system of dif-

ferential equations:

$$P_j(t) = -(\lambda_j + \mu_j + c)P_j(t) + \mu_{j+1}P_{j+1}(t) + \lambda_{j-1}P_{j-1}(t) \qquad (2.7)$$

$$j = 0,1,2,\ldots,\ell-2 ,$$

and

$$P_{\ell-1}(t) = -(\lambda_{\ell-1} + \mu_{\ell-1} + c)P_{\ell-1}(t) + \lambda_{\ell-2}P_{\ell-2}(t) . \qquad (2.8)$$

Note that equation (2.8) excludes any probabilities which involve ℓ or more elements in a failure state because the system operation logic requires that that state must not have occurred before the time t.

The solution of equations (2.7) and (2.8) can be obtained by using a suitable numerical method. In the cases that the failure rates and repair rate are constants, the Laplace transform method can be effective. Some solutions for low order values of ℓ are summarized as below:

Case 1: m-out-of-m system

$$P_0(t) = e^{-(\lambda_0 + c)t}.$$

Case 2: (m-1)-out-of-m- system

$$P_0(t) = A_{1,1}e^{s_1 t} + A_{1,2}e^{s_2 t}$$

$$P_1(t) = A_{2,1}e^{s_1 t} + A_{2,2}e^{s_2 t}$$

where s_1 and s_2 are the roots of the characteristic equation

$$(s+\lambda_0+c)(s+\lambda_1+\mu_1+c) - \lambda_0\mu_1 = 0$$

and $\quad A_{1,1} = (s_1+\lambda_1+\mu_1+c)(s_1-s_2)^{-1}$

$$A_{1,2} = -(s_2+\lambda_1+\mu_1+c)(s_1-s_2)^{-1}$$

$$A_{2,1} = \lambda_0(s_1-s_2)^{-1}$$

$$A_{2,2} = -A_{2,1} .$$

Case 3: (m-2)-out-of-m system

$$P_i(t) = \sum_{j=1}^{3} A_{i+1,j}e^{s_j t} , \qquad i=0,1,2 ,$$

where s_j, $j = 1,2,3$, are the roots of the characteristic equation

$$s^3 + (w_0+W_1+W_2) s^2 + (W_0W_1+W_1W_2 +$$

$$W_0W_2-W_3-W_4) \, s + (W_0W_1W_2-W_0W_3-W_2W_4) = 0$$

and $$W_0 = \lambda_0+c$$

$$W_1 = \lambda_1+\mu_1+c$$

$$W_2 = \lambda_2+\mu_2+c$$

$$W_3 = \lambda_0\mu_1$$

$$W_4 = \lambda_1\mu_2 \; .$$

The coefficients, $A_{i,j}$, are given by

$$A_{i,j} = g_i(s_j)Q^{-1}(s_j), \qquad i = 1,2,3; \quad j = 1,2,3 \; ,$$

where

$$g_1(s) = (s+W_1)(s+W_2)-W_4$$

$$g_2(s) = \lambda_0(s+W_2)$$

$$g_3(s) = \lambda_0\lambda_1$$

and $Q(s_j)$ is the derivative of the characteristic equation evaluated at $s = s_j$.

Denote by T_ℓ the elapsed time for the system to pass from a state of no element down to a state of ℓ or more elements down for the first time, then the system survival probability, $P_s(t)$, for a specified time duration is given by

$$P_s(t) = P\left\{T_\ell > t \mid D(0) = 0\right\} .$$

The survival probability may be obtained directly from the solution of equations (2.7) and (2.8). It is clear from equation (2.6) that $P_s(t)$ is given by

$$P_s(t) = \sum_{j=0}^{\ell-1} P_j(t) . \tag{2.9}$$

The mean time to system failure can be directly obtained by

$$MTTF = \sum_{j=0}^{\ell-1} \int_o^\infty P_j(t)dt . \tag{2.10}$$

It may also be obtained by evaluating the derivative of the Laplace transform solutions of Equations (2.7) and (2.8) and setting s=0. For the above three cases, the mean times to system failure are given by

Case 1:

$$E(T) = (\lambda_0 + c)^{-1} .$$

Case 2:

$$E(T_2) = (\lambda_0 + \lambda_1 + \mu_1 + c) \left[(\lambda_0 + c) (\lambda_1 + \mu_1 + c) - \lambda_0 \mu_1 \right]^{-1}.$$

Case 3:

$$E(T_3) = \left[W_1 W_2 + \lambda_0 W_2 - \lambda_1 W_4 \right] \left[W_0 W_1 W_2 - W_0 W_4 - W_2 W_3 \right]^{-1}.$$

An alternative method for computing T_ℓ utilizes a recursive calculation. The idea is that T_ℓ may be expressed as the sum of independent first-passage times, first from zero down to one down, then from one down to two down, finally, from $\ell-1$ to ℓ. It may be shown that

$$P\left\{ T_\ell > t \,|\, D(0) = 0 \right\} = P\left\{ T_\ell^* > t \,|\, D(0) = 0 \right\} e^{-ct} \qquad (2.11)$$

where T_ℓ^* is the system survival time of a simple birth and death process or equivalently, a special case of Model type I with $c=0$. The Laplace transform of equation (2.11) can therefore be expressed as

$$\int_0^\infty e^{-st} P\left\{ T_\ell > t \,|\, D(0) = 0 \right\} dt = \int_0^\infty e^{-(s+c)t} P\left\{ T_\ell > t \,|\, D(0) = 0 \right\} dt \qquad (2.12)$$

$$= (s+c)^{-1} \left\{ 1 - E\left[e^{-(s+c)T_\ell^*} \right] \right\};$$

see Feller [5] for the latter expression. In other words, the transform of system time to failure, T , under both chance failure and common mode failure influences can be expressed in terms of the transform for time to failure under chance failure influence alone. In addition, it may be shown that [5]

$$E\left[e^{-(s+c)T_\ell^*}\right] = \prod_{i=0}^{\ell-1} \phi_i(s+c) \qquad (2.13)$$

and that ϕ_i satisfies the following recursion relation,

$$\phi(s+c) = \frac{\lambda_i}{\lambda_i+\mu_i+s+c-\mu_i\phi_{i-1}(s+c)} , i = 1,2,3... , \qquad (2.14)$$

where

$$\phi_0(s+c) = \frac{\lambda_0}{\lambda_0+s+c} .$$

Thus the transform of $P\left\{T_\ell > t;c\right\}$ can be derived, and inverted either algebraically or numerically, using a suitable computer code. The mean time to system failure is

$$E\left[T_\ell;C\right] = \frac{1 - E[e^{-cT_\ell^*}]}{c}; \qquad (2.15)$$

the latter expression may be evaluated by use of (2.13) and subsequent formulas, including the recursion (2.14). This method becomes
useful in evaluating the mean-time-to-failure when the system is
highly redundant. Table 2 shows some results for a one-out-of-three
system with different constant values of common mode failure rate.

Table 2.   Survivor Probabilities $P_s(t)$ For a one-out-of-three System
            of Model Type I.

$$\mu = 1.0, \qquad \lambda = 0.01$$

Time	C = 0.0	C = 0.0001	C = 0.001
0	1.000000	1.000000	1.000000
10	.999975	.998976	.990025
20	.999946	.997948	.980146
30	.999917	.996922	.970365
40	.999888	.995896	.960682
50	.999859	.994872	.951095
100	.999713	.989766	.904578
200	.999422	.979632	.818258
500	.998550	.949850	.605651
1000	.997098	.902211	.366812
1500	.995648	.856962	.222159
2000	.994200	.813982	.134550
3000	.991310	.734381	.049354
4000	.988429	.662564	.018104
5000	.985557	.597770	.006641
10000	.971318	.357328	.000044
MTTF	3.5×10^5	10×10^3	10×10^2

3. Models for Characterizing Environment Changes

Model Type II. Since Model 1 treats any common mode as being catastrophic, it may not satisfactorily represent common, but less
stringent, environmental stresses, human errors, or the effects of
maintenance procedures. Model II, a random environment model, is
proposed as being a possibly more realistical in assessing the
effects of these mild changes on the system performance.

3.1. Model Description. A random environmental system model
may be specified as follows:

A system is made up of m functionally identical components
or subsystems. Each component has a chance failure rate
$\lambda(i)$, and repair rate $\mu(i)$, provided the environment is in
state i; here i = 1,2. Furthermore, the environment
changes in accordance with a Markov process; $\alpha(i)$ is the
rate at which the environment leaves state i, so $(\alpha(i))^{-1}$
is the expected duration time for state i. Conditional
upon the environmental state i, the number of components
down is again a Markov process. The system operates
successfully until ℓ units out of m are down simultaneously.

The above model assumes that a system operates with "normal"
failure and repair rates $\lambda(1)$, $\mu(1)$ for a period of time which is exponentially distributed with parameters $\alpha(1)$. Then, owing to the

occurrence of an operating environment change, these parameter values
are assumed to give way to others, namely $\lambda(2)$, and $\mu(2)$. These
"stress associated" values are perhaps such that $\lambda(2) >> \lambda(1)$, and
$\mu(1) > \mu(2)$. The latter values prevail for an exponentially distributed time of parameter $\alpha(2)$, after which the environment changes and
the parameters shift again to $\lambda(1)$, $\mu(1)$, and so the process
continues. Model type II has been introduced in mathematical
ecology [6], and in certain queueing studies [7,8]. They are also
appropriate in the present context, supplementing the severe shock
models.

3.2. First Passage Time and Survival Probabilities. The
steady-state system availability of this model may not be a useful
measure of system performance because of the presence of stressed
failure rate. The mean-time-to-failure and system survival probabilities for a specified period of time may be more indicative of the
system effectiveness.

In order to describe the process state at time t, the values of
two state variables are needed: $D(t)$ describes the number of units
down at t, while $X(t)$ describes the environmental state. The
possible values of $D(t)$ are $0,1,2,\ldots m$, and those of $X(t)$ are 1 and
2. Usually $X(t) = 1$ will mean the environmental state is normal or
benign, while $X(t) = 2$ will mean that the environment is stressful,
perhaps for natural or man-related reasons. Failures that occur
during such times may be attributed to common mode initiators. Table

3 shows the transition probabilities during the time interval

(t, t+dt) for a system which is in state j and the environment

$X(t) = i$, where the failure rate and repair rate are defined the same

as in the previous section.

In order to study the system survival probabilities, dif-

ferential equations for the model can be derived in the same manner

as for the model type I. Denote the state prob-

abilities, $\left\{ P_j (i,t,u) \right\}$, as below:

$$P_j(i,t;u) = P\left\{ D(t) = j, \ 0 \le D(t') \le \ell-1, \ \text{all } t' \le T, \ X(t) \right.$$

$$\left. = i \,|\, D(0) = 0, \ X = u \right\}$$

$$j = 0,1,2,\ldots,\ell-1;$$

$$i,u = 1,2, \text{ and}$$

where

$$X(t) = \begin{cases} 1 & \text{if environment in state 1 at t,} \\ \\ 2 & \text{if environment in state 2 at t.} \end{cases}$$

Thus, $P_j(i,t;u)$ is the probability that j units are down at t, there

have been no more than $\ell-1$ down simultaneously up to time t, and the

environmental state is i (i=1, or 2). All of this depends upon the

initial conditions $D(0) = 0$ and $X(0) = u$.

Table 3. Transition States and Transition Probabilities for
                  Model Type II.

     State Transition   Probability

From D(t) = j, X(t) = i to D(t+dt) = j+a, X(t+dt) = i $\lambda_j(i)dt$

                              D(t+dt) = j-1, X(t+dt) = i $\mu_j(i)dt$

                              D(t+dt) = j,   X(t+dt) = i $1-[\lambda_j(i)+\mu_j(i)+$

                                                                          $\alpha(i)]$ dt

From D(t) = j, X(t) = 1 to D(t+dt) = j,   X(t+dt) = 2 $\alpha(1)dt$

From D(t) = j, X(t) = 2 to D(t+dt) = j,   X(t+dt) = 1 $\alpha(2)dt$

where j=0,1,2,........,m and i = 1,2.

Based on the model formulation it may be shown that the state
probabilities satisfy the following system of differential equations
(the initial conditions may be suppressed):

$$P_j(1,t) = -\left[\lambda_j(1)+\mu_j(1)+\alpha(1)\right]P_j(1,t)+\mu_{j+1}(1)P_{j+1}(1,t)$$

$$+ \lambda_{j-1}(1)P_{j-1}(1,t)+\alpha(2)P_j(2,t) \tag{3.3}$$

$$P_j(2,t) = -\left[\lambda_j(2)+\mu_j(2)+\alpha(2)\right]P_j(2,t)+\mu_{j+1}(2)P_{j+1}(2,t)$$

$$+ \lambda_{j-1}(2)P_{j-1}(2,t)+\alpha(1)P_j(1,t).$$

For j=0,1,2,......, ℓ-2, while the pair of equations for j=ℓ-1 remain
the same except that the terms $\mu_\ell(1)$ $p_\ell(1,t)$ and $\mu_\ell(2)$ $P_\ell(2,t)$ are

removed. For the constant failure and repair rates, the system of linear differential equations (3.3) can be solved by the Laplace transform method. The system survivor probability, $P_s(t)$, for a specified time duration t is, again, given by

$$P_s(t) = P\left[T_\ell > t \mid D(0) = 0, X(0) = U\right] = \sum_{j=0}^{\ell-1} \left[P_j(1,t) + P_j(2,t)\right]$$

where U is specified by the initial conditions which may well be an important factor to the systems effectiveness. For all practical purposes U can be assumed to be 1; that is, all the components initially are in the normal state. When the parameters in the model are time dependent, numerical methods may have to be used to solve for P_j .

To evaluate the mean-time-to-failure, the method of recursion discussed for model type I may be generalized to handle this random environment model. Because of the complexity of the method, only a brief outline of the recursion procedure is given in the appendix; a detailed derivation of the method is to be presented in a special technical report [9]. The method of recursion has been applied in this paper to the random environment model. The mean time to system failure for a three-component and a five-component model are shown in Tables 4 and 5, respectively. It is noticed that the initial state of the environment is the predominant factor affecting the MTTF. The

effect of environmental change decreases as the redundancy reduces,

that is for a k out of m system the environmental change becomes less

significant as k increases. Note that when $\lambda(1) = \lambda(2)$ the model is

equivalent to a simple reliability calculation without any common

mode effect.

Table 4.   Mean time to system failure of a three-component system
                for the random environment model, Model Type 2.

$\lambda(1) = 0.01,$ $\mu(1) = \mu(2) = 1.0,$ $\alpha(1) = 0.001,$ and $\alpha(2) = 1.0$

Operation logic	X(0)	(2) = 0.01	(2) = 9.00	(2) = 99.0
One-out-of-three	3.5×10^5		1226	1016
			230	19
Two-out-of-three	1750		673	640
			59	5
Three-out-of-three	33.0		32.2	32.2
			1.2	0.1

Table 5. Mean time to system failure of a five-component system
          for the random environment model, Model Type 2.

$\lambda(1) = 0.01$, $\mu(1) = \mu(2) = 1.0$, $\alpha(1) = 0.001$ and $\alpha(2) = 1.0$

Operation logic	X(0)	$\lambda(2) = 0.01$	$\lambda(2) = 9.00$	$\lambda(2) = 99.0$
One-out-of-five	1	9×10^7	1317	1024
	2		318	4
Two-out-of-five	1	5×10^6	1158	1013
	2		159	13
Three-out-of-five	1	7×10^4	1058	980
	2		89	8
Four-out-of-five	1	545	359	354
	2		17	2
Five-out-of-five	1	20	20	20
	2		0.5	0.5

4. <u>Discussion</u>. The proposed models could perhaps be applicable to quantify the design base safety analysis. For instance, the common mode failure rate, C, of the model I may be used to represent the arrival rate of a design base earthquake, or the occurrence rate of a design base tornado missile or maintenance error in the system analysis to quantify the effects of those devastating natural phenomenon on the system performance. Since the model I assumes the catastrophic failure in the presence of a common mode initiator, the results calculated from this class of models could be overly conservative and, at best, be used as a lower bound of the system performance. The class of random environment models, model II, may be more adequate in characterizing the behavior of the common mode failure initiators and may provide a realistic assessment of system performance. However, the data base required in the application of the model II may be extensive and, sometimes, unavailable. Using this class of models for the parameter sensitivity analysis supplemental to the result obtained from the model I may be satisfactory for the design safety analysis.

It is worthwhile to mention that these two models may also be useful to analyze systems which consist of nonidentical components, having different failure and repair rates. Furthermore, for Model I, the common mode failure rate could be made to be state dependent. That is, the magnitude of the common mode failure rate could depend on the number of components which are still in operation. This feature might be useful when the common mode failure initiators are induced by physical parameters, say, pressure or temperature.

Attempts have been made to evaluate these models for some simple time dependent failure and repair rates using a predictor-corrector numerical scheme. It appears that a numerical unstability makes solution difficult for the cases studied. It is believed that the variation of $P_j(t)$ with time induces numerical difficulties. It would appear that the coefficient matrix of the differential equations may be ill-conditioned. Further investigation seems warranted.

Finally, for the examples described the effects of common mode failure on system effectiveness may be a reduction of several orders of magnitude in the mean-time-to-system failures (Tables 2, 4 and 5). When the system is highly redundant, common mode failures can thus potentially become the controlling failure mechanism, as is implicitly indicated in Tables 4 and 5 by comparing the two one-out-of-n values; because of this, further increase in redundancy may not be justifiable (in this case one-out-of-three is as good as one-out-of-five).

Appendix

A-1. Method of Recursion.

Let $G_k(i,j,s)$ be the Laplace transform, thus

$$G_n(i,j;s) = E\left[e^{-sT_n}; X(T_n) = j | D(0) = n, X(0) = i\right] , \qquad \text{(A-1)}$$

$n=0,1,2,\ldots$ Here X represents the state of the environment process: $X=1$, or $X=2$. T_n represents the first-passage time from n units down to $n=1$ down. $G_n(i,j;s)$ is the Laplace transform of the time to pass from n down, the environment being in state i, at some initial instant $t=0$, to $n+1$ down, the environment being in state j. It may be shown that

$$G_n(i,j;s) = \frac{\lambda_n(i)}{d_n(i)} \ell_i(j) + \frac{\mu_n(i)}{d_n(i)} \sum_{k=1}^{2} G_{n-1}(i,k;s)G_n(k,j;s) \qquad \text{(A-2)}$$

where

$$d_n(i) = \lambda_n(i) + \mu_n(i) + \alpha(i) + s \qquad \text{(A-3)}$$

and the indicator function

$$\ell_i(j) = \begin{cases} 1 \text{ if } i=j , \\ 0 \text{ if } i \neq j . \end{cases}$$

Examination of (A-2) reveals that, given n, and i, two simultaneous linear equations may be solved to yield G_n in terms of G_{n-1}; a beginning is made by solving

$$G_0(1,j;s) = \frac{\lambda_0(1)}{d_0(1)} \ell_1(j) + \frac{\alpha(1)}{d_0(1)} G_0(2,j;s)$$

(A-4)

$$G_0(2,j;s) = \frac{\lambda_0(2)}{d_0(2)} \ell_2(j) + \frac{\alpha(2)}{d_0(2)} G_0(1,j;s), \quad j=1,2 ,$$

and the solutions then provide inputs to the equations for G_1, and so forth. Finally, define

$$P_n(i,j;s) = E\left[e^{-sT}n; X(T_n) = j|D(0) = 0, X(0) = i\right],$$

(A-5)

the Laplace transform of the first-passage time to n+1 from zero, assuming that $X(0) = i$, and that $X(t_n) = j$. Then by the Markov character of the process,

$$P_n(i,j;s) = \sum_{k=1}^{2} P_{n-1}(i,k;s)G_n(k,j;s)$$

(A-6)

or, in matrix notation,

$$\underset{\sim}{P}_n\underset{\sim}{P}_{n-1}\underset{\sim}{G}_n = \underset{\sim}{G}_0\underset{\sim}{G}_1\cdots\underset{\sim}{G}_n.$$

(A-7)

In order to calculate the transform of time to system failure, compute

$$\underset{\sim}{P}_{\ell-1}\underset{\sim}{I} = \underset{\sim}{G}_0\underset{\sim}{G}_1\cdots\underset{\sim}{G}_{\ell-1}\underset{\sim}{I} \tag{A-8}$$

where $I = (1,1)^T$ is a column vector with unit entries; the ith row of the above is equal to the desired $E \left[e^{-sT} \Big|_\ell D(0) = 0, X(0) = i \right]$.

A-2.   Mean Time to System Failure.

In order to compute the mean time to system failure, one may differentiate in matrix chain rule fashion with respect to s at s=0 through (A-8).

$$E\left[T_\ell | D(0) = 0, X(0) = i\right] = -\underset{\sim}{P}_{\ell-1}\underset{\sim}{I} = -\sum_{j=0}^{\ell-1} \left(\prod_{i=0}^{j-1} \underset{\sim}{G}_i\right) \underset{\sim}{G}_j \left(\prod_{i=j+1}^{\ell-1} \underset{\sim}{G}_i\right). \tag{A-9}$$

The derivative terms, $\underset{\sim}{G}$, are obtained by differentiating through the recursion (A-7), setting s=0, and solving the resulting recursion formulas.   Detailed formulas will appear in a later report.

A-3.   Probability of System Failure During j^{th} Environmental State (e.g., Severe or Benign).

Equation (A-8) yields the probability that the system fails during either of the two environmental states.   Simply put s=0; then the element i the ith row, j^{th} column of $\underset{\sim}{P}_{\ell-1}$ equals the probability that a system that begins with D=0 eventually reaches D=ℓ, and does so while experiencing environmental state j.

ACKNOWLEDGEMENTS

The authors wish to express appreciation to Dr. G. S. Lellouche for his suggestions and comments throughout the preparation of this report.

REFERENCES

[1] G. E. APOSTOLAKIS, The Effect of a Certain Class of Potential Common Mode Failures on the Reliability of Redundant Systems, Nucl. Engrg. and Design, 36(1976).

[2] D. P. GAVER, Failure Time For a Redundant Repairable System of Two Dissimilar Units, IEEE Trans. on Reliability, R-13(1964), pp. 14-22.

[3] R. HARRIS, Reliability Applications of a Bivariate Exponential Distribution, Operations Research, 16(1968) pp. 18-27.

[4] S. KARLIN, A First Course in Stochastic Processer, Academic Press, New York, 1966.

[5] W. FELLER, An Introduction to Probability Theory and Its Applications, II(1971), 2nd Edition, John Wiley and Sons, New York.

[6] ROBERT M. MAY, Stability and Complexity in Model Ecosystems, Princeton Univ. Press, Princeton, NJ, 1973.

[7] U. YECHIALI and P. NAOR, Queueing Problems With Heterogeneous Arrivals and Service, Operations Research, 19(1971), pp. 722-734.

[8] P. PURDUE, The M/M/i Queue in a Markovian Environment, Operations Research, 22(1974) pp. 562-569.

[9] B. CHU and D. GAVER, EPRI Special Technical Report, (to be published).

Topic 4
METHODOLOGY I

DISCUSSION BY THE EDITORS

Topic 4
Methodology I

The first three papers in this section present methods which are either based upon previously performed reactor plant risk analyses or presuppose that such an analysis has been performed in at least a gross fashion. All three papers illustrate uses of a risk analysis beyond those which might be directly attached to the risk analysis itself. The remaining papers deal with specific problem areas in a risk assessment; i.e. reliability assessment of the containment for a pressurized water reactor and analysis of accidental transients.

The first paper, by M. Maekawa, W. E. Vesely, and N. C. Rasmussen, contains a method for obtaining a set of equations relating the public risk in potential nuclear reactor accidents to variables such as population distribution and radioactive releases. Thus changes in the risk curve can be determined for postulated changes in the variables. Choosing the variable of interest as population density, the authors present and solve a hypothetical site selection problem. A hypothetical radioactive release problem, involving an iodine cleanup system, is also formulated.

In the second paper, G. R. Burdick, D. M. Rasmuson and J. Weisman, illustrate how to construct models describing availability and costs associated with redundancy and diversity allocations for nuclear systems and, through use of a random number technique, how to use the models to construct functions of cost versus system availability. The functions provide curves which approximate the best cost obtainable for a given level of availability. The method is extended to the case where a gross risk analysis has been performed and acceptable levels of risk have been defined for plant operation. In this case, the cost versus availability functions are used in an optimization procedure which performs trade-offs of cost and availability among safety systems. The solution set of unavailabilities from this problem are used as constraints in an integer optimization problem, the solution of which determines the systems optimal configurations. The authors point out additional design applications of the cost versus availability curves. Although the examples used to illustrate the methods are taken from a simplified liquid metal reactor design, the approach is general and hence applicable to other systems as well.

The third paper describes a procedure for establishing the ranking of research and development goals through the use of probabilistic risk assessments. The authors, M. I. Temme and K. A. El-Sheikh, partition consequences, determined from a risk assessment, into categories and assign event sequences to each consequence category. Three measures of importance are defined which an analyst could use to select the most important category. Using probability as a ranking criterion, the most important event or event sequence is found for the most important category. The procedure is thus a systematic technique for determining and focusing attention on those areas where R & D action could be applied to achieve significant risk reduction.

The fourth paper, by G. I. Schuëller, presents a concept for the reliability assessment of the containment of a boiling water reactor. The paper discusses the statistical nature of various external hazards such as earthquakes, pressure waves, aircraft crashes, as well as internal load conditions such as the LOCA. The hazards are considered rare events in a discussion of a key problem which is the establishment of a relation between the frequency of occurrence of the hazards and their intensity. A numerical example is presented to illustrate the proposed methodology.

The fifth paper, by J. Amesz, S. Garriba, and G. Volta, establishes a theoretical framework for computing probability distributions of loads and residual strength in the primary cooling system of nuclear reactors. Two cases are considered. One case deals with the design stage where designers must forecast system performance based on available data; the second case treats the operational stage where information is available about the states of the system and its components. The developed methods are sophisticated and general in nature.

AN APPLICATION OF RISK ANALYSIS: FUNCTIONAL RELATIONSHIPS OF NUCLEAR RISKS†

MITSURU MAEKAWA*, WILLIAM E. VESELY** AND NORMAN C. RASMUSSEN*

Abstract. A method is developed for deriving an explicit set of
equations relating the public risk in potential nuclear reactor accidents to more basic variables, such as population distribution and
radioactive releases. The methodology consists of two steps. The
first step involves parametrically fitting the risk distributions of
frequency versus consequence. The second step involves relating the
distribution parameters to the basic variables of interest. Regression
techniques are used for this second step.

The methodology is demonstrated on examples based on the results
of the Reactor Safety Study (WASH 1400) (1). The calculated distributions of early fatalities in nuclear reactor accidents and the historical records of fatalities in hurricanes, tornadoes, earthquakes and
dam failures are examined to determine an appropriate family of parametric distributions. From these examinations, the Weibull distribution is found to be appropriate for all of the examined events.

A set of equations is then derived which relates the population
distribution and the parameters of the Weibull distributions for early
fatalities from PWR accidents. The derived equations are useful in
analyses of population effects on risk. Regression equations relating
the parameters to the characteristics of radioactive releases are also
derived. The derived equations are useful for evaluating release
effects on risk.

1. Introduction. The recently released Reactor Safety Study (1)

estimated the risk to the public for a 100 reactor nuclear power indus-

---

† This paper is based on the Ph.D. thesis of Mitsuru Maekawa, "A
Method of Risk Analysis of Nuclear Reactor Accidents", Department of
Nuclear Engineering, MIT, July 1976. Funded by NRC contract AT(49-29)
0263.

* Massachusetts Institute of Technology, Cambridge, MA 02139.
** U.S. Nuclear Regulatory Commission, Washington, DC 20555.

try in the United States. The study considered six risks associated
with hypothetical reactor accidents. These risks were early fatalities,
early injuries, latent cancers, latent thyroid injury, genetic effects,
and property damage. One of the principal results of the study was a
risk curve for each of these effects in the form of a complementary
cumulative probability function.

The generation of the risk curves was accomplished using a complex
model (2) which related the particular consequence to the basic variables. The basic variables considered were the population density, the
magnitude of the radioactive release, the atmospheric stability, the
wind speed, the mixing height, the type of precipitation if any, and
the effectiveness of evacuation. To generate the risk curves given in
WASH 1400 distributions of possible values for each of the basic variables were approximated by histograms. The actual calculation considered 96 different population densities, 14 different radioactive releases, and 90 different weather categories. Since all combinations of
these variables were in essence calculated the process required over
120,000 different accidents to be evaluated. This was a lengthy calculation even on a very large, fast computer.

This paper describes a simplified analytical method for generating
these risk curves for hypothetical reactor accidents. The analytical
method has been developed in detail for risk curves for early fatalities. By investigating various families of distributions, the Weibull
distribution is found to adequately fit the risk curves. A regression

analysis is then used to express the Weibull coefficients in terms of basic input variables. Once the relationship between the Weibull coefficients and the basic input variables has been determined it is possible to use this analytical relation to determine the changes in the risk curve for any postualted change in the basic variables. Examples are given for changes in population density and for changes in design of the radioactive iodine removal system.

2. Selection of the Weibull Distribution. There was no a priori reason for selecting any particular distribution to represent the risk curves. Based on risk curves of historic data four candidate distributions were chosen, the exponential, the gamma, the Weibull and the log normal. These candidate distributions were then tested by fitting them to the United States fatality risk curves for hurricanes, earthquakes, dam failures, tornadoes, and to three different nuclear risk curves generated by the WASH 1400 consequence model.

Each of the candidate distributions were fitted to the seven risk curves and the residual mean square deviations were calculated. In all cases the Weibull distribution gave either the smallest or next to the smallest deviation.

As a result of the close fit to all the curves studied, the Weibull distribution of the following form was chosen.

$$(1) \quad f(x) = \alpha \cdot (\frac{\beta}{\eta}) \cdot (\frac{x-x_o}{\eta})^{\beta-1} \cdot \exp\left[- (\frac{x-x_o}{\eta})^{\beta}\right]$$

where

$$x \geq x_o, \ x_o \geq 0, \ \alpha > 0, \ \beta > 0 \text{ and } \eta > 0$$

Its moments are of the form:

(3) $M_1 = \alpha \cdot \eta \cdot \Gamma (1 + \frac{1}{\beta})$

(4) $M_2 = \alpha \cdot \eta^2 \cdot \Gamma (1 + \frac{2}{\beta})$

(5) $M_m = \alpha \cdot \eta^m \cdot \Gamma (1 + \frac{m}{\beta})$

where $\Gamma(\cdot)$ is the Gamma function.

3. Procedure for the Regression Analysis. Having selected the Weibull distribution, the next step is to derive the equations that relate the Weibull parameters to the basic variables that affect the consequences of nuclear reactor accidents. There are six fundamental steps in this process.

1) Identification of the basic variables

2) Selection of the dependent variables

3) Determination of data to be used in the regressions

4) Formulation of the regression equations

5) Estimation of the regression constants

6) Determination of the adequacy of derived equations

3.1. Identification of Basic Variables. As a result of
sensitivity studies carried out with the WASH 1400 consequence
code it was known that population density, radioactive release
magnitude, and evacuation procedures had a strong influence on the
risk curves. The meteorological conditions varied somewhat from
site to site but had a less marked effect on the risk curves. For
the purposes of the analysis the population density and the release
magnitude were specifically chosen as the basic variables of interest.

3.2. Selection of Dependent Variables. There are a number of
possible dependent variables that might be used to relate the basic
variables to the fitted distribution. They include among others:

1) Scale factor, shape factor, and normalization
   constant of the fitted distribution

2) Risk moments about a specific magnitude of
   consequence

3) Slope of the complementary cumulative distri-
   tion at a given consequence magnitude.

For the analysis the first two moments, M_1 and M_2, plus the normalization constant α were chosen as the values describing a particular Weibull. The values of η and β can then be estimated from the moments using the following equations.

(6)
$$\frac{[\Gamma(1 + \frac{1}{\beta})]^2}{\Gamma(1 + \frac{2}{\beta})} = \frac{M_1^2}{M_2\alpha}$$

(7)
$$\eta = \frac{M_1}{\alpha\Gamma(1 + \frac{1}{\beta})}.$$

3.3. Assembling of Needed Data. In general the data needed for the regression analysis may be obtained from historical records or from a calculational model. In the case of nuclear reactor accidents no historical data is available so the calculational model of WASH 1400 was used to generate data for a variety of postulated accident situations. The examples discussed later will illustrate how the specific data was used.

3.4. Formulation of Equations to Relate Regressor and Dependent Variables. The general regression equation is:

(8)
$$y = h(z_1, z_2, \ldots, z_m \,|\, \tau_1, \ldots, \tau_k) + \varepsilon$$

where ε is a random error variable, y is the dependent variable, the z's the regressor variables, and the τ's the unknown regression constants. In practice it is desirable to limit the number of constants to a relatively small number to reduce the overall calculation complexity.

The significance of adding additional unknown constants is tested using an F test (3).

3.5. Test of the Adequacy of the Derived Equations. The following criteria were used to investigate the adequacy of the derived equations.

1) The F value can be used to measure the significance of additional variables. These observed
   significance levels can be used to determine the most
   suitable equation from several possibilities.

2) There should be no systematic biases in the fitting,
   i.e., no systematic underestimations or overestimations.

3) The risk distribution derived using the assumed basic
   variable relations should closely fit the empirical
   risk distribution derived from the data.

4. Regression Analysis for Population Distribution.

4.1. Population Distribution. The population density is known to have a significant impact on the consequences. Following the format of WASH 1400 the population as a function of position is defined using a polar coordinate system centered at the reactor site.

The population in the kth annular segment in sector j can be expressed as

(9)
$$N_{jk} = \int_{r_k - \Delta r_k/2}^{r_k + \Delta r_k/2} n_j(r)\,dr$$

where

$$n_j(r) = \int_{\frac{\pi}{8}} n(r,\theta)d\theta$$

and $n(r,\theta)$ is the population as a function of position. In this example the radial distance is divided in 34 segments.

4.2. Regression Data and Analyses. The data base consisted of the population distributions within 500 miles of the 68 sites on which the first 100 United States reactors will be built. The population data was obtained from the 1970 Census Bureau data (4). This data included a wide variation in total population ranging from over 100 million in the highest case to about 7 million in the lowest case.

The WASH 1400 consequence code was used to calculate fatality risk curves for a PWR with typical eastern river valley meteorology for each of these 68 population distributions. A Weibull distribution was then fitted to each of these risk curves. The first and second moments and the normalization constants were then determined for each of the 68 population cases. This became the basic data for the remaining analysis process.

Using the data base, the next step was to formulate a model to relate the dependent variables M_1, M_2, and α to N_{jk}, the populations in the annular segments. For a given trial, the consequence code will calculate the number of fatalities occurring for a particular value of the radioactive release, a particular weather state, a particular evacuation speed, a given wind direction, and a specific population. The fatalities, x, in a given trial can be mathematically expressed as

(10) $\qquad x = \int_r A(r) \cdot n_j(r) \cdot dr.$

The variable $A(r)$ is the number of fatalities per unit population at a given radial distance and $n_j(r)$ is the population density per unit radius in the wind direction j.

A typical calculation at a given site considers 90 different weathers, 9 different radioactive releases, 3 evacuation speeds and 16 different population distributions, each with a known probability. Each trial gives a number of fatalities, x, and a probability of occurrence. These results are combined to produce a frequency distribution of accident probability versus magnitude.

The first moment of the frequency distribution over all the trials is the expected value of x noted by $E(x)$:

$$M_1 = E(x).$$

Since the 16 wind directions are treated as equally likely in the code, M_1 can be expressed as:

(11) $\qquad M_1 = \frac{1}{16} \sum_j \int E[A(r)] \cdot n_j(r) dr.$

Where the expectation is now for a given direction it is useful to define a function $a(r)$ as:

(12) $\qquad a(r) = \frac{1}{16} \cdot E[A(r)].$

Since M_1 is the annual expected number of fatalities, $a(r)$ is just the expected fatalities per individual per year at distance r. In this paper $a(r)$ will be called the first transfer function.

In a like manner the second risk moment M_2 is the expected value

of x^2. From Eq. (10), M_2 can be expressed as:

(13) $M_2 = E[x^2] = \frac{1}{16} \sum_j \int_r \int_{r'} E[A(r) \cdot A(r')] \cdot n_j(r) \cdot n_j(r') dr dr'$.

The second transfer function $b(r,r')$ is defined as:

(14) $b(r,r') = \frac{1}{16} E[A(r) \cdot A(r')]$

giving M_2 as:

(15) $M_2 = \sum_j \int_r \int_{r'} b(r,r') \cdot n_j(r) \cdot n_j(r') dr dr'$.

The quantity $b(r,r')$ can be interpreted as the probability that

individuals at r and r' will both be killed in the same accident.

The final parameter characterizing the Weibull is the normaliza-

tion constant α and is the probability per year that an accident will

kill one or more people. The probability can be expressed as:

(16) $\alpha = E[H(\int_{d_j}^{\infty} A(r) \cdot n_j(r) dr)]$

where the step function $H = 0$ for $x = 0$ and $H = 1$ for $x > 0$, and d_j

is the closest distance to the site that people inhabit. From Eq. (16)

the third and final transfer function $c(r)$ is defined as:

(17) $\alpha = \sum_j [c(r)]_{r=d_j}$.

4.3. Evaluation of the Transfer Functions.   From execution of the consequence code, the values of $a(r)$ at specific radial distances k are plotted versus r in Figure 1. The value of $a(r)$ was estimated by taking the number of fatalities in a given radial segment divided by the population in the segment. Three different candidate functions were fitted to the data. The two exponential forms given in Figure 1 gave the best fits. The partial F statistic showed that the two added constants in the second form were statistically significant. However, a careful analysis of the differences in the values of the first risk moment showed that the improvement obtained by using two added constants was small compared to the modeling uncertainty in the consequence code calculation so the simpler form $a_1 \exp(-a_2 r)$ was used. Fitting analyses were also done with candidate functions for $b(r,r')$ and $c(r)$. In each case based on statistical analyses and code considerations the chosen functions were:

$$b_1 \exp[-b_2 \cdot (r+r')] \cdot \exp[-b_3 \cdot |r-r'|] \text{ and } c_1 \exp[-c_2 \cdot r], \text{ respectively.}$$

5.   Potential Applications of Regression Results.

5.1. A Hypothetical Site Selection Problem. Consider a straight section of a river along which four cities are located, as illustrated in Figure 5. The population density in these cities is assumed to follow a Gaussian distribution with the σ's indicated. The total population of each city is also indicated. The problem is to find where a reactor (PWR) could be located along line AD so that the first risk moment would not exceed the average of the first 100 reactors of 4.5×10^{-5}/reactor/year. In this example we assume that eastern river valley weather is appropriate. This problem will be formulated in one dimension; however, there is no reason a more generalized two dimensional problem could not be solved by the same procedure.

If we let r be the direction along the line connecting the cities and ζ be a direction normal to it, then the population density in the region can be expressed by summing the contributions from four overlapping Gaussian distributions, each of the form

$$(18) \qquad \rho(r,\zeta) = \frac{N_T}{\sqrt{2\pi}\ \sigma_R} \exp\left[-\frac{(r-R)^2}{2\sigma_R^2} - \frac{\zeta^2}{2\sigma_R^2}\right].$$

The corresponding $n_T(r)$ population in an annulus per unit r is approximated as:

$$n_T(r) = \int_{-\infty}^{\infty} \rho(r,\zeta)d\zeta$$

$$(19) \qquad = \frac{N_T}{\sqrt{2\pi}\ \sigma_R} \exp\left[-\frac{(r-R)^2}{2\sigma_R^2}\right].$$

Adding the contribution of all four cities

$$(20) \qquad n_T(r) = \sum_\ell \frac{(N_T)_\ell}{\sqrt{2\pi}\ (\sigma_R)_\ell} \exp\left[-\frac{(r-R_\ell)^2}{2(\sigma_R)_\ell^2}\right]$$

where ℓ refers to the different cities.

The first risk moment is:

$$M_1 = \int_0^\infty a(r) \cdot n_T(r)dr$$

$$= \int_0^\infty a_1 \cdot \exp\left[-a_2 \cdot r\right] \cdot \left\{ \sum_\ell \frac{(N_T)_\ell}{\sqrt{2\pi}\,(\sigma_R)_\ell} \exp\left[-\frac{(r-R_\ell)^2}{2(\sigma_R)_\ell^2}\right] \right\} dr$$

$$= \sum_\ell a_1 \cdot (N_T)_\ell \cdot \exp\left[-a_2 \cdot R_\ell + \frac{a_2^2 \cdot (\sigma_R)_\ell^2}{2}\right]$$

(21)

$$\cdot \int_0^\infty \frac{1}{\sqrt{2\pi}\,(\sigma_R)_\ell} \exp\left\{-\frac{[r - R_\ell + a_2^2 \cdot (\sigma_R)_\ell^2]^2}{2(\sigma_R)_\ell^2}\right\} dr.$$

If $R_\ell > 2(\sigma_R)_\ell + a_2 \cdot (\sigma_R)_\ell^2$

then M_1 is given approximately by

(22) $\qquad M_1 = \sum_\ell a_1 \cdot (N_T)_\ell \cdot \exp\left[-a_2 \cdot R_\ell + \frac{a_2^2 \cdot (\sigma_R)_\ell^2}{2}\right].$

Using the values of a_1 and a_2 from Figure 1 it is possible to plot the first risk moment as a function of r as shown in Figure 6. The first risk moment is smaller than the risk criteria for r between 13 and 16 miles; thus, a plant can be located anywhere between 13 and 16 miles and meet the specified conditions.

In addition to the first moment the second moment and the
normalization constant could also be used if more detailed risk
criteria (controlling the shape of the fatality distribution) were
imposed. With explicit expressions for the transfer functions $a(r)$,
$b(r,r')$ and $c(r)$ it is, therefore, possible to investigate a wide
range of different applications with rather simple calculations. To
calculate a curve of the type shown in Figure 6 with the WASH 1400
consequence code would, on the other hand, require a large number of
long and expensive computer runs.

5.2. A Hypothetical Radioactive Release. In the previous discussion the basic variable of interest was population density. A
similar procedure with radioactive release as a basic variable was
carried out for a BWR. Because of space limitations a full description of the process will not be given here but only the results.

It was found that the transfer functions relating risk characteristics to release characteristics could be expressed by exponential
forms similar to those used for the population   In this case
expressions were developed relating the constants a_1, a_2, b_1, b_2, b_3,
c_1, and c_2 to variables which chararcterized the radioactive release.
These variables included the time of release t_d, the height of release
h, the energy of release E and a parameter ψ called the effective
source which was a sum of the release fraction of each isotope weighted
by its effectiveness for producing dose to the critical organ.

The regression and fitting analysis yielded relations of the following form.

$$a_1 = 7.73 \times 10^{-2} \cdot E^{-.53} \cdot h^{-.46} \cdot \psi^{.40} \, ,$$

$$a_2 = 2.93 \cdot t_d^{.23} \cdot E^{.059} \cdot \psi^{-.98} \, ,$$

$$b_1 = 4.16 \times 10^{-2} \cdot h^{-.27} \cdot E^{-.39} \, ,$$

$$b_2 = 1.75 \cdot h^{.043} \cdot t_d^{.19} \cdot E^{.12} \cdot \psi^{-.99} \, ,$$

$$b_3 = 1.45 \cdot \psi^{-.52} \, ,$$

$$c_1 = 8.63 \times 10^{-2} \cdot h^{-.37} \cdot t_d^{-.65} \cdot E^{-.65} \cdot \psi^{.93} \, ,$$

$$c_2 = 2.43 \cdot h^{-.080} \cdot \psi^{-1.02} \, .$$

These expressions can be used to explicity relate risk characteristics to release variables.

As an example of a use of the above results we consider the effect on the first risk moment of the efficiency of an iodine cleanup system. The particular effect of the iodine cleanup system is to reduce the value of the effective source ψ.

The effect on the first risk moment is shown in Figure 7. From this curve and a given criteria on the risk moment the acceptable values of the iodine cleanup system efficiency could be obtained.

6. Conclusion. An analysis procedure has been developed using the WASH 1400 consequence code that permits the risks moments to be explicitly expressed in terms of basic controlling variables. Rela-

tively simple expressions have been obtained for the population distribution effects and the release effects as they influence early
fatalities. With these relatively simple analytical relations a
wide variety of postulated conditions can be investigated without
repeating the laborious WASH 1400 calculations.

The analytical relation approach also gives a better understanding of the influences of the various physical phenomena and can also
serve as a basis for decision-making. Since the two moments and the
normalization constant define the Weibull distribution the entire
curve of probability versus consequence is obtained once these three
parameters are obtained. Examples are given which show how
the method can be applied to a typical siting problem and an evaluation of the effectiveness of an iodine removal system.

REFERENCES

(1) U.S. NUCLEAR REGULATORY COMMISSION, Reactor Safety Study -- An
     Assessment of Accident Risks in U.S. Commercial Nuclear
     Power Plants, WASH 1400 (NUREG-75/014). Washington:
     U.S. Nuclear Regulatory Commission, October 1975.

(2) ibid., Appendix 6

(3) See, for example, N.R. Draper and H. Smith, Applied Regression
     Analysis, John Wiley & Sons, Inc. 1977.

(4) U.S. DEPARTMENT OF COMMERCE, Master Enumeration District List
     with Coordinates. Washington: U.S. Department of Commerce,
     1970.

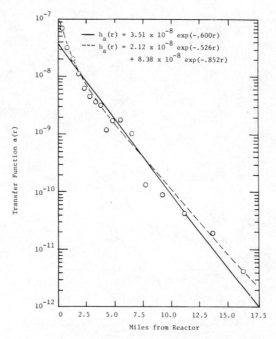

Figure 1. Transfer Function a(r) for PWR Accidents

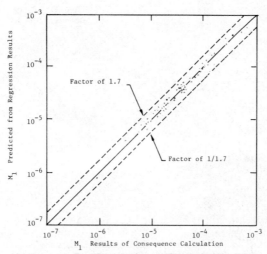

Figure 2. Test of the Regression Results of the
First Risk Moment for $a(r) = a_1 \cdot \exp(-a_2 \cdot r)$

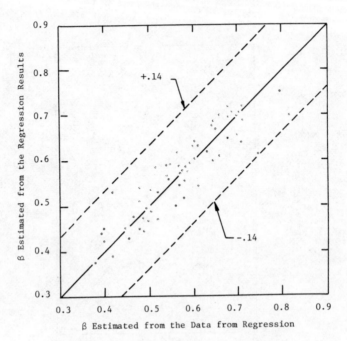

Figure 3.   Test of the Regression Results for
the Weibull Shape Factor

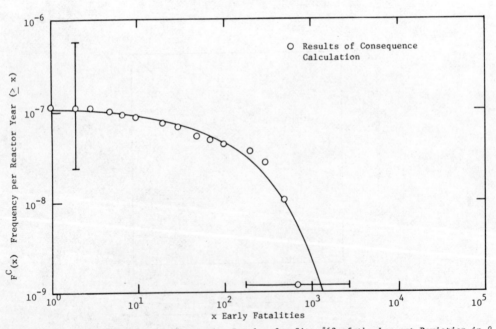

Figure 4.   Test of the Regression Results for Site #63 of the Largest Deviation in β

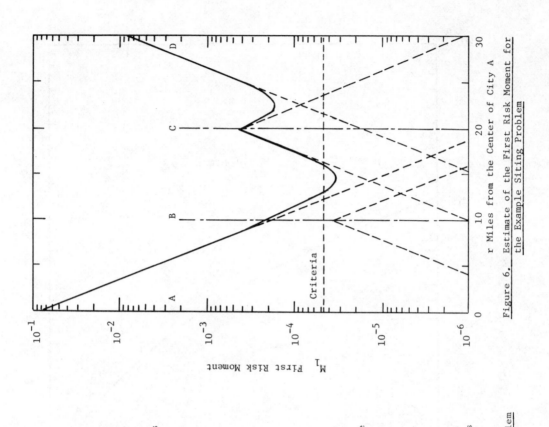

Figure 6. Estimate of the First Risk Moment for the Example Siting Problem

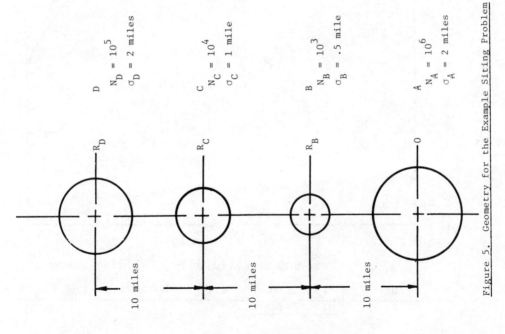

Figure 5. Geometry for the Example Siting Problem

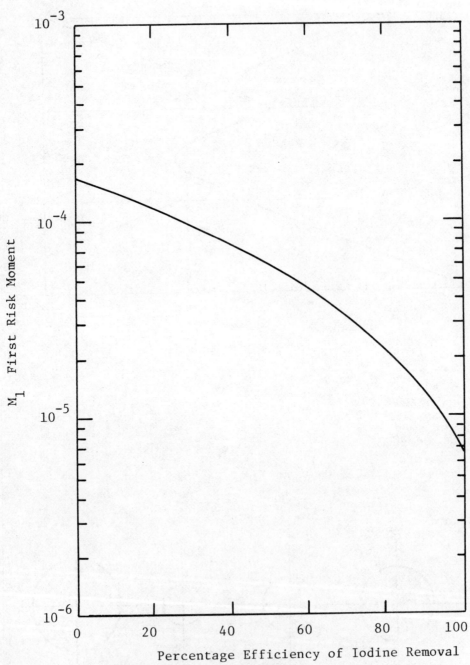

Figure 7.   Effect of the Iodine Removal on the First
Risk Moment

PROBABILISTIC APPROACHES TO ADVANCED
REACTOR DESIGN OPTIMIZATION†

G. R. BURDICK*, D. M. RASMUSON* AND J. WEISMAN**

Abstract. Approaches to advanced reactor design which combine
probabilistic methods with system optimization techniques are proposed. In recognition that the best available design and manufacturing methods have been applied to component design and manufacture,
it is suggested that significant improvement of nuclear systems performance is possible only through modification of the systems configuration. Thus, the models used in the optimization procedures are
those for redundancy and diversity allocation or models which provide
goals for such allocation. It is shown that diversity and redundancy
allocation may be improved through the construction and use of functions which represent an estimate of the best cost obtainable for a
given level of system availability.

The use of functional relationships between cost and availability
is extended to the case where acceptable levels of risk have been
defined for plant operation. In this case, the cost functions for the
various safety systems are used to form an overall objective function.
Societal risk expressions are formulated in terms of safety system
unavailabilities. Construction of the cost objective function and
risk expressions constitutes completion of the top-level problem. A
second-level problem is then solved which performs trade-offs of cost
and availability among the safety systems. The second-level problem
solution is a set of availability goals and cost guidelines which can
be used by designers in solution of third-level problems which are
determinations of the actual safety system configurations.

This paper suggests a method whereby the economic risk may be
minimized for a given level of societal risk. The authors make no
attempt to define what might be an acceptable level of societal risk.

---

*EG&G Idaho, Inc., Idaho Falls, ID 83401. †This research was supported
by the U. S. Energy Research and Development Administration under
Contract EY-76-C-07-1570. **Department of Chemical and Nuclear Engineering, The University of Cincinnati, Cincinnati, OH 45221.

395

1. Introduction. In the nuclear industry as a whole, and in the liquid metal fast breeder reactor (LMFBR) project in particular, each hardware component is being fabricated in essentially the best manner now known. Within the technology available, the reliability of individual components cannot now be significantly improved. As further research indicates better ways to design components, designers will incorporate those indicated changes. However, more reliable components cannot now be specified for use in proposed advanced reactor systems since the technology for producing such components is not available.

In view of the foregoing, the only means to achieve design improvement in plant safety and productivity is by intelligent choice of both the types of systems and subsystems to be used in the design and the configurations of these systems. In order to make such intelligent choices, those responsible for the decisions must have some way of assessing what functional reliability is realistically achievable from the various systems they must choose among. Moreover, if several systems can be made to achieve equal levels of availability through (1) use of different numbers and sizes of components in series and/or parallel (i.e. redundancy allocation) or (2) use of alternative means to accomplish a given purpose (i.e. diversity choices), it is necessary to determine which system can be made to do so at the least cost. Clearly, the ability to make such determinations would be most valuable if applied during the earliest possible stage of the design process.

If such determinations could be made, and made well, for the designs of prototypical nuclear power plants, the design effort would be substantially reduced on subsequent plants of that same type.

There is no intention to imply that design decisions are not now being made intelligently. However, given a task to be performed by some system, there are a number of types of systems that will perform that task. There also are a large number of possible, workable, reliable configurations for each type. Granted, common sense and good engineering judgment can be used to eliminate a great many candidates from the field, but there will still remain a large number of possibilities to choose from. This paper proposes new applications of probabilistic methods and system optimization techniques which can improve the design decision making process.

The methods to be described require basic systems designs as starting points. A basic design is meant to describe a system that is devoid of redundant and diverse components, loops, and modules. Unavailability logic models can be constructed from the basic designs. Mathematical expressions, containing integer variables indicating diversity and redundancy choices, can then be derived from the logic models. Cost functions can also be constructed in terms of the integer variables. The foregoing procedure is described in Section 2 through use of a much simplified LMFBR system.

Section 3 illustrates how the expressions derived in Section 2 may be used to generate functions which provide estimates of the best cost obtainable for a given level of system availability. Several possible uses of these functions are discussed.

Additional uses of the cost functions are described in Section 4, where it is assumed that some numerical level of acceptable societal risk for plant operation has been defined. The cost functions are used to develop a method for performing cost and availability trade-offs among various safety systems.

2. Modeling the Basic System. Consider the basic LMFBR system of Figure 1. For illustrative purposes, a very simple shutdown heat

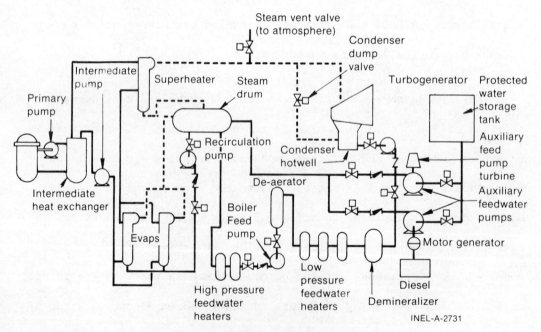

Fig. 1 Basic LMFBR System

removal system (SHRS) is included. The SHRS consists of two subsystems: (1) a condenser dump valve in conjunction with the normal
return path to the steam drum and (2) a steam vent valve, a protected
water storage tank and a line to the steam drum through either a steam
or a motor generator driven pump. (Assume that pump diversity here is
a requirement imposed on the design by regulation.) For simplicity,
the SHRS was the only safety system included.

Several additional simplifying assumptions were made in the
development of the logic models for this example. Some of these are:

(1) Turbogenerator failure was neglected.

(2) Sodium pump drive motors are not needed during shutdown
     heat removal (SHR).

(3) No credit is taken for natural circulation.

(4) Makeup feedwater is not required during normal operation.

(5) Three loops are required for power generation.

(6) One loop is required for SHR.

(7) All components have constant failure rates.

(8) Probability of SHR malfunction is dominated by its unavail-
     ability (i.e. probability of SHR failure during its mission
     time is negligible).

(9) Components considered for parallel redundancy are assumed to
     have a 30-year (plant) lifetime.

Additional assumptions are implicit in the logic models of Figure 2.

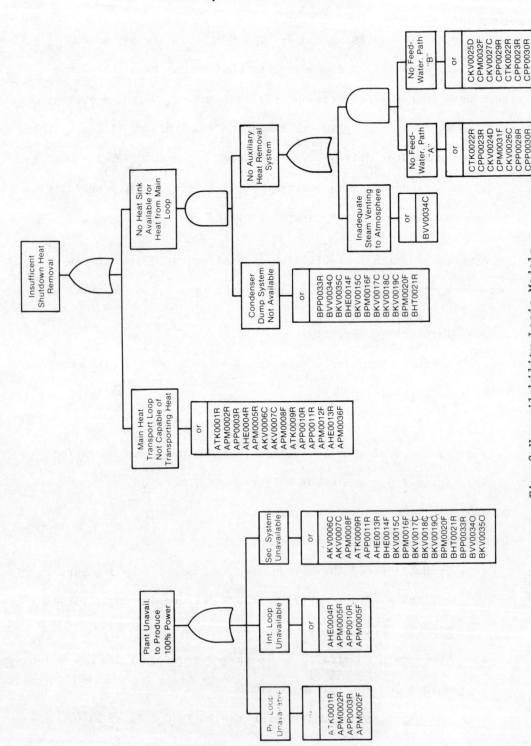

Fig. 2 Unavailability Logic Models

EGG-A-1467

Some questions which can be answered within the framework of a constrained optimization procedure are: (1) How many heat transport loops should there be? (2) How many pumps per loop are required? (3) How many steam vent and condenser dump valves are necessary? (4) How many hotwell pumps and boiler feed pumps should be installed? (5) How many recirculation pumps are required? For a more inclusive basic design (i.e. one that included containment, a shutdown system, an overflow heat removal system, etc.) similar additional questions could be asked.

Optimizing total plant cost subject to a constraint on safety system unavailability requires the construction of logic models for SHRS and plant unavailability. A constraint expression and cost objective function can then be derived from these logic models. In what follows, the $\lambda\tau$ approximation [1] was used for component and module unavailabilities and intersection terms were ignored for simplicity; thus, the unavailability expressions provide upper bounds on systems unavailabilities [1]. Redundant components were considered to be in cold standby although hot and tepid [2] standby situations could have been treated as well. For an alternate optimization approach, with application to a light water reactor plant, see Reference 3.

In the logic models for plant unavailability and SHRS unavailability of Figure 2, eight-symbol alphanumeric identifiers, similar to those used in Reference 4, have been used to label the various

BURDICK, RASMUSON AND WEISMAN

failure events in the logic models. Table I is a sample fault summary
sheet an analyst might use to assemble the necessary failure and
repair data for the failure modes appearing in the logic models. Cost
data could be gathered and also tabulated. Because the purpose of
this section is to illustrate a procedure, via a simple example,
accuracy of the data was not paramount and a complete listing of the
data used was deemed unnecessary.

TABLE I

SAMPLE FAULT SUMMARY ENTRIES

			Primary Failure		
Event Name	Event Component	Failure Mode	Failure Rate λ (10^6 hr)	Fault Duration τ (hr)	$\lambda\tau$ (10^{-6})
APP0003R	Primary Na Loop Piping	Rupture, Leakage	0.63	9000	5670
AHE0004R	Intermediate Heat Exchange	Rupture, Leakage	0.38	72	27.4
APM0005R	Intermediate Na Pump	Rupture, Leakage	0.2	37	7.4
AKV0006C	Steam Generator Isolation Valve (Solenoid)	Transfer Closed, Rupture	1	50	50
AKV0007C	Steam Generator Isolation Valve (Solenoid)	Transfer Closed, Rupture	1	50	50
APM0008F	Recirculation Pump	Rupture, Fails OFF	14	37	518

From the appropriate logic model of Figure 2, the following function approximating system unavailability to produce 100% power, $\bar{Q}_p$,
was obtained:

$$(1) \quad \bar{Q}_p = C_1 \, [C_2 + U_1(U_1^*)^{(N_1-1)} + U_8(P+I) + U_5(U_5^*)^{(P-1)}$$

$$+ \, U_6(U_6^*)^{(I-1)}]^{(L-2)} + C_3 + U_2N_2 + U_3N_3 + U_4(U_4^*)^{(N_4-1)}$$

$$+ \, U_7(U_7^*)^{(N_5-1)}$$

where

$$C_1 \quad = \quad \begin{array}{l} 3 \text{ when } L = 3 \text{ for } 1/3 \text{ logic} \\ 6 \text{ when } L = 4 \text{ for } 2/4 \text{ logic} \end{array}$$

L = number of heat transport loops (3 or 4)

P = number of primary sodium pumps/loop

I = number of intermediate sodium pumps/loop

N_1 = number of recirculation pump modules/loop

N_2 = number of steam vent valves in parallel

N_3 = number of condenser dump valves in parallel

N_4 = number of boiler feed pump modules

N_5 = number of hotwell pump modules

U_i = unavailabilities of components

U_i^* = unavailabilities of standby components

C_i = unavailabilities of fixed portions of the design,

i ≠ 1.

For additional details, see Reference 5.

From the remaining logic model of Figure 2, the following function approximating SHRS unavailability, $\overline{Q}_s$, was derived:

$$(2) \quad \overline{Q}_s = \left[C_4 + U_1 (U_1^*)^{(N_1 1)} + U_8 (P+I) + U_9 (U_9^*)^{(P-1)} + U_{10} (U_{10}^*)^{(I-1)} \right]^L$$

$$+ [U_{11}^{N_3} + U_4 (U_4^*)^{(N_4-1)} + U_7 (U_7^*)^{(N_5-1)} + C_5](C_6 + U_{12} N_2)$$

$$+ C_7 U_2^{N_2}.$$

Numerical values of the unavailabilities can be calculated, using
component failure mode information as illustrated in Table I, for
each C, U, and U*. By use of the values of $\overline{Q}_p$ so generated and the
tabulated cost estimates, an objective function can be formed as
follows:

$$(3) \qquad Ob = K_p \overline{Q}_p + \sum_{i=2}^{5} K_i N_i + L (K_6 P + K_1 N_1 + K_7 I + K_8)$$

where the coefficient of $\overline{Q}_p$ is determined by the cost of downtime per day with total cost taken over a 30-year plant lifetime. The last term represents total cost of the fixed numbers of components per loop. Expressions (2) and (3) can be used in a straightforward optimization procedure using any one of several techniques. This was, in fact, done at the Idaho National Engineering Laboratory (INEL). Failure rates, mean downtimes, and cost data were input to Equations (2) and (3) and constraints were put on the integer variables as follows:

$$(4) \qquad\qquad\qquad 3 \leq L \leq 4$$

$$(5) \qquad\qquad\qquad 1 \leq P \leq 3$$

$$(6) \qquad\qquad\qquad 1 \leq I \leq 3$$

$$(7) \qquad\qquad 1 \leq N_i \leq 3; \; i = 1,\ldots,5.$$

When a constraint of 10^{-6} on SHRS unavailability was assumed, the solutions appearing in Table II were obtained using the code ARSTEC [6]. ARSTEC is the INEL version of an adaptive random search code used by Campbell and Gaddy [7] to optimize a light water reactor system using reliability considerations. No significance should be attached to these solutions other than the fact that ARSTEC was able to solve

TABLE II

SAMPLE PROBLEM SOLUTIONS

	N_1	N_2	N_3	N_4	N_5	I	L	P
Optimum SHRS Cost	2	1	1	1	1	1	3	1
*Optimum Plant Cost No. 1	2	1	1	2	2	1	3	1
**Optimum Plant Cost No. 2	1	1	1	2	2	1	4	1

*Cost/day of plant downtime = $ 85,000
**Cost/day of plant downtime = $500,000

such problems successfully. This was determined by a comparison of

the ARSTEC solutions with solutions for all possible values of the

integers. This was found to be the case with several reformulations

of the problem.

Table II illustrates that the solutions are sensitive to cost

data. This indicates that a configuration which is optimal for a

small demonstration plant may not be optimal for large commercial

plants.

3. <u>Obtaining Functions of Cost versus Unavailability</u>. As formulated, only integer variables appear in Equations (1), (2), and (3).
By using a Monte Carlo approach to randomly select a large number of
integer combinations (system configurations), the costs and unavailabilities may be evaluated for a large number of configurations. When
the results of these calculations are plotted in terms of cost and
unavailability, a cloud of points will be obtained. Lower bounding
curves can then be constructed for the cloud of points. The bounding
curves provide the designer with functions which approximate the best
cost obtainable for a level of availability at least as good as that
indicated.   The cloud of points and curve of Figure 3 were obtained

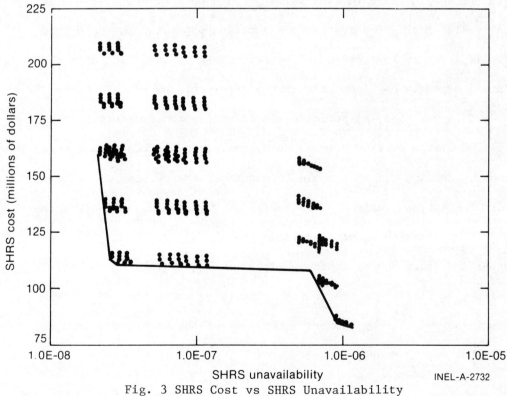

Fig. 3 SHRS Cost vs SHRS Unavailability

INEL-A-2732

from the sample problem of the preceding section. The points were not
randomly selected but, due to the small number of integer variables in
the sample problem, are actually plots of cost versus unavailability
for all possible integer combinations. Linear interpolation can be
used to obtain the curves; however, other methods are also available.
For example, spline interpolation could be used to provide differen-
tiable functions of cost versus unavailability.

Plots and curves, such as those of Figure 3, contain information
which can be used to advantage in system design. For example, a set
of bounding curves might indicate that for a modest additional cost a
significant increase in availability could be achieved over the original design goal. The system configuration providing the described
availability at low cost could be determined by examination of the
data previously obtained, assuming a record of the data is kept. If a
record is not kept, an optimization is necessary to determine the
desired system configuration. An optimization procedure (alternate to
that of Section 2) that could be used for this purpose is to transform the integer variables to binary (0-1) variables and use an implicit enumeration scheme [8] to locate the desired configuration.
Upper and lower bounds could be placed as constraints on unavailability to facilitate the search. The curve could then be used to
check the solution by comparing the cost of the solution configuration
with the cost given by the curve.

It would also be possible to derive Equations (1), (2), and (3) for a number of types of systems which are candidates for performance of a given task. Cost versus availability curves can be generated for each of these candidates. The curves could be compared to perform a preliminary weeding out of less desirable systems (i.e. those with unsatisfactory cost versus availability characteristics). The remaining systems could then be subjected to more refined selection processes which might involve optimization subject to a specific availability goal. Alternately, the cost versus availability curves could indicate that none of the systems considered could achieve a specific availability goal. This would suggest that further research and development would be needed to devise alternative systems for the performance of the function in question.

By accurately scaling component, loop, and other costs, the objective function (3) can be modified to describe different size plants. The resulting families of curves could be used to establish system availability goals as a function of plant size.

4. Optimization Involving Constraints on Risk. The concept that a specified level of risk of operation of nuclear power plants may be useful in the design process of such plants is not new. F. R. Farmer [9] discussed this idea in a paper in 1967. Schleicher [10] further investigated the area in 1972. Estimation of total system reliability characteristics from those of subsystems is also an area which has

been previously investigated [11,12,13]. This section combines the
latter idea, the idea of the use of risk constraints in design, and
the cost versus unavailability functions for subsystems discussed in
the previous section, to develop a new approach to total system optimization. In principle, the optimum design of a large system
subject to risk constraints should be solvable by the approach described in Section 2. The risk expressions are functions of the
safety system availabilities which are in turn functions of the
integer variables describing the system configuration. The problem is
thus a nonlinear integer programming problem as was the one solved
in Section 2. However, in any realistic modeling of a large system,
the total number of integer variables required to describe its configuration will be large. While available programming techniques
perform satisfactorily with small and moderate numbers of integer
variables, satisfactory solutions are generally not attainable with
reasonable computing times when the number of integer variables is
large. To avoid this difficulty, the problem is decomposed into
several smaller problems, each of which is of a size which can be
readily solved by available techniques.

It is assumed that the overall system consists of a set of
"weakly interacting subsystems". Subsystems are said to be weakly
interacting when the configuration of any one subsystem affects the
availability of that subsystem but not the availability of any other

subsystem. Subsystems interact only to the extent that they all affect overall availability and risk. In many large systems most, if not all, subsystems will be weakly interacting. When the subsystems are weakly interacting, we may determine system behavior by considering each subsystem individually.

The proposed technique is a three-level process. In the first level, cost versus unavailability curves are determined, as described in Section 3, for each of the subsystems of interest. The cost versus unavailability functions are then used as input to a second-level problem where overall system cost is minimized subject to acceptable constraints on risk. The risk expressions will contain products of unavailabilities or unit complements of unavailabilities. If all subsystems may be considered weakly interacting, this second-level problem will be a nonlinear programming problem with all continuous variables. Such problems are readily solvable by a variety of techniques.

The solution to the second-level problem represents the optimum trade-off between availability and cost among the various safety systems involved. Moreover, the availabilities are consistent with the overall safety objectives of the plant. The second-level solution therefore provides a set of availabilities and cost guidelines to be used in the third-level problems by safety system designers as they determine the actual configuration of the various safety systems.

4.1. Problem Formulation. (Much of this section also appears in Reference 14 and is similar to the formulation introduced in Reference 15.) Before the risk expressions can be formed, there are several events that must first take place. Therefore, the following assumptions are made.

(1) An acceptable level of societal risk has been defined.

(2) A basic LMFBR design has been constructed. This basic design is devoid of redundancies and is composed of basic descriptions of those systems known to be required for safety and operational functions.

(3) A gross reliability analysis has been performed. That is, event sequences have been identified from loss of various safety functions following selected initiating events.

(4) Consequences, given particular accident sequences, have been calculated.

(5) Accident sequences are assumed to be mutually exclusive and exhaustive.

Risk constraints can then be formulated as follows:

(8) $$Pr[f > X_j] = \sum_{i=1}^{n} Pr[f > X_j | AS_i] Pr[AS_i] \le k_j \ , \ j = 0, \ldots, N$$

where

$Pr[\ ]$ = probability that bracketed quantity will occur

f = number of fatalities

n = number of distinct accident sequences

AS_i = particular accident sequences

k_j = given acceptable levels of risk

N = number of risk constraints.

The risk expressions, Equation (8), would be required to produce values in the acceptable risk region under the curve of Figure 4.

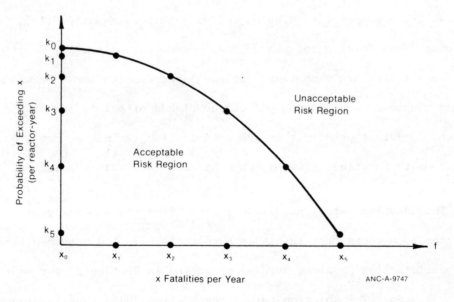

Fig. 4 Risk Criteria Curve

The quantities $\Pr[f > X_j | AS_i]$, by assumption 4, are known. The $\Pr[AS_i]$ are functions of the unavailabilities of the systems appearing in the various accident sequences. That is:

$$(9) \qquad\qquad \Pr[AS_i] = f(\overline{Q}_1, \overline{Q}_2, \ldots \overline{Q}_m)$$

where the $\overline{Q}_\ell$ ($\ell = 1, \ldots, m$) are safety system unavailabilities. What is now required is an objective function of the form:

$$(10) \qquad\qquad Ob = g(\overline{Q}_1, \overline{Q}_2, \ldots, \overline{Q}_m)$$

which is to be minimized subject to the constrained risk expressions (8).

4.2.Problem Solution. The objective function, Equation (10), can be obtained by forming the sum of cost versus unavailability functions for the various safety systems. Spline interpolation could be used to obtain the lower bounding curves on the clouds of points for each system. Gradient or direct search optimization techniques combined with penalty functions [16] may then be used to obtain a solution.

The solution set of unavailabilities, from the second-level problem, can be used as constraints in a third-level, redundancy-allocation optimization problem, such as described in Section 2, for each safety system. The solution set of costs, from the second-level problem, can be used as a check on the final costs of the configurations determined by solution of the third-level problem.

5. Conclusions. The tasks of making R&D and design decisions, as applied to conceptual advanced reactor systems, are extremely difficult ones; this is in part due to the seemingly endless number of possible alternatives the decision-maker is forced to choose among. This paper has described some probabilistic methods, coupled with optimization techniques, that offer some promise of providing tools which may have utility as aids in the selection of system types, and systems configurations, which fit best in the overall plant design.

The sample problem, used to illustrate the methods, was quite simplified. Effective use of the described methods would require much larger models for the more complex systems involved. For risk constrained total system optimization, dependencies among systems would limit the number of weakly interacting subsystems which could be isolated. The remaining top level problem might then be a mixed integer programming problem of a size which is still difficult to solve. However, the problem will be significantly smaller than the original problem and effective solution procedures will be more easily achieved.

Acknowledgment. The authors thank J. L. vonHerrmann and J. R. Wilson, EG&G Idaho, Inc., for their assistance with the sample problem. J. H. Carlson, F. X. Gavigan, and J. D. Griffith, U. S. Energy Research and Development Administration, are gratefully acknowledged for their interest and support.

REFERENCES

[1] J. FUSSELL, How to Hand Calculate System Reliability and Safety
     Characteristics, IEEE Trans. on Rel., 24 (1975), pp. 169-
     174.

[2] A. POLOVKO, Fundamentals of Reliability Theory, Academic Press,
     New York, 1968.

[3] J. WEISMAN and A. HOLZMAN, Optimal Process System Design Under
     Conditions of Risk, Ind. Eng. Chem. Process Des. Develop.,
     11 (1972), pp. 386-397.

[4] Reactor Safety Study, NUREG-75/104, USNRC (1975), available from:
     NTIS, U.S. Dept. of Commerce, 5825 Port Royal Road, Spring-
     field, VA 22161 USA.

[5] G. BURDICK, D. RASMUSON and J. WILSON, An Investigation Con-
     cerning a Risk-based Approach to Design of Liquid Metal
     Fast Breeder Reactor Plants, RE-S-76-174 (1975), available
     from: Safety Assessment Branch, Division of Reactor De-
     velopment and Demonstration, U.S. ERDA, Wash., D.C. 20545.

[6] D. RASMUSON and N. MARSHALL, ARSTEC: A Computer Program for
     Solving Nonlinear, Mixed-Integer, Optimization Problems
     Using the Adaptive Random Search Technique, RE-S-76-180
     (1976), available from: Argonne Code Center, 9700 South
     Cass Avenue, Argonne, IL 60439 USA.

[7] J. CAMPBELL and J. GADDY, Optimization of a Nuclear Reactor
     Primary Cooling System, Presented at the 68th Annual Meeting
     of AICHE, Los Angeles, CA (November 1975), available from:
     Prof. J. Gaddy, University of Missouri - Rolla, Rolla, MO
     65401.

[8] A. GEOFFRION, Integer Programming by Implicit Enumeration and
     Balas' Method, SIAM Rev., 9 (1967), pp. 178-190.

[9] F. R. FARMER, Reactor Safety and Siting: A Proposed Risk Crit-
     erion, Nuclear Safety, 8 (1967), pp. 539-548.

[10] R. SCHLEICHER, Stochastic Decision Making Applied to Nuclear
     Reactor Safety, Paper No. 72-20, Cornell Energy Project
     (August 1972), available from: Center for Environmental
     Management, Cornell University, Ithaca, NY 14850.

[11] Z. BIRNBAUM and J. ESARY, Modules of Coherent Binary Systems, SIAM J. of Appl. Math., 13 (1965), pp. 444–462.

[12] E. KRISHNAMURTHY and G. KOMISSAR, Computer-Aided Reliability Analysis of Complicated Networks, IEEE Trans. on Rel., 21 (1972), pp. 86–89.

[13] A. SHOGAN, Sequential Bounding of the Reliability of a Stochastic Network, Op. Res., 34 (1976), pp. 1027–1044.

[14] G. BURDICK, D. RASMUSON, and S. DERBY, A Risk-based Approach to Advanced Reactor Design, IEEE Trans. on Rel., (To appear, August 1977).

[15] S. DERBY and O. GOKCEK, Advanced Safety Analysis Eighth Quarterly Report, GEAP-14038-8 (1976), available from: TIC, P. O. Box 62, Oak Ridge, TN 37830.

[16] B. GOTTFRIED and J. WEISMAN, Introduction to Optimization Theory, Prentice Hall, Englewood Cliffs, NJ, 1973.

A PROCEDURE FOR THE USE OF PROBABILISTIC RISK ASSESSMENTS TO ESTABLISH RANKING OF RESEARCH AND DEVELOPMENT GOALS

M. I. TEMME AND K. A. EL-SHEIKH*

Abstract. A procedure for Ranking Research and Development goals on the basis of their importance to risk is described. Using information developed in a typical LMFBR risk assessment, rules for selection of the most important events or groups of events (e.g., event sequences) are defined. The procedure is illustrated through its application to an example, using a First Cycle Risk Assessment of FFTF.

1. Introduction. An often-expressed reservation as to the value

of probabilistic risk assessment of LMFBR is that there is insufficient

data to carry out a credible analysis. Proponents of the probabilistic

approach contend that risk assessments can be carried out at virtually

any level of detail commensurate with the availability of data, and

that the risk assessment itself provides a useful structure in which

further information needs can be identified in order of importance.

A systematic approach to the establishment of R&D priorities is des-

cribed in this paper. The illustrative case for the procedure is a

limited risk assessment of two initiating events in the FFTF reactor

[3]. While this serves to illustrate the procedure, general con-

clusions regarding R&D needs would require a complete risk assessment

---

*General Electric Co., Fast Breeder Reactor Dept., Sunnyvale, Calif.
This paper describes work performed under U.S. ERDA Contract
EY-76-C-03-0893, Task 13.

covering all initiating events of significance.

The procedure for selection of important events is described in Section 2. Section 3 discusses the application, and inferences of R&D work which would affect the significant probabilities of the FFTF First Cycle Assessment [3].

2.   The Procedure. The three major steps: Risk assessment, ranking of contributors to risk, and identification of information needs are described in Sections 2.1 through 2.3.

2.1. Risk Assessment. To be compatible with the procedure for ranking R&D needs, the probabilistic risk assessment model must contain these generic elements:

1.   A mutually exclusive set of all accident initiating events which contribute significantly to the measure of risk,

2.   Assessment by fault-tree or comparable analysis of initiating event probabilities,[1]

3.   Development of a finite set of event sequences associated with each initiating event. Assignment of a probability

---

[1] While some investigators have quantified risk in terms of accident and/or consequence frequencies, we have chosen the complementary cumulative probability distribution of consequence as the risk measure. Consistent with this measure, initiating event probabilities - specifically the probabilities that the events will occur once in a year - are used. Where needed, the second and sucessive occurrences of the same events would be treated following the same logic.

and an ultimate consequence (e.g., number of fatalities)
to each sequence,

4. Definition of "consequence intervals" - finite ranges of the
    consequence parameter - such that each event sequence can
    be assigned to a single consequence interval,

5. Uncertainty evaluation of each sequence probabilities.

A risk assessment procedure along the lines of the Reactor Safety
Study [1] fits these requirements. A comparable procedure for LMFBR
has been developed and used in several applications [2, 3, 4, 5]

A "First Cycle" (i.e., first iteration) risk assessment of FFTF
[3] is used to illustrate the R&D ranking procedure. The analysis
considers just two accident initiating events, thus it does not meet
requirement 1, foregoing. The possible effect of this on conclusions
regarding R&D needs is considered in the paper (Section 3). Also, it
is noted that the consequence parameter of the First Cycle Analysis is
Site Boundary Dose, rather than health effects. Conceivably, a risk
assessment which does address health effects (e.g., acute fatalities,
latent cancer fatalities) would produce a different ranking of R&D
needs. Therefore, the conclusions reached here regarding R&D priorities bear further evaluation beyond the scope of the paper.

The risk model is delineated in Figure 1. It contains the following elements:

1. Initiating Event Definition and Probability:

The two initiating events considered are [3]:

A1: Uncontrolled withdrawal of a control rod at the maximum possible mechanical speed of 9.8"/min. The probability of this event was estimated by fault tree analysis with uncertainty in the probability estimate primarily due to lack of information on operating procedures.

A2: A pipe break at the inlet nozzle of one primary HTS loop. The probability of this event was estimated by judgment based on sodium leak data extrapolation models, analytical models, and LWR data. Uncertainty in the probability estimate was due to scatter in the information base.

2. Reactor Shutdown System (RSS) Response Definition and Probability:

The potential responses of the reactor shutdown system (reactor scram and heat transport systems) were developed using event trees such as that of Figure 2. Six system event sequences (SES) are defined for each initiating event. The conditional probability of failure-to-scram was assigned by judgment with consideration of system

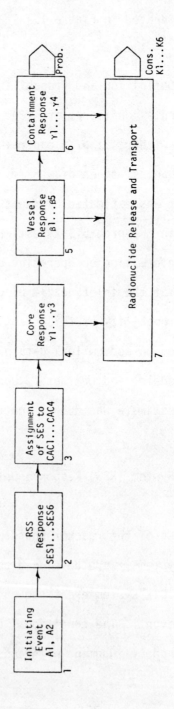

FIGURE 1. FFTF FIRST CYCLE RISK ASSESSMENT MODEL

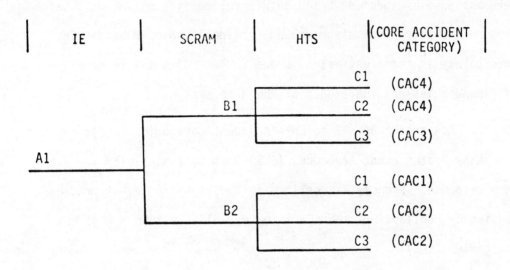

IE	SCRAM	HTS	(CORE ACCIDENT CATEGORY)

A1

	B1	C1	(CAC4)
		C2	(CAC4)
		C3	(CAC3)
	B2	C1	(CAC1)
		C2	(CAC2)
		C3	(CAC2)

IE: Initiating Event

HTS: Heat Transport System

A1:   Uncontrolled Rod Withdrawal

B1:   Successful Scram

B2:   Failure to Scram

C1:   Full Power Heat Removal
      Capability

C2:   Shutdown Heat Removal
      System (SHRS) Functioning

C3:   SHRS Failed

CAC1: Transient Overpower
       W/O Scram (TOP)

CAC2: Transient Under-
       cooling at Power
       (TUC)

CAC3: Loss of Shutdown Heat
       Removal Capability
       (LSHRS)

CAC4: Safe Shutdown (SS)

FIGURE 2

Event Tree for Accident Initiator A1

dependencies. The conditional probabilities of the HTS configurations were assigned by judgment based on HTS reliability estimates obtained from reliability block diagrams and consideration of dependencies. Uncertainty in these estimates is due to lack of detailed modeling of dependencies and inadequacy of the data base.

3. Assignment of SES to Core Accident Categories (CAC):

Many system event sequences (SES) lead to essentially the same core response. Economy is realized in the remaining event tree expansion by "collapsing" the SES into four different Core Accident Categories (CAC):

CAC1: Transient overpower of 10¢/sec W/O Scram (TOP),

CAC2: Transient undercooling due to loss of HTS full power flow capability W/O Scram (TUC),

CAC3: Loss of the shutdown heat removal system capability (LSHRS),

CAC4: Safe shutdown (SS).

Table 1 lists the SES contributions to each Core Accident Category, including probability uncertainty estimates.

4. Core Response:

Three core response categories are defined:

α_1 = 0 - 1% core damage

α_2 = 1 - 10% core damage

TABLE 1

ASSIGNMENT OF SYSTEM EVENT SEQUENCES TO CORE ACCIDENT CATEGORIES

CORE ACCIDENT CATEGORY

TOP (CAC1)	TUC(CAC2)	LSHRS (CAC3)	SS (CAC4)
A1 B2 C1	A1 B2 C2	A1 B1 C3	A1 B1 C1
3×10^{-9}*	10^{-9}	10^{-11}	A1 B1 C2
$(2 \times 10^{-10} -$	$(10^{-10} - 10^{-8})$	$(5 \times 10^{-14} - 2 \times 10^{-9})$	A2 B1 C2
$4 \times 10^{-8})$**			10^{-3}
	A1 B2 C3	A2 B1 C1	$(10^{-4} - 10^{-2})$
	3×10^{-17}	10^{-15}	
	$(10^{-18} - 10^{-15})$	$(10^{-17} - 10^{-13})$	
	A2 B2	A2 B1 C3	
	10^{-14}	10^{-12}	
	$(5 \times 10^{-17} - 2 \times 10^{-12})$	$(2 \times 10^{-15} - 5 \times 10^{-10})$	

TOTAL PROBABILITY/YEAR OF THE CORE ACCIDENT CATEGORY

3×10^{-9}	10^{-9}	10^{-11}	10^{-3}
$(2 \times 10^{-10} -$	$(10^{-10} - 10^{-8})$	$(5 \times 10^{-14} - 2 \times 10^{-9})$	$(10^{-4} - 10^{-2})$
$4 \times 10^{-8})$			

A1 = Uncontrolled rod withdrawal

A2 = Pipe break at inlet nozzle

B1 = Successful scram

B2 = Failure to scram

C1 = Full Power Heat Removal Capability

C2 = Shutdown heat removal system (SHRS) functioning

C3 = SHRS failed.

 * = Probability estimate

** = Range of probability estimate uncertainty.

α_3 = 10 - 100% core damage

Conditional probabilities of these categories given a Core Accident Category were assigned by judgment based on SAS [6] and MELT [7] code analyses. Uncertainties in probability estimates were assigned by judgment.

  5. Vessel Response:

Five Vessel Response Categories are defined:

  β_1 = No vessel damage,

  β_2 = Minor vessel head damage with no vessel melt-through,

  β_3 = Moderate vessel head damage with no vessel melt-through,

  β_4 = Moderate vessel head damage and late vessel melt-
        through,

  β_5 = Large vessel head damage and early vessel melt-through.

Conditional probabilities of these categories given a Core Accident Category and a Core Response Category were assigned by judgment based on VENUS [8] and REXCO [9] code analyses. Uncertainties in probability estimates were assigned by judgment.

  6. Containment Response:

Four Containment Response Categories were defined:

  γ_1 = No containment failure,

  γ_2 = Late containment failure by overpressurization,

  γ_3 = Early containment failure by overpressurization,

γ_4 = Containment isolation failure.

Containment isolation failure probability was obtained by fault tree analysis with uncertainty primarily due to inadequacy of data. The overpressurization failure probability estimate and its uncertainty were assigned by judgment.

7. Radionuclide Release and Transport:

The 30-day whole body dose at 4.5 miles for sequences formed from Core Accident, Core Response, Vessel Response, and Containment Response categories was based on parametic radiological evaluations using the CACECO [10], HAA [11], and COMRADEX [12] codes. Six Consequence Categories covering the resulting range of 30-day whole body dose were defined as:

$$K_1 = 0 - 10^{-3} \text{ Rem}$$
$$K_2 = 10^{-3} - 10^{-2} \text{ Rem}$$
$$K_3 = 10^{-2} - 10^{-1} \text{ Rem}$$
$$K_4 = 10^{-1} - 1 \text{ Rem}$$
$$K_5 = 1 - 10 \text{ Rem}$$
$$K_6 = 10 - 10^{2} \text{ Rem}.$$

Sequences leading to each of these Consequence Categories are shown in Table 2. Table 2 also contains the probability of each sequence and the uncertainty in each probability estimate.

TABLE 2

ASSIGNMENT OF SEQUENCES TO CONSEQUENCE CATEGORIES

Consequence Category
(30 Day Whole Body Dose at 4.5 Miles (Rem))

	$0-10^{-3}$ K1	$10^{-3}-10^{-2}$ K2	$10^{-2}-10^{-1}$ K3	$10^{-1}-1$ K4	$1-10$ K5	$10-10^{2}$ K6
	SS 10^{-3} $(10^{-4}-10^{-2})$	TOP α2 β2 γ1 3×10^{-9} $(2 \times 10^{-10} - 4 \times 10^{-8})$ TUC α3 β2 γ1 9×10^{-10} $(9 \times 10^{-11} - 9 \times 10^{-9})$ TOP α3 β 2 γ1** LSHRS α3 β2 γ1** LSHRS α2 β2 γ1** TUC α2 β2 γ1**	TUC α3 β3 γ1 10^{-10} $(10^{-11}-10^{-9})$ LSHRS α3 β3 γ1** TOP α3 β3 γ1**	TUC α3 β4γ 2 2×10^{-11} $(10^{-12}-10^{-11})$ LSHRS α3 β4 γ2 9×10^{-12} $(4 \times 10^{-14} -8 \times 10^{-9})$	TOP α2 β2 γ4 10^{-11} $(10^{-12}-10^{-10})$ TUC α3 β5 γ3** TUC α3 β2 γ4** LSHRS α3 β4 γ4** LSHRS α3 β5 γ3**	TUC α3 β3 γ4 5×10^{-13} $(5 \times 10^{-14} -5 \times 10^{-12})$ TUC α3 β4γ4** TUC α3 β5 γ4** LSHRS α3 β3 γ4** LSHRS α3β4γ4** LSHRS α3 β5γ4**
Probability/Year	10^{-3} $(10^{-4}-10^{-2})$	4×10^{-9} $(3 \times 10^{-10} -5 \times 10^{-8})$	10^{-10} $(10^{-11} -10^{-9})$	3×10^{-11} $(4 \times 10^{-13} -4 \times 10^{-9})$	10^{-11} $(10^{-12}-10^{-10})$	5×10^{-13} $(5 \times 10^{-14} - 5 \times 10^{-12})$

**Probabilities negligible compared to other sequences under the same column.

2.2. Ranking of Contributors to Risk. The risk contribution of any event* (E_i) is measurable by examination of its effect on the total risk curve. Specifically, if sequences which do not contain E_i are omitted, a lower risk curve is obtained. Thus a basis for ordering events according to their importance (i.e., risk contribution) is provided. The same logic would apply to ordering of groups of events, such as the event sequences shown in Table 2. The concept is illustrated in Figure 3.

In the comparison of any two events or event sequences, there are two possible situations, both illustrated in Figure 3. In the first case, the probability associated with one event dominates that associated with the other event at all levels of consequence. This situation (called "stochastic dominance" in Decision Theory) is illustrated in the comparisons of CAC1 to CAC3 and CAC2 to CAC3. If as illustrated in the comparison of CAC1 to CAC2, stochastic dominance is not the case, selection of the more important risk contributor requires additional criteria.

At this point, measures of preference ("utilities") are introduced. Many choices are available, but all have in common that they attempt to quantify the decision maker's risk aversion -- his diminishing willingness to accept risk of increasingly larger

---

* Examples of "events" are: TOP (Transient Overpower With Failure to Scram), β_2 (Minor Vessel Head Damage), γ_4 (Containment Isolation Failure)

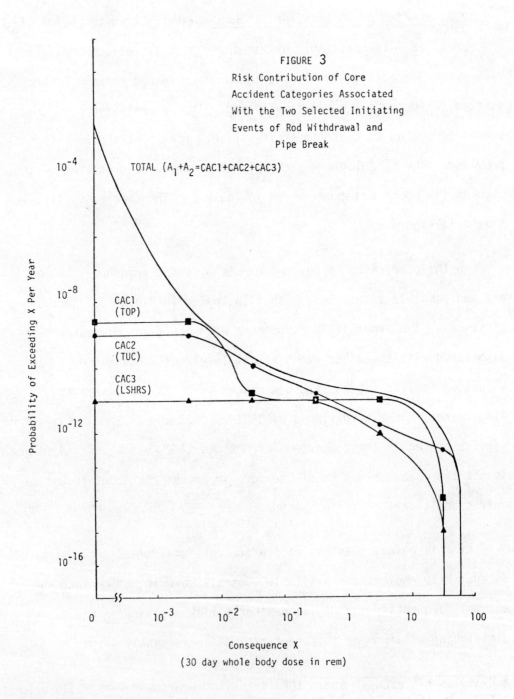

FIGURE 3
Risk Contribution of Core
Accident Categories Associated
With the Two Selected Initiating
Events of Rod Withdrawal and
Pipe Break

TOTAL ($A_1+A_2=$CAC1+CAC2+CAC3)

CAC1
(TOP)

CAC2
(TUC)

CAC3
(LSHRS)

Probability of Exceeding X Per Year

Consequence X
(30 day whole body dose in rem)

consequences. In this study, we experimented with three different measures.

Numerical evaluations are simplified by dealing with complementary cumulative probability distributions in discrete form, rather than in the continous form illustrated in Figure 3. Therefore, the problem is further considered in terms of the six "Consequence Intervals" of Table 2. The most important Consequence Interval was selected according to each of three importance measures:

(1) An expected value measure, defined as the product $P_i K_i$; where P_i and K_i are the probability and consequence of Consequence Category i,

(2) A risk measure with consequence aversion, defined as $P_i K_i^n$, where each consequence is weighted by the factor K_i^{n-1}, with n greater than 1,

(3) An extreme expected value measure, defined as the product of Max $\{P_i\}$ and K_i where Max $\{P_i\}$ is the upper bound estimate of the probability of K_i.

The first two measures reflect varying degrees of risk aversion. An expected value criterion places no weight at all on large consequences -- a "one-in-a-million" chance of 1 fatality would weigh equally with a "one-in-a-billion" chance of 1000 fatalities. The second measure (of which the first is just a special case) reflects reluctance to accept large consequences. In the trial application, it was found that the rankings of consequence intervals remained un-changed for $1 \leq n < 1.3$. The last measure places greater importance on the uncertainties of probability estimates. This type of measure might be used if demonstration of meeting a design goal was an objective.

Following determination of the most important consequence inter-val according to one of the three measures, the most important event (or event sequence) is found by using probability as a ranking measure. Nominal probability estimates are used in conjunction with importance measures 1 and 2, while upper bound probabilities are used with im-portance measure 3.

2.3. Identification of R&D Needs. The ranking technique defines events or event sequences most important to the risk. If the risk is unacceptable, it can be modified by a combination of design changes, research and development. These actions affect the risk contribution of a given event (or sequence) in two ways:

1. Design changes attempt to reduce the nominal probabilities of events, and

2. Research and Development reduces uncertainties which in turn either reduces bounding probability estimates or modifies nominal probability estimates.

The design approach is fairly straightforward. Modifications to increase redundancy or diversity in safety systems fall in this category. So-called "consequence mitigating" features do so as well. For example the probabilities of $\gamma 2$ and $\gamma 3$ (containment failure by over pressurization) can be reduced by the addition of venting and cooling systems.

Research and Development includes the acquisition of data on processes, components and system behavior, and improvement of risk assessment models (including more detailed definition of the design itself). These (R&D) activities reduce the uncertainties associated with inputs to the risk assessment. The resulting nominal estimates of event probabilities may then shift either upward or downward. Thus R&D will produce better information about risk, but does not by itself guarantee reduction of risk.

Emphasis on one or the other of the above plans is decided by such considerations as the direct cost penalty, impact on the plant schedule, and potential to resolve licensing issues. A major factor in the decision is the technical risk (i.e., chance of success) of either an attempt to change the design or an R&D effort. None of these factors has been addressed in our studies to date.

The identification of areas of research and design features to alter the probabilities (and uncertainties) associated with important

events is -- at least in this application -- a less structured process
than the one used for event selection. We relied on intuition gained
from performance of the risk assessment itself. It seems desirable
to develop a more structured approach to this part of the evaluation,
so that the results would be reproducible.

3.   Application to the First Cycle Risk Analysis of FFTF.

3.1. Ranking of Contributors to Risk. The risk contributions of
Core Accident Categories are shown in Figure 3. CAC1 (TOP) and CAC2
(TUC) both dominate CAC3 (LSHRS) in these results, but a choice
between CAC1 and CAC2 would require application of the consequence
importance measures described in Section 2. The bias introduced by
limiting the risk assessment to just two initiating events will have
an effect here. Although the analysis has not been carried out, our
knowledge of the risk assessment leads us to expect that CAC3 (LSHRS)
would dominate the other Core Accident Categories if all events were
included.

With these reservations noted, the risk assessment considering
only events A1 and A2 suffices as a concrete example through which
the procedure for ranking R&D needs is illustrated. The remaining
discussion will consider this limited risk assessment as though it
were complete. We chose to rank the accident sequences rather than
single events.

None of these sequences is stochastically dominant therefore the most important Consequence Categories were first selected according to the importance measures of Section 2.

The Expected Value Measure results in Consequence Category K5 as most important. The Risk Aversion Measure with n = 1.3 results in K6 as most important. The Extreme Value Measure results in K4 as most important. Rather than eliminate two of these by choosing one importance measure over the others, the dominant sequences in all three categories (K4, K5 and K6) are determined.

The dominant Sequence leading to K4 is:

K4-1. LSHRS $\alpha 3 \beta 4 \gamma 2$: loss of the shutdown heat removal system after successful scram (LSHRS), whole core damage ($\alpha 3$) moderate vessel head damage and late vessel melt-through ($\beta 4$), and late containment failure by overpressurization ($\gamma 2$).

Uncertainty in the sequence K4-1 leads to the high ranking of K4 based on the Extreme Value Measure. Uncertainty in the LSHRS probability estimate is the major contributor to this uncertainty.

The dominant sequence leading to K5 is:

K5-1. TOP $\alpha 2 \beta 2 \gamma 4$: transient overpower with failure-to-scram (TOP), partial core damage ($\alpha 2$), minor vessel head damage ($\beta 2$), and containment isolation failure ($\beta 4$).

The probability of sequence K5-1 is the major contributor to the high ranking of K5 based on the Expected Value Measure.

The dominant sequence leading to K6 is:

K6-1.   TUC $\alpha3\beta3\gamma4$: TUC with whole core damage ($\alpha3$), moderate
        vessel head damage ($\beta3$), and containment isolation fail-
        ure ($\gamma4$).

3.2.   Identification of R&D Needs

3.2.1. Classification of R&D Actions. Application of the ranking
procedure in Section 3.1 focuses attention on the issues of:

(1) Contributions to the bounding probability of Sequence K4-1
     (LSHRS $\alpha3\beta4\gamma2$), due largely to uncertainty in estimation of
     the probability of LSHRS,

(2) Contributions to the nominal probability of sequence K5-1
     (TOP $\alpha2\beta2\beta3\gamma4$),

(3) Contributions to the nominal probability of sequence K6-1
     (TUC $\alpha3\beta3\gamma4$).

Actions which address these issues are categorized by Line of
Assurance (LOA) [13], defined as follows:

LOA1:   prevention of accidents,

LOA2:   limitation of core damage,

LOA3:   containment of core debris,

LOA4:   attenuation of radiological products.

Within each LOA, R&D needs are further classified into the

following three areas:

1. Basic Data and Phenomena,

2. Methods and Criteria Development, and

3. Design Strategy.

Conclusions related to LOA1 are summarized under Section 3.2.2. Space limitation prevents the inclusion in this paper of results for other LOA, but the interested reader will find such discussion in Reference 3.

3.2.2. R&D Needs Under LOA1. Probabilities of interest in LOA1 are those leading to the Core Accident Categories LSHRS, TOP and TUC. Examination of Table 1 suggests in turn consideration of three System Event Sequences:

A1B1C3: Uncontrolled withdrawal of a control rod A1, with successful scram B1, and SHRS failure C3. This sequence leads to LSHRS (K4-1),

A1B2C1: Uncontrolled withdrawal of a control rod A1, with failure to scram B2, and continued full power heat removal capability C1 (K5-1),

A1B2C2: Uncontrolled withdrawal of a control rod A1, with failure-to-scram B2, and pump coastdown C2 (K6-1).

Considerations of the above sequences and definition of the ranking measures used lead to the following R&D goals:

● Reduce uncertainty in the probability estimate of A1B1C3,

• Reduce probabilities of A1B1C3, A1B2C1 and A1B2C2.

The following discussion addresses means of accomplishing these goals. Reduction of uncertainty in the probability estimate of A1B1C3 is discussed under Basic Data and Phenomena, and Methods and Criteria Development. Reduction of the probability of A1B2C2 is discussed under Design Strategy. Due to the limited scope of the First Cycle risk assessment and the large uncertainty associated with the probability estimate of C3, no recommendations of design strategy to reduce the probabilities of A1B1C3 and A1B2C1 are made at this time.

3.2.2.1. R&D Needs - Basic Data and Phenomena. Uncertainty in the probability estimate of the uncontrolled withdrawal of a control rod (A1) is due to lack of information on the operating procedure for the FFTF [3].

Uncertainty in the SHRS failure probability estimate (C3) is attributed to deficiencies in the reliability model used in the First Cycle Analysis [3] and the lack of verification by test or experience at the system level. The model used does not specifically account for common mode failures or propagation of failures between loops.

3.2.2.2. R&D Needs - Methods and Criteria Development. The following R&D needs are identified:

1. SHRS Failure Propagation Analysis: The probability of

SHRS failure, P(C3), was estimated on the basis that any
failure within the SHRS is equivalent to failure of the
entire system. This probability could be reduced by code
development to analyze structural and functional failure
propagation in the SHRS. The output of this analysis
should be directed towards providing criteria for adequate
characterization of system failure to identify the importance of:

    a.   structural failures under various flow conditions,

    b.   failures of snubbers and check valves,

    c.   loop and pump cavitation,

    d.   sodium freezing.

2.   Common Mode (Cause) Failure Analysis: Bounding estimates
of P(B2) and (C3) could be reduced by development of quantitative tools to provide systematic analysis of common
mode failure. The rigidity of causal approaches such as
fault tree analysis needs to be weighed against acceptability of uncertainty through such techniques as sensitivity analysis.

3.   Data Transfer Procedures: There is a need for procedures
to transfer information at the component and system levels
from more mature technology areas, e.g. LWR applications.

3.2.2.3. R&D Needs - Design Strategy. The TUC accident A1B2C2
occurs as a result of tripping the HTS pumps by the scram signal
while failing to insert the control rods because of failures in the
scram system mechanical subsystem (SSMS). These conditions are also
possible for other initiating events.

This TUC entry mode could be avoided if the pump trip signal
were delayed until the neutron flux started to decay. This suggests
blocking the HTS pump trip signal by a flux signal such as flux/de-
layed flux or flux/flow. A logic to perform this function is within
the state of the art and could provide 2 to 3 orders of magnitude
probability reduction for this TUC entry mode.

Another strategy would be to consider probability reallocation
among accident initiators and the scram system electrical subsystem
to render the SSMS failure unimportant. There is a need for further
investigation of these strategies to evaluate their pay off.

4. Closure. The foregoing describes a simple procedure to
single out important information needs by systematic examination of
risk assessment results.

Illustration of the method by its application to a limited risk
assessment of FFTF shows that it is workable, and that it can produce useful information. The general conclusions drawn from this
illustration are not profoundly surprising. It is well known that

uncertainty in LMFBR risk predictions is influenced by inadequate
modelling of common mode failures, and that the reliability of shutdown heat removal systems is a primary focus of design to produce
high shutdown reliability. The value of the method lies in the fact
that it systematically focuses on the issues important to risk, thus
providing the means to determine importance rankings when the analysis
is too complex to rely on intuitive conclusions.

In the example application, the more important R&D goals in each
category (i.e., Research, Methods, Design) and within each line of
Assurance (LOA) were identified. Development of preference criteria
and models based on optimal allocation of resources (as suggested in
Section 2.3) would provide a basis for further narrowing the emphasis
of R&D efforts. For example, quantitative measures which reflect both
the desirability and the chance of success of R&D in LOA1 as compared
to LOA3 could be used as a basis for choice between these categories.

REFERENCES

[1] U. S. NUCLEAR REGULATORY COMMISSION, Reactor Safety Study - An
     Assessment of Accident Risks in U.S. Commercial Nuclear
     Power Plants, WASH-1400 (NUREG-75/014), October 1975.

[2] K. G. FELLER, Method for Assessing the Risk of an LMFBR Loss-of-
     Flow With Failure to Scram Accident, General Electric Co.
     GEFR 00036 (L), February 2, 1977.

[3] K. A. EL-SHEIKH, Risk Analysis Application to FFTF - First Cycle,
     General Electric Co. GEFR 00066, August 1977.

[4] K. A. EL-SHEIKH, ET AL., Risk Analysis Application to FFTF -
     Second Cycle, General Electric Co., GEFR 00029, March 1977.

[5] CRBRP SAFETY STUDY, CRBRP-1, March, 1977.

[6] F. E. DUNN, ET AL., The SAS3A LMFBR Accident Analysis Computer
     Code, Argonne National Laboratory ANL/RAS 75-17, April 1975.

[7] ALAN E. WALTAR, ET AL., Melt-III, A Neutronics, Thermal Hydrau-
     lics Computer Program for Fast Reactor Safety Analysis -
     Volume I, Hanford Engineering Development Laboragory, HEDL-
     TME 74-47, December 1974.

[8] J. F. JACKSON AND R. B. NICHOLSON, VENUS - II: An LMFBR Dis-
     assembly Program, Argonne National Laboratory ANL-7951,
     September 1972.

[9] Y. W. CHANG AND J. GVILDYS, REXCO-HEP: A two-Dimensional Com-
     puter Code for Calculating the Primary System Response in
     Fast Reactors, Argonne National Laboratory ANL/RAS-75-11,
     March 1975.

[10] R. D. PEAK AND D. D. STEPNEWSKI, Computational Features of the
     CACECO Containment Analysis Code, Hanford Engineering
     Development Laboratory, HEDL-SA-922, May 1975.

[11] R. S. HUBNER E. V. VAUGHN AND L. BAURMASH, HAA-3 User Report,
     Atomics International, AI-AEC-13038.

[12] J. M. OTTER AND P. A. CONNERS, Description of the COMRADEX-III
     Code, Atomics International, TI-001-130-053, June 30 1975.

[13] J. D. GRIFFITH, Reflections on the Recriticality Conference at
     Argonne National Laboratory, Nuclear Safety, 18(1970), pp.
     45-52.

A CONCEPT FOR THE RELIABILITY ASSESSMENT OF THE CONTAINMENT OF A PWR

GERHART I. SCHUËLLER*

Abstract. A concept for the reliability assessment of the Containment of a PWR based on classical statistics is presented. The statistical nature of various external hazards such as earthquakes, external pressure waves, aircraft crashes, as well as internal load conditions such as the LOCA are discussed. The key problem, which is the establishment of the relation between the time-dependent frequency of occurrence of the hazards and their intensity, is discussed in light of their rare event property. For this reason the Poisson process is used as prediction model. The intensity distributions are modeled by asymptotic distributions of the Fisher-Tippett type. Various failure modes such as fracture and the yield condition and their influence on the design are discussed. The work concludes with a numerical example in which the reliability of the steel hull of a sample containment structure under a large LOCA is calculated.

*Institut für Bauingenieurwesen III of the Technische Universität München, 8 Munich 2, Germany. This research is supported by the Bundesministerium für Forschung und Technologie under Contract RS-201.

1. __Introduction__. During its design life the containment structure of a pressurized water reactor (PWR) may be exposed to various man-made or natural hazards. Since,in case of an accident, the vapour container is the last barrier to prevent the release of radioactive pollution and the consequent contamination of the surrounding site, its reliability should be given particular attention. For a credible reliability assessment sound prediction methods with regard to their frequency of occurrence and their intensity distributions are essential. In addition the material behaviour, i. e. the probability density functions of strength depending on the respective rate of strain caused by the load history, is an important factor for a realistic reliability analysis.

Generally the containment structure consists of two parts:

(a) of the steel hull, which may be a cylinder closed on top by a spherical dome or a spherical shell and which has to remain leaktight under all credible loading conditions.

(b) of a reinforced and/or prestressed concrete shell structure which is supposed to protect the hull against external hazards.

The sketch of a typical structure to be analysed herein is shown in Fig.1. In this paper primary emphasis will be given to the analysis of the steel hull under internal load conditions.

Presently, the design of containment structures is carried out using existing deterministic codes. Uncertainties in the predictions of load and material properties are recognized in these codes and believed to be covered by providing "ample" safety-factors, which are expected to take care of these uncertainties. However, no direct relation between these factors and the reliability of the structures to be analysed is given [1]. The establishment of this relation for containment structures under load conditions will be shown. By doing this, one of the major problems appears to be the inclusion of the fracture mechanical aspects. Finally it should be stated that the probabilistic structural analysis is consistent with the modern systems analysis approach used in risk analysis of nuclear reactors as shown for example in [2,3].

2. The Reliability Concept. The basic principles of the classical structural reliability analyses have been presented by Freudenthal et al. [1]. Some aspects in connection with the design of structural and equipment design for nuclear power plants are given in ref. [4-9].

The hazards to the containment may develop from either internal or external causes or by a combination of both. The basic question is first to identify and then to find proper models to forecast the hazards which may jeopardize the integrity of the structural system and secondly to develop a realistic frequency of occurrence - intensity relation.

It is now generally recognized, that these variables and models are associated with statistical uncertainty. Moreover the material properties used to calculate the structural resistance show statistical scatter as well. The combination of all these uncertainties with respect to space and time is accomplished by using a reliability model such as the one suggested here.

In this context, the first aspect to be resolved is the prediction of the frequency with which a certain hazard, e. g. an earthquake, aeroplane crash, LOCA etc. is likely to occur. To answer these ques-

tions the theory of stochastic processes has to be applied. Relevant
models have to be chosen for the various types of loads on the basis
of statistical fit procedures and physical reasoning. It has been
shown previously [6] that the Poisson process is for many load cases
the most appropriate model, as these extreme loads are considered
rare events.

The second question to be answered is the probability distribution of the load intensity, once the hazard event has occurred. For
example, given an earthquake, what is the distribution of the intensity? Several types of theoretical probability distributions may
be fitted to sample data, which may also be obtained by simulation
procedures. The acceptance condition and preference of a particular
distribution is, again, based on statistical and physical reasoning.
It should be noted that the distributions are unique to each individual site.

Third, the question of modeling the resistance in probabilistic
terms remains, as material properties generally show considerable
scatter. The acceptance criteria for the various models are similar
to those as stated for the loads.

Once the probability distributions of the loads are known, the
structural failure probability under a single load application may
be easily determined by the well known convolution integral

(1) $$p_f = \int_{-\infty}^{\infty} F_R(x) \cdot f_S(x) \, dx$$

where $F_R(x)$ is the cumulative distribution function of the resistance
and $f_S(x)$ the probability density function of the load. It should
be noted that Eq. (1) is based on the assumption of stochastic inde-

pendence between the loads and the resistance.

The probability of failure within a given period of time, however, includes the frequency of occurrence of the hazard as described above. The failure probability $F_L(t)$ may be expressed as [1]

(2) $$F_L(t) = 1 - \left[\int_{-\infty}^{\infty} p_r(t) \left[F_S(x) \right]^r f_R(x) \, dx \right]$$

where $p_r(t)$ is the probability of occurrence of a particular hazard event which may be described by appropriate stochastic models. The probability of surviving this event or the conditional reliability - conditioned on a particular hazard event category - of a structure for its design life $[0,t]$ is then obviously

(3) $$Z_L(t) = 1 - F_L(t)$$

The analysis has to be repeated for each of the hazard category to be considered. Simultaneous occurrence of two hazard events, i. e. two rare events, may safely be neglected. It should be pointed out, however, that the occurrence of a particular event might influence the occurrence probability of another event. For example, due to the occurrence of an earthquake, the likelihood of the occurrence of a LOCA might be increased. This problem, however, will not be considered here.

3. Description of Loads.

3.1. General Remarks. With regard to the previous discussion, man-made and environmental hazards may be considered as rare events. It was shown in ref. [6] that based on physical reasons inherent in the derivation of this process, the Poisson process is most suitable to model the frequency of occurrence of these events. The probability of occurrence of r hazard events with the period $[o,t]$ is

(4) $$p(r \mid \mu,t) = \frac{(\mu t)^r e^{-\mu t}}{r\,!}$$

where μ is the mean rate of occurrence of events to be statistically
estimated from historical or simulated data.

With regard to the choice of suitable probability distributions
to model the load intensities and the structural resistance, one should
should refer to the definition of the safety factor which is the ratio
between an extremely low resistance and an unexpected extremely high
load. By this definition therefore the only physically germane and realistic distribution functions are those of the extreme values of the
population of the variables describing the load and the resistance. It
has been shown by Gumbel [10] that this leads consequently to the application of the asymptotic distribution of the Fisher-Tippett type.

There are three types of external hazards which are required to be
considered for design of the containment structure: (a) impact due to
aircraft crashes, (b) external pressure due to explosion in the vicinity
of the plant and (c) earthquakes. While (a) and (b) are categorized as
man-made hazards, the load type (c) is a natural hazard. An additional
load case, the tornado or hurricane, may be considered to be covered
by load case (b); missile impacts from objects generated by these storms
are covered by (a).

Internal load conditions may be caused by transients such as loss
of off-site power, loss of heat sink, steamline break, loss of load,
etc., or the loss of coolant accidents (LOCA).

3.2. External Hazards.

3.2.1. Aircraft Crash. In Germany this load case has to be considered for design for all cases, while in some other countries it has
to be taken into account only in the vicinity of airports. It is quite
obvious that the risk of an aircraft crashing on a nuclear power plant
depends strongly on the density of population and the aviation activities of a country. Sütterlin [11] claims that for the Federal Republic
of Germany a mean rate of occurrence of 10^{-6}/year and plant has to be
expected. Military planes are considered to be most hazardous to the
plants.

For the impact-response determination a military fighter of the
"Phantom" type has to be used. The impact velocity is 215 m/s and the
total mass of the plane is 20t. The time history of the impact of the
impact of the aircraft as suggested by the Institut für Reaktorsicher-
heit (IRS) [12] is shown in Fig. 2. The maximum impact is to be expected after 40 msec. The IRS is proposing a smoothing of the curve as
indicated by the dash-dotted line. However, it seems to be more realistic to recognize the statistical scatter of the input function as
indicated in the figure and to determine the load effect distribution,
i.e. the load intensity S as random variables $f_S(s)$.

3.2.2. External Pressure Wave.* The question of endangering a
nuclear power plant by chemical explosives is, again, a question of the
density of the population and furthermore a question of potential
sources of explosives. A time history of a gas deflagration as suggested
in ref. [12] is shown in Fig. 3. As the volume of the gas, its con-

---

* Sabotage is not considered in this paper.

centration, i.e. type and the distance of the explosion cannot be predicted deterministically, the load intensity may also be considered a
random variable. The load diagram suggested by the IRS might serve as
mean value function. At this time sufficient statistical information is
not available with respect to a reliable estimation of the mean rate of
occurrence of this type of hazard. It is believed, however, that the
occurrence rate is about the order of magnitude of a plane crash on a
plant.

3.2.3. Earthquakes. Considerable more information is available
regarding the quantification of this type of risk [5,12-16]. Like the
other types of loads earthquake occurrence frequencies are modeled by
using the Poisson process [17] as well. Since the earthquake activities
change with geographical regions, the mean rate of occurrences have to
be determined according to a regionalization concept. Similar attempts
have already been made for determining the structural design wind
velocity.

Fig. 4 shows the intensity distribution in terms of ground accelerations for two different sites. The rather large difference between
the two curves emphasizes the influence of the difference of the tec-
tonics of the ground. Peak accelerations are determined with regard to
the accepted exceedance probabilities and the time histories to be used
for the structural analysis are then scaled accordingly.

3.3. Internal Loading Conditions. No attempt has been made to
elaborate on the systems analysis aspect in detail. For this purpose one
should refer to ref. [18] where the details of the determination of the
probability density functions of the loadings following a Loss of
Coolant Accident (LOCA) - as used herein - are described.

The LOCA - an important consideration in safety analysis - involves a major loss of high-pressure reactor coolant and results in a drastic, sudden and very complex change of system conditions. The LOCA may also be a core melt initiating event. The containment structure is supposed to hold the release of steam and gases for a period of time during which pressure and temperature increases to a certain peak and then decrease eventually to atmospheric conditions. The pressure decrease is accomplished by condensation heat transfer and the emergency core cooling system. The LOCA is considered in this paper the sample internal loading case which has been treated in the numerical example.

Using the ZOCO computer code [19] the pressure distribution as a function of time for the LOCA with full availability of the emergency safety feature can be calculated (see ref. [18]). The result of this calculation is shown in Fig. 5. It can be seen clearly that the maximum pressure built up is reached already within 10 sec after initiation of the accident. Furthermore, the analysis performed by Kafka and Augustin [18] has shown that the pressure distribution is subject to statistical variation, if one considers the LOCA with different systems operational. This contains the special case of all systems operational as well. The consideration of the core melt problem with steam explosion is beyond the scope of the paper. The mean value function as well as the probability distribution of the pressure are indicated in the figure.

The frequency of occurrence, i.e. the annual mean rate of occurrence of such an initiating event is indicated in ref. [2] to be 10^{-4} for a large LOCA and 10^{-3} for a small one. The frequency of occurrence of the core melt following a LOCA is part of the German Risk Study, the results of which are not yet available.

Finally, it should be mentioned that the calculations have to be repeated for other internal loading situations.

4.   Structural Analysis.   Static and dynamic analysis is required for the calculation of the load effects caused by the loads described in the previous section. While the static cases of the cylindrical and the spherical shell are rather simple problems, the calculation of the stress distributions of shells with holes usually requires the application of approximate methods. If the classical shell theory is applied, the approximation by Geckeler [20] might be used. However, as high speed digital computers are available in most cases the Finite Element method [21] is generally utilized. The dynamic analysis for the various stochastic loads is performed by using either the time history method or the response spectrum method. For this purpose the structure is idealized to a lumped mass system and both linear and nonlinear responses are then computed. Material and geometric nonlinearities may be considered. If the calculations are carried out in the time range, a statistical analysis of the response function yields the data for the probability distribution of the load effect. Multipurpose computer codes [22,23] are generally used to analyse containment structures.

5. Failure Criteria.   It should be pointed out clearly that - particular with reference to the steel hull - two failure criteria, which are independent of each other, govern the design of the containment structure. There is on one hand the ultimate load criterion and on the other the fracture criterion [24].

The ultimate load criterion is generally well known and is therefore reviewed only briefly. If one considers for example a spherical shell and assumes ideal homogeneous material properties, the yield condition of the entire system can be calculated as follows:

(5)
$$p_1 = (2\delta/R) \cdot \sigma_y$$

where δ is the wall thickness, R the radius of the shell and σ_y the yield stress. Eq. (5) is based on the von Mises yield condition recognizing the triaxial state of stress. It should be noted that according to the equation above the condition of failure of the structure is attained when unrestrained plastic flow sets in and the dimensions of the structure therefore change rapidly. Hence Eq. (5) also expresses an instability condition. Since the yield stress σ_y is a random variable because of statistical scatter of the material properties, the yield pressure, p_1, is also to be considered a random variable. However, due to the air locks and the piping the spherical shell is perforated by holes of circular shape. The stress concentrations around these holes may be determined by finite element analysis. For this case the above equation can be written as follows:

(6)
$$p_2 = (2\delta/R) \frac{1}{C} \cdot \sigma_y$$

where C is the stress concentration factor. In this case only the "yield pressure" of the local range (around the hole) is determined, while the previous equation determines the global yield pressure.

Since the materials which are actually used do not satisfy the assumptions as described above, another criterion - which is independent of the yield condition - is needed. For this purpose fracture mechanical aspects are considered. This criterion accounts for small cracks and other damage of the steel and welds which may occur although stiff quality control regulations are being applied. The utilization of the Griffith equation - which is valid only for brittle materials - results for the spherical shell into the following expression:

(7) $$p_3 = K_{Ic} \, (2\delta/R) \cdot \frac{1}{F\sqrt{c}}$$

where K_{Ic} is a parameter depending on the type of material and its crack propagation property (dimension: $(kp/cm^2) \cdot \sqrt{cm}$), F is a shape factor and 2c is the crack or flaw length. In the above equation K_{Ic} expresses the statistical uncertainties inherent in the material. Finally, the case of cracks in the vicinity of stress concentrations is investigated. Similar to Eq. (6) the stress concentration factor C is introduced. The following equation results from Eq. (7):

(8) $$p_4 = K_{Ic} \cdot (2\delta/R) \, \frac{1}{F\sqrt{c}} \cdot \frac{1}{C} \, .$$

6. Numerical Example. In order to exemplify the application of the proposed methodology a numerical example is presented. Since this example is only meant as an illustration, it should be kept in mind that the actual design procedure needs considerable more detailed analysis than has been shown here. However, it is believed that some of the significant aspects are indicated here.

This example is concentrated on the evaluation of the reliability of the steel hull of the containment structure under the overpressure load caused by a LOCA. As a sample structure the containment shown in Fig. 1 is analysed. The structure consists of a spherical hull with a diameter of 56.0 m. Its wall thickness is 0.029 m, around the holes of the staff and material locks it is 0.034 m. The type of material which has been used for construction is a FG 47 WS special type steel. This material is generally utilized for containments in Germany.

For modeling the structural resistance, the Fisher-Tippett-Type III distribution of the smallest values - which is also called Weibull distribution - has been used. From actual material test data the parameters

of the distribution have been estimated. This resulted in scale para-
meter of u = 0.091 kp/mm^2 and a shape parameter α = 41.7 $\left[\text{kp/mm}^2\right]^{-1}$.
The coefficient of variation of this distribution is therefore 3%. The
scale and shape parameters for the fracture related resistance
K_{Ic} are u = 0.103 kp/mm and α = 20.0 $\left[\text{kp/mm}^2\right]^{-1}$ respectively. This
yields a coefficient of variation of 6.2%. Referring to section 3.3:
the parameters of the pressure distribution caused by a LOCA are
0.0445 kp/mm^2 for the scale parameter, u, and 1348 kp/mm^2 $^{-1}$ for the
shape parameter α. This results in a coefficient of variation of 2.12%.
The load distribution has been modeled by using the Fisher-Tippett-
Type I distribution of the largest values, which is also called the
Gumbel-distribution.

Utilizing Eq. (1) and Eq. (2) through (8) the failure probabilities, p_f - i.e. the probability of failure under a single load application - are calculated for yield and fracture failure conditions. The
spherical shell with and without holes has been analysed. A stress concentration factor of 1.7 resulted from an extensive Finite Element
analysis [25]. The results are listed in Table 1 and should be read in
connection with Fig. 6. For the fracture failure condition a crack
length of 2 c = 2 cm is assumed.

The mean rate of occurrence μ of a large LOCA - which is in this
example assumed the initiating event of the pressure build up - is
chosen to be 10^{-4}/year. Utilization of Eq. (2) and (3) yields then the
failure probabilities $F_L(t)$ with respect to the design life of the
structure. For a design life of 30 years the results are listed in
Table 1.

   7. Discussion of Results and Conclusions. The analysis reveals
a number of interesting aspects which deserve some attention. First of
all it becomes clear, that an analysis, in which the various types of
uncertainties as described in section 1 are treated in a consistent
manner, is feasable. A possible solution of the rather difficult problem
of deriving a frequency of occurrence intensity relation has been presented. The method seems very convenient as the amount of numerical
effort which is required for the performance of the analysis remains
in acceptable limits. In order to demonstrate the analysis a numerical
evaluation of the failure probability of the steel hull under a large
LOCA is carried out. Both failure mechanisms, i.e. the yield and the
fracture condition are analysed with respect to their contribution to
the risk of failure. Given the maximum crack length - which might not
be detected despite of stiff quality control - it is possible to show
in terms of risks which mechanism governs the design. From Table 1 it
may be seen that, assuming a crack length of 2 cm in the hull, the
probability of occurrence of the fracture failure condition is about
six orders of magnitude higher than that for the yield failure. In other
words, the risk increases from $2.8 \cdot 10^{-15}$ to $1.9 \cdot 10^{-9}$. It is interesting
to note that in the area of stress concentrations this difference is
smaller than one order of magnitude. Attention should be drawn to the
fact that the failure probabilities of the hull without perforations
describe the global failure chance, while the values for the area around
the holes define a local chance of failure, which does not necessarily
lead to a total failure, i.e. "explosion" of the shell. With regard to
the general order of magnitude of the probability of failure under a
single load application p_f, it should be mentioned that this is an
"extreme" failure probability for it is based on extreme value dis-

tributions, i.e. extreme populations. If the calculations were based on the distributions of the total populations, their computational value would decrease considerably. For the hull without holes (fracture condition) the number would decrease from $6.4 \cdot 10^{-8}$ to $2.5 \cdot 10^{-12}$. The analysis shows clearly the need for the quality control of the material and the welds as the failure probability increases considerably around stress concentrations. Furthermore, it should be pointed out that the analysis allows an optimization of the acceptable crack length, i.e. the quality control. For this purpose, for example, the crack length parameter c, which has been used in Eq. (7), may be varied so that it results in connection with Eq. (1) in the same failure probability as one would obtain by using Eq. (5). In other words the risk of failure under fracture and yield is of the same value.

It should be pointed out again that melt down in conjunction with large LOCA or large LOCA with simultaneous occurrence of a seismic event, explosion or aircraft crash has not been considered. Neither the situation where the leak rate of the container is in excessive of design prior to the accident.

In conclusion, it might be stated that the approach put forward here enables the engineer to make rational comparisons between several types of risks. Moreover, by using this method a comparison of risks of various nuclear plants built at different sites is possible.

TABLE OF NOMENCLATURE

$F_R(x)$ .... cumulative distribution function of resistance R

$f_S(x)$ .... probability density function of the load S

p_f .... probability of failure under single load application

$p_r(t)$ .... probability of occurrence of a particular hazard event

L .... design life

$F_L(t)$ .... failure probability with regard to the design life

$Z_L(t)$ .... reliability with regard to the design life

r .... number of hazard events

μ .... mean rate of occurrence of events

δ .... wall thickness

R .... radius of spherical shell

σ_y .... yield stress

p_y .... yield pressure ($p_1 \div p_4$)

C .... stress concentration factor

2c .... crack (flaw) length

K_{Ic} .... parameter for crack propagation

F .... shape factor

REFERENCES

[1]   A.M. FREUDENTHAL, J.M. GARRELTS and M. SHINOZUKA, The Analysis
       of Structural Safety, J. Struct. Div., Proc. ASCE, Vol. 92,
       No. ST1, Feb., 1966, pp. 267 - 325.

[2]   REACTOR SAFETY STUDY, Wash 1400, U.S. NRC, Oct. 1975

[3]   A.E. GREEN and A.J. BOURNE, Reliability Technology, Wiley-
       Interscience 1972.

[4]   A.M. FREUDENTHAL and G.I. SCHUÉLLER, Risikoanalyse von
       Ingenieurtragwerken, Rep. No. 25, Konstr. Ingenieurbau,
       Berichte, Ruhr-Universität Bochum, Aug. 1976, pp 7 - 96.

[5]   S. ZENDEHROUH and M. SHINOZUKA, A Structural and Equipment
       Design Methodology for Nuclear Power Plants, Techn. Rep.
       No. 8, Columbia Univ., April, 1976, p. 149.

[6]   G.I. SCHUËLLER and R.F. SCHWARZ, Some Aspects of the
      Reliability-Based Design of Reactor Containment Structures,
      Journ. Nucl. Engr. Des., Vol. 37, No. 2, May, 1976,
      pp. 299 - 305.

[7]   A.M. FREUDENTHAL, Reliability of Reactor Components and Systems
      Subject to Fatigue and Creep, J. Nucl. Engr. Des., Vol. 28,
      No. 2, Sept. 1974, pp. 196 - 217.

[8]   A.M. FREUDENTHAL, Extreme Value Risk Analysis in the Structural
      Design of Reactor Components, Journ. Nuclear Engr. Des.,
      Vol. 37, No. 2, May, 1976, pp. 179 - 181.

[9]   G.I. SCHUËLLER, On the Structural Reliability of Reactor Safety
      Containments, J. Nuclear Engr. and Design 27, 1974,
      pp. 426 - 433.

[10] E.J. GUMBEL, Statistics of Extremes, Columbia Univ. Press,
      New York, N.Y., 1958.

[11] L. SÜTTERLIN, Zur Auslegung kerntechnischer Anlagen gegen Ein-
      wirkungen von außen, Teilaspekt: Betrachtungen über das
      Risiko bei Flugzeugabsturz auf ein Kernkraftwerk (Zwischen-
      bericht); IRS-W-12, March, 1975.

[12] K. DRITTLER, Technisch physikalische Modelle für äußere Ein-
      wirkungen und Ableitung der Lastannahmen, Tagungsber.,
      Schutz von Kernkraftwerken gegen äußere Einwirkungen,
      IRS-T-27, April, 1975.

[13] C.A. CORNELL, Engineering Seismic Risk Analysis, Bull. Seism.
      Soc. of America, Vol. 58, No. 5, Oct., 1968, pp. 1583 - 1606.

[14] B. ISACKS, J. OLIVER and L.R. SYKES, Seismology and New Global
      Tectonics, J. Geophys. Res., Vol. 73, No. 18, Sept., 1968,
      pp. 5855 - 5899.

[15] N.M. NEWMARK, J.A. BLUME and K.K. KAPUR, Seismic Design Spectra
      for Nuclear Power Plants, J. Power Div., Proc. ASCE,
      Vol. 99, No. PO2, Nov., 1973, pp. 287 - 303.

[16] R.L. SHARPE, Systems Aspects of Seismic Design for Nuclear
      Plants, J. Power Div., Proc. ASCE, Vol. 99, No. PO1,
      May, 1973, pp. 175 - 192.

[17] J.R. BENJAMIN, Probabilistic Models for Seismic Force Design,
      J. Struct. Div., ASCE, Vol. 94, No. St5, Paper 5950, 1968.

[18] P. KAFKA and W. AUGUSTIN, Magnitude and Probability of Internal
      Load Behaviour of the Containment of a PWR, 4th Int. Conf.
      Struct. Mech. React. Tech., Sess. J., San Francisco,
      Aug., 1977.

[19]   D. BROSCHE, ZOCO V, A Computer Code for the Calculation of Time-
       and Space- Dependent Pressure Distributions in Reactor
       Containments, Nuclear Eng. and Des. 23 (1972), pp. 239 - 272.

[20]   K. GIRKMANN, Flächentragwerke, Springer-Verlag, Wien, 1946.

[21]   O.C. ZIENKIEWICZ, The Finite Element Method in Engineering
       Science, McGraw-Hill, London 1971.

[22]   K.-J. BATHE, E.L. WILSON and F.E. PETERSON, SAP IV, A Structural
       Analysis Program for Static and Dynamic Response of Linear
       Systems, Univ. Calif., Berkeley,Report Nr. EERC 73-11,
       June, 1973.

[23]   K.-J. BATHE, E.L. WILSON and R.H. IDING, NONSAP, A Structural
       Analysis Program for Static and Dynamic Response of Non-
       linear Systems, Struct. Mat. Res. Rep. No. UC SESM 74-3,
       Univ. of Calif., Feb., 1974.

[24]   A.M. FREUDENTHAL, Introduction to the Mechanics of Solids,
       John Wiley & Sons, Inc., New York, 1966.

[25]   J. BAUER and G.I. SCHUÉLLER, PRW Containment: Structural
       Reliability Assessment under Internal and External Loading
       Conditions, 4th Int. Conf. Struct. Mech. React. Tech.,
       Sess. J, San Francisco, Aug., 1977.

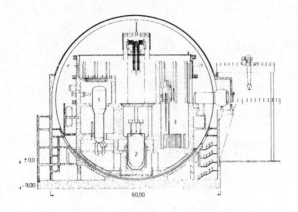

FIG. 1 : SCHEMATIC SKETCH OF THE CONTAINMENT OF A PWR

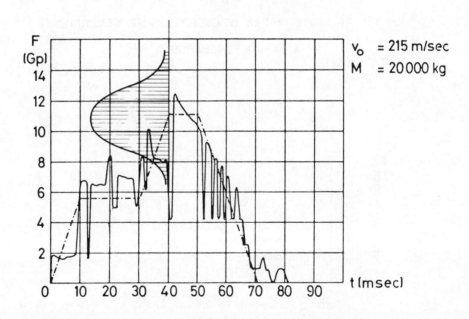

FIG. 2: TIME HISTORY OF IMPACT OF AN AIRCRAFT (TYPE "PHANTOM") CRASH

ON A STIFF PLATE [12]

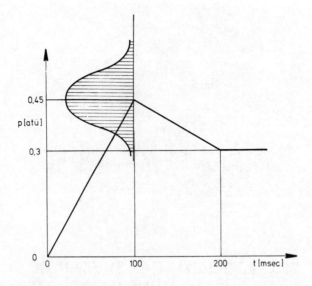

FIG.3 : TIME HISTORY OF OVERPRESSURE GENERATED

BY A GAS – DEFLAGRATION [12]

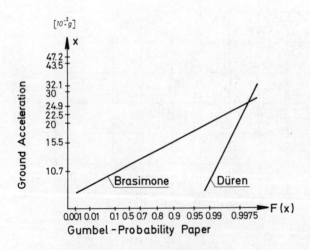

FIG.4 : SAMPLES OF EARTH QUAKE ACCELERATION DISTRIBUTIONS

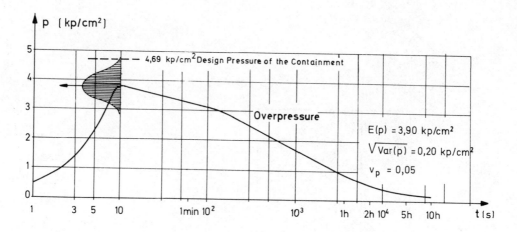

FIG.5: PRESSURE DISTRIBUTION IN THE STEEL HULL FOLLOWING A LARGE LOCA [18]

FAILURE CONDITIONS

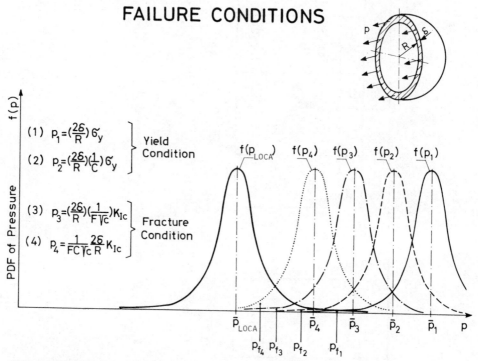

FIG. 6: FAILURE CONDITIONS OF THE STEEL HULL

SCHUËLLER

Failure Condition	Shell	Load		Resistance			p_{f_i}	$F_L(t)$**
		u	α	Eq.	u	α		
Yield	no holes	0.0445	1348	(5)	0.091	41.7	$9.3 \cdot 10^{-14}$	$2.8 \cdot 10^{-15}$
Yield	holes	0.0445	1348	(6)	0.0535	43.0	$9.8 \cdot 10^{-4}$	$2.9 \cdot 10^{-5}$
Fracture*	no holes	0.0445	1348	(7)	0.103	20.0	$6.4 \cdot 10^{-8}$	$1.9 \cdot 10^{-9}$
Fracture*	holes	0.0445	1348	(8)	0.0609	20.0	$2.5 \cdot 10^{-3}$	$7.5 \cdot 10^{-5}$

*) crack length 2c = 2 cm   **) design life L = 30 years

Table 1: Results of Numerical Analysis

PROBABILISTIC ANALYSIS OF ACCIDENTAL TRANSIENTS IN NUCLEAR POWER PLANTS

JAN AMESZ*, SERGIO GARRIBBA** AND GIUSEPPE VOLTA*

Abstract. The aim of this work is to establish a theoretical frame for computing probability distribution of loads and residual strength in the primary cooling system of nuclear power reactors. Random point processes provide a sufficiently general frame for describing operational and accidental load transients. Once failure modes have been ascertained and a failure hypothesis has been established, strength appears represented by a filtered process and reliability can be in principle computed. Information concerned with failures and performance characteristics of components allows us to infer the statiatical structure of the random processes. The case is displayed where advantage is taken of all information which is available prior to the design. The case is then considered where use is made of additional data obtained through operation practice of the system.

1. Introduction. Estimation and evaluation of nuclear reactor safety run into two recurrent problems which seem should deserve more attention than they received thus far. First, there is the problem of providing a satisfactory description of the load history. By load we mean any type of external demand or any combination of stress whose deployment may influence the performance of the nuclear system and its component parts. Second, there is the problem of finding a measure for accumulated

---

* CEC - Joint Research Centre, 21020 Ispra (Varese) Italy

** CESNEF, Nuclear Engineering Department, Politecnico di Milano, 20133 Milano - Italy.

damage and making an appraisal of residual life. It is apparent that the
solution of the first problem is a prerequisite and a condition for the
solution of the second one. Attention will be mainly focused on structural (or mechanical) components. Needless to say, nearly all considerations and findings of the present study can be easily transferred to
any other type of component or item from the nuclear primary cooling
system.

Loads, as far as their time behaviour is concerned, can be station-
ary or transient. In their turn, transients (loads) are classified as
normal, operational, and accidental. All transients develop in time randomly and a record has the appearance of a realization of a stochastic
process (Figure 1). Our interest will be placed in operational and accidental transients only, for these transients can have a considerable
magnitude and contribute to accumulation of progressive damage. Furthermore, it has to be pointed out that generalized carrying capacity of components is affected by loads and by load-activated damaging mechanisms.
In this respect, the description of damage should also be viewed as a
realization of a random process.

Until now designers solved the problem of the description of transients through the consideration of lists where supposedly all types of
transients are classified and expected numbers of occurrences are accounted for [5] . These lists match with design codes and regulations commonly accepted in the practice [1-3] . However, such a procedure leads us to
disregard completely the statistical characteristics of the transients.
Possible frequency and magnitude fluctuations are in fact ignored. In
other words, this approach is equivalent to a deterministic treatment of
the problem. Conservative assumptions apply to load history, while damage
and deterioration effects are evaluated according to some unquantifiable
worst case (upper envelope) hypothesis or similar suppositions. Instead,
the complete statistical description would grant the appreciation of the
real level of safety. At the same time there will be the possibility of
correct calibration of the engineered safety devices.

The present study originates from the idea that marked point proces-

ses may offer an adequate representation for time sequence and size of
operational and/or accidental transients. Methods to infer the statistical structure of these random processes are then needed. Our concern will
be with two situations. There is the situation which occurs at the design
stage, where the designer should make forecasts of the performance of the
system and decisions must be made grounded on the available information
about components data, failure rates, repair rates and so on (prior analysis). Conversely, there is the situation which occurs at the operation
stage where new information about the states of the system and its components has become available (posterior analysis). A final step towards a
satisfactory probabilistic treatment is the question of damage accumulation. It will be shown how damage can be reasonably modelled as a response
to (points of) the marked point process which is taken to represent load
history. The interest of this filtered process model seems due to the wide
variety of wear and damaging phenomena for which it provides a mathematical frame.

   2. Representation of load history in terms of marked point process.
2.1. We assume that loads in a nuclear reactor are described through a
succession of random transients which arrive at random times. During the
life of the plant $T = [t_o, t_F]$ the time distribution of the transients
is determined by time arrivals of the corresponding initiating events
$E_j \in \underline{E}$. The denumerable set $\underline{E} = \{E_1, \ldots, E_j, \ldots\}$ can be divided into two
parts $\underline{E}^{(i)}$ and $\underline{E}^{(e)}$ depending on events being of internal or external
origin, respectively. Internal events $\underline{E}^{(i)}$ are produced either by the
failure of some device and component or by some operational occurrence
within the plant. Example are failure to work of components, failure of
the electrical supply system, tests and so on. External events $\underline{E}^{(e)}$ derive
from interactions of the external environment with the plant. Examples
are earthquake, flood, sabotage and the like. The shape and the magnitude
or size (height and extent) of the transients will depend upon the states
of a variety of control and consequence limiting systems at the time the
initiating event happens.

Without introducing any significant restriction in the development of the theory we will henceforth refer essentially to loads of internal origin. Moreover let us suppose that elements $E_j \in \underline{E}^{(i)} = \{E_1^{(i)}, \ldots, E_j^{(i)}, \ldots, E_J^{(i)}\}$ are mutually exclusive and represent all possible (internal) occurrences.

<u>2.2.</u> Let $\underline{q} = \{q_1, \ldots, q_1, \ldots, q_L\}$ stay for a set of parameters which uniquely determine the generalized load characteristics of the system (i.e. pressure, temperature, neutron flux, chemistry of the coolant and so on). The first problem consists of determining the marginal representation of a generic component $q \equiv q_1$. Let $q_o \equiv q_{1_o}$ denote the normal operating condition (base-load condition). A particular experimental record of transients may be expressed in the form

$$q(t) = q_o + \sum_j^{1,\ldots,J} \sum_k^{1,\ldots,K_j} Q_{jk} q_{jk}(t,t_{jk}) \qquad (2.1)$$

where Q_{jk} is the height of the transient and q_{jk} is its shape function. Admittedly

$$q_{jk}(t_{jk},t_{jk}) = 1 \quad \text{and} \quad \lim_{|t-t_{jk}| \gg 0} q_{jk}(t,t_{jk}) = 0 .$$

By j one identifies the transient produced by the initiating event E_j, $j = 1,\ldots,J$. Subscript k assigns numbers to transients of the same type or class which are consecutive in time. Superscript i has been omitted for sake notational simplicity.

If one hypothesizes that pulses are not strongly shaped by the developing damage, the transients of the same type, say j, will present similar shapes and equation (2.1) can be rewritten as

$$q(t) = q_o + \sum_j^{1,\ldots,J} \sum_k^{1,\ldots,K_j} Q_{jk}\, q_j(t-t_{jk}). \qquad (2.2)$$

Assuming the additional hypothesis that time intervals between two successive transients is large if compared with the duration of each transient,
the effect of the accidental loads on the plant can be expressed analyti-
cally as

$$q(t) = q_o + \sum_{j}^{1,\ldots,J} \sum_{k}^{1,\ldots,K_j} Q_{jk}\, \delta(t-t_{jk}), \qquad (2.3)$$

where

$$\delta(t-t_{jk}) = 1, \quad \text{if} \quad t = t_{jk}, \quad \delta(t-t_{jk}) = 0 \text{ otherwise.}$$

In the statistical treatment of (2.1) it is supposed that $q_1(t)$ has associated a point process in time for times after t_o. That is a point process
$\tilde{t}_j = \{t_j : t \in T\}$, where instants $t_{j1} < \ldots < t_{jk} < \ldots < t_{jK}$ are a random sequence of times. Furthermore, if $N_j(T)$ is the number of events E_j having
epochs in the interval of time $[t_o, t_F]$, one may also introduce the
associated stochastic counting process $\tilde{N}_j(t) = \{N_j(t) : t \in T\}$.

Now consider that each occurrence time t_{jk} from the basic random
point process $\tilde{t}_j$ has connection with a vector mark $\underline{Q}_{jk}$ which contains the
information regarding the height of the pulse and its shape. Thus, the
description of $\tilde{q}_1(t)$ appears modelled through a marked point process. In
principle $\underline{Q}_{jk}$ may underline a continuum of possible values in a nondenu-
merable mark space $\underline{Q}$. There is no requirement that the marks be identically distributed and mutually independent. Nor there are requirements
that they be independent of the point process they label.

We are interested in evaluating the conditional probability
$\lambda(t,\underline{Q}_i)\, dt\, d\underline{Q}_j$ of there being a point in $[t, t+dt]$, with mark in
$[\underline{Q}_j, \underline{Q}_j + d\underline{Q}_j]$ given the number of points in $[t_o,t]$ their occurrence times,
and their marks. To this purpose we define the random process
$\tilde{\lambda}(t,\underline{Q}_j) = \{\lambda(t,\underline{Q}_j) : t \in T, \underline{Q}_j \in \underline{Q}\}$ where

$$\lambda(t,\underline{Q}_j) \equiv \begin{cases} \lambda_j(t,\underline{Q}_j;0) & \text{for } t_o \leqslant t \leqslant t_{j1} \,, \\ \\ \lambda_j(t,\underline{Q}_j;\ N_j(t);\ t_{j1},\ldots,t_{jk-1},\ \underline{Q}_{j1},\ldots,\underline{Q}_{jk-1}) \\ \\ \qquad \text{for } t_{jk-1} < t \leqslant t_{jk} \end{cases} \qquad (2.4)$$

We term this process the intensity process for the k-th mark and the j-th initiating event [6]. It is a left-continuous function of t.

2.3.   A few examples of processes of practical interest which may be included as special cases in this marked point process model are now offered.

(i) Generalized compound Poisson process. Points occur in a compound Poisson process as a Poisson process and marks are assigned independently to each point. If $\lambda(t)$ is the intensity of the nonhomogeneous Poisson process and if marks are continuous random variables with probability density function $p(\underline{Q}_j)\ d\underline{Q}_j$, then $\lambda(t,\ \underline{Q}_j)$ has the product form

$$\lambda(t,\ \underline{Q}_j) = \lambda_j(t)\ p(\underline{Q}_j). \qquad (2.5)$$

Marked points can be interpreted as occurring in a multidimensional space $T \times \underline{Q}$ [6]. Points have a time co-ordinate $t_{jk} \in T$ and a mark-space co-ordinate $\underline{Q}_{jk} \in \underline{Q}$. Therefore one is led to define the number of points $P(dt,\ d\underline{Q}_j)$ which lie in the region $[t,t+dt] \times [\underline{Q}_j,\ \underline{Q}_j + d\underline{Q}_j]$. This number results itself in a time-space Poisson process $\tilde{P}(dt,\ d\underline{Q}_j)$ whose intensity is (2.5). The relation with counting process $\tilde{N}_j(t)$ holds

$$\tilde{N}_j(t) = \int_{t_o}^{t} \int_{\underline{Q}} \tilde{P}(dt,d\underline{Q}_j) \qquad (2.6)$$

Furthermore, if the mark accumulator process $\underline{\tilde{m}}_j = \{\underline{m}_j(t) : t \in T\}$ is defined by

$$\widetilde{\underline{m}}_j = \overset{1,\ldots,K_j}{\underset{k}{\Sigma}} \underline{Q}_{jk} \tag{2.7}$$

then $\widetilde{m}_j$ can be obtained through the functional relation

$$\widetilde{\underline{m}}_j = \int_{t_o}^{t} \int_{\underline{Q}} \underline{Q}_j \ \widetilde{P}(dt, \ d\underline{Q}_j) \tag{2.8}$$

(ii) Ceneral self-exiting point process. If marks are assigned independently to the points of the process, point processes depend only upon t, $N_j(t)$, and times $t_{jk-m}, \ldots, t_{jk-1}$ with $m \geqslant 1$. The particular position m 1 identifies renewal process as the best explored case. It is of interest to consider the circumstance where a point process is the result of the superposition (or pooling) of many independent self-exciting and inhomogeneous point processes. A limit theorem then holds which states that if component processes are reasonably sparse the resulting process converges in distribution to a Poisson process [7]. This theorem seems to have some resemblance with the central limit theorem for sums of random variables and to explain why Poisson processes may be reasonably used to approximate a variety of practical situations.

(iii) Doubly stochastic Poisson process. This process arises when intensity parameter (2.4) is function of the random process $\widetilde{q}$. It can be observed that a doubly stochastic Poisson process is a particular type of self-exciting point process. Statistical properties can then be obtained as a direct consequence of this observation. There appear to be a number of physical situations where doubly stochastic Poisson processes provide useful mathematical models. For instance, a doubly stochastic Poisson process would allow to represent possible uncertainties in parameters estimation and their time evolution immediately in the design stage of the system.

3. Inference of statistics from Random processes of transients in practical cases.

3.1. The application of marked point processes has interest both for design and operation of the nuclear reactor. In the stage of design the objective is to predict the statistical structure of the process moving from the knowledge of data concerned with failure and performance characteristics of all components. Essentially, this prediction corre-sponds with the attempt to provide the designer with a simulation of the behaviour of the system. Basic tools and techniques are of two types. Fault-tree analysis and truth tables (or event-trees as a substitute) allow us to determine all situations of interest and to perform frequency computations. On the other hand, there is the need of computer codes which describe the thermohydraulic transient behaviour of the system. Simplified examples of this approach are the analysis of the overpressure transients in a PWR (Figure 3) [8] and the study of the temperature tran-sients in a LMFBR (Figure 4 and 5) [9] .

Conversely, in the stage of operation, the statistical structure of the load process can be inferred through a more standardized approach. Aim is at the confirmation of predictions already expressed during the design stage. This confirmation concerns input data as well as the simu-lative models that were assumed as a basis for computation. An experiment of this type is now being carried out at Kernkraftwerke Obrigheim PWR (Figure 6 and 7) [4].

3.2. In view of this experience, it has found of great help to develop a group of methods and techniques which may serve at stage of design to infer the structure of the random process of loads. It is admitted that this process can be modelled in terms of a point process. Steps are coordinated according to the block diagram of Figure 2. A short description follows.

Step (i). Define a set of initiating events $\underline{E}' = \{E'_1, \ldots, E'_j, \ldots, E'_J\}$. Check is the set is complete (or exhaustive) for it must contain all situations which may happen in the system.

Step (ii). For each initiating event $E'_j \in \underline{E}'$ characterize the corresponding point process $\tilde{t}_j$. The process is admittedly Poisson-like and its intensity function $\lambda_j(t)$ is inferred from a fault-tree having the initiating event E_j', as its top event. Given a number of initiating events $\in \underline{E}'$ the question arises of their statistical independence. The degree of correlation can be checked by means of fault-tree analysis based on the use of list processing techniques [10]. Then it is possible to arrange the primitive set $\underline{E} = E_1, \ldots, E_j, \ldots, E_J$ containing all initiating events having the property of completeness and mutual independence.

Step (iii) and (iv). Associate mark Q_j to each time instant of point process $\tilde{t}_j$. A specific hypothesis which helps to reduce the computational effort required by these steps is mutual independence of the statistics of initiating events E_j with respect to the statistics of states A_{cs} of control and safety subsystems (or interventions). This independence can also be checked by a combined fault-tree analysis. Should independence hold, the conditional probability distribution of a point in $[t, t + dt]$ with mark in $[Q_j, Q_j + dQ_j]$ is expressed by a product form (2.5).

Step (v) and (vi). Determine p.d.f. $p(Q_j)$. Now, if step (v) is afforded along a deterministic approach, one finds that a single value of Q_j is associated with each combination of E_j with A_{cs}. Thus statistics of combinations produces the statistics of Q_j. Indeed this is the type of assumption that has been made in the applications [8,9]. Should step (v) be afforded according to an entirely probabilistic approach, the identification of $p(Q_j)$ requires a statistical treatment of step (v) and the random nature of transients should than be explored.

Now step (vi) is accomplished through application of fault-tree analysis of each combination of E_j with A_{cs}. A result of this search could be possible pooling of various initiating events. The two prerequisites needed in order to apply the limit theorem for superposition of point processes seem fairly well satisfied by accidental transients. Therefore the Poisson process seems to provide a reasonable description under rather comprehensive circumstances.

Pooling of marks can be also considered. For instance, this happens in all cases where relevance of transients stems from their height and shape considerations are comparatively less important.

Step (vii). Create histograms for the expected number of transients vs a set of discrete values for mark Q_j.

A final remark is that uncertainties in input data may originate a doubly stochastic Poisson process. The case has been considered by Amesz et al. [8]. The basic technique for such investigation has been a fault-tree analysis apt to accept input data given by means of their p.d.f.'s [11].

3.3. Transient analysis performed at operation stage basically entails a problem of characterization of time series. Data have to be extracted from documents that have a normal use along purposes other than transients investigation. Particularly, sources of data are the various records required for plant control and operation [4]:

- operator diary,
- monthly report,
- yearly report,
- failure and switch computer records,
- failure history computer records,
- failure information sheets,
- pressure registrations.

Classification of failures proceeds according to the groups of initiating events:

. electrical (e),
. mechanical (M),
. testing, of installations (T),
. spurious (S),
. operator error (O),

• meteorological (W),

• unknown (U),

as shown by Figure 7. The results obtained seem to confirm that a compound Poisson process can be used to produce a fully acceptable mathematical representation.

4. Representation of damage accumulation in terms of filtered point process.

4.1. Under many circumstances failure of a structure or of a system is the result of gradual accumulation of damage. Examples can be found in the estimation of residual life of nuclear components. Carrying capacity appears influenced by load history and by operating characteristics. For instance, fatigue may induce variations in strength properties. Other variations may be activated by effects of corrosion and nuclear radiation. Usually, given the variety of loads and their interactions, practical situations are characterized by the coexistence and superposition of more failure modes. Then suppose that all failure modes have been ascertained and that failure criterion (or hypothesis) H is expressed by an analytic form of the type

$$\underline{h} = \underline{h}(q) > \underline{\alpha} , \qquad (4.1)$$

where $\underline{h} = \{h_1, \ldots, h_m, \ldots, h_M\}$ is a generalized damage accumulation function and $\underline{\alpha} = \{\alpha_1, \ldots, \alpha_M\}$ is a set of constants. In complete generality loads are stochastic processes which develop during the life of the system T as well as in the space domain of interest $\underline{x}$. For each $\tilde{q}_1 \in \tilde{q}$ it may be written $\tilde{q}_1 = \{q_1(t,\underline{x}) : t \in T, \underline{x} \in \underline{X} \}$ and $\underline{h} = \underline{h} \, (\tilde{q})$ will also result a stochastic multidimensional process.

Now as far as reference is made to (4.1), the reliability of the system conditional upon H is expressed by the probability of absence of $\underline{\alpha}$-level crossings in the interval T and domain X

$$R(T,\underline{X} \mid H) = \bigcap_{\forall t \in T, \underline{x} \in \underline{X}} \{\underline{h}(\tilde{q}) < \underline{\alpha} \}. \qquad (4.2)$$

Thus reliability computation appears to be stated mathematically through a problem of (predictive) filtering. There is the need of knowing some statistical properties of process $\widetilde{h}$. In its turn $\widetilde{h}$ is the result (output) of a specified transformation of the signal $\widetilde{q}$ (input).

To find answers one should move from easier-to-handle schemes. In this context it will be assumed that computation can be confined to a specific point in the space domain $\underline{x}^* \in \underline{X}$ and some reduction procedure is allowed leading to the consideration of scalar variables q and h instead of vectors q and h, respectively.

A single model then occurs if q is a marked point process and response h can be expressed as the superposition of separate responses to each marked point (t_k, Q_k). More complicated physical phenomena may also be described by removing the superposition hypothesis and imposing the additional structure of Markov processes.

4.2. We will first consider the case where $\widetilde{h} = \{h : t \in T\}$ is a filtered point process generated by a superposition of point responses. For $t \geqslant t_o$ the process has the form

$$h = \begin{cases} h_o & \text{if } N(t) = 0, \\ \sum\limits_{k}^{1,\ldots,N(t)} h(t, t_k, Q_k) & \text{if } N(t) > 0. \end{cases} \qquad (4.3)$$

$h(t, t_k, Q_k)$ is the k-th weighting function or the impulse-response at time t to the k-th marked point. More particularly, we term $\widetilde{h}(t)$ a filtered Poisson process when $\{q : t \geqslant t_0\}$ is a compound Poisson process. In this case $\{N(t) : t \geqslant t_0\}$ is a Poisson counting process, and the marks Q_k are mutually independent, independent of $\widetilde{N}(t)$, identically distributed, vector-valued random variables. A specialization of (4.3) is the mark accumulator process (2.7).

It may be of interest to remark that if process $\widetilde{h}$ is non-decreasing and $p(h,t)$ stays for the p.d.f. of the statistical variable h at time t, relation (4.2) takes the form

$$R(T \mid H) = \prod_{\forall t \in T} \{h(q) < \alpha \} = \int_{-\infty}^{\alpha} p(h, t_F) \, dh.$$

Two problems for which filtered point processes provide reasonable models follow.

(i) Fatigue failures. Let us assume that damage caused by a given cycle of stress is independent of the state of the structure at a given time and of the preceding history of the loading. Under this assumption Miner's rule for cumulative fatigue damage states that the increment in the measure h for a number n_k of cycles with stress amplitude Q_k equals

$$h_k = \frac{n_k}{N(Q_k)},$$

where $N(Q_k)$ is the limiting number of cycles under load cycles of type k. Summing the damage we obtain the failure condition ($\alpha = 1$)

$$h = \sum_k^{1, \ldots, N(t)} \frac{n_k}{N(Q_k)} \geqslant 1. \qquad (4.4)$$

In terms of (4.3) it is

$$h(t, t_k, Q_k) = \frac{1}{N(Q_k)}. \qquad (4.5)$$

(ii) Crack propagation. In the crack propagation stage of fatigue it is recognized that determinants of crack growth are the cumulative damage ahead of the advancing crack and the non-cyclic progressive deformation ahead of the crack root. The idea of progressive growth can be formulated in macromechanical terms through the equation

$$\frac{dL}{dN} = C(Q \sqrt{L})^m, \qquad (4.6)$$

where L is the instantaneous crack length, N is the number of cycles, Q is the range of stress intensity factor. C and m are constants specific to the material where crack propagates. A cumulative mechnism can be expressed as follows:

$$h \equiv L = 1,\ldots, \underset{k}{\overset{N(t)}{\sum}} L_k,$$

where

$$h = \begin{cases} h_o = L_o & \text{initial crack length} \\ h_k = L_k = C(Q_k{}^{0,\ldots,k-1} \underset{r}{\sum} L_r)^m & r = 0,1,\ldots,k-1. \end{cases} \tag{4.7}$$

4.3. A more comprehensive filtered process is based on a model whose response is generally not a superposition of individual effects. This process will be referred to as a point process driven Markov process. It satisfies the differential form

$$dh(t) = a(t,h)dt + \int_Q b(t,h,Q)P(dt,dQ) \tag{4.8}$$

with the initial condition $h_o = h(t_o)$. When P(dt, dh) is associated with a Poisson process, process underlied by (4.8) may be termed a Poisson driven Markov process. Then stochastic differential equation (4.8) can be viewed as defining a nonlinear transformation of the Poisson process P into the process h. In cases where random loads occur beyond the elastic limit equation (4.8) may serve to represent the distribution for residual deformation at the end of an interval[t_o,t] . Another applica-tion is related to creep under ramdomly varying stress. The differential dh =h (t +dt)-dh(t) would denote an infinitesimal increment in deformation which occurs during [t, t+dt]. If during this interval no variations in load occur in the incident marked point process, P(dt,dh)= 0 and the increment in h results from process a (t,h)dt (base-creep only). If a point having a mark (load) Q does occur during [t,t+ dt], the increment in h is a (t,h)dt +b(t,h,Q) which is dominated by b(t,h,Q).

5. Conclusion. The approach to the analysis of reactor transients which has been presented seems to allow satisfactory handling of random load history and life evaluation of nuclear components. When load history is treated as a (marked) point process many problems are considerably simplified. The point process assumption lends particular significance to a chapter of statistics whose application to nuclear safety assessment might be productive. More particularly, results obtained would find a direct use in nuclear plant design and operation. A number of difficulties remain to be solved. Two are of immediate relevance for our research programs: a) At this time analysis of complex systems can be hardly afforded owing to the large quantity of information (and data) to handle. Simplified methods should be searched for. b) Calculation of transients (step (v) of Figure 2) is performed essentially on the basis of deterministic computer programs. Instead, a coherent statistical approach would require full recognition of their random nature.

Acknowledgment. Authors are indebted to Framatome and Kernkraftwerk Obrigheim for the help offered in data collection and interpretation. Valuable suggestions for the use of marked point process has been given by M.R. Leadbetter.

REFERENCES

[1] Nuclear Power Plant Components. Section III. ASME Boiler and Pressure Vessel Code; ASME, New York, N.Y. 1974.

[2] Case 1592. Class 1 Components in Elevated Temperature Service. Section III. Cases of ASME Boiler and Pressure Vessel Code. ASME, New York, N.Y. (Approved by Council April 29, 1974).

[3] Application de la réglementation des appareils à pression aux chaudières nucléaires à eau, Journal Officiel de la Republique Française (le 26 février 1974).

[4] J. AMESZ, G.F. FRANCOCCI, Validation of Probabilistic Transient Analysis by Comparison with Historical Data of the Obrigheim PWR, Euratom JRC Technical Note, Engng. Division, Restricted Distribution, Euratom JRC, Ispra, Italy, March 1977.

[5] Failure Analysis and Failure Prevention in Electric Power Systems, EPRI NP-280, Electric Power Research Institute, Palo Alto, Cal., November 1976.

[6] D.L. SNYDER, Random Point Processes, J. Wiley & Sons, New York, N.Y., 1975.

[7] E. CINLAR, Superposition of Point Processes, in Stochastic Point Processes Statistical Analysis, Theory, and Applications, P.A.W. Lewis (ed.), J. Wiley & Sons, New York, N.Y. 1972.

[8] J. AMESZ, C. COUDERT, G. VOLTA, Analyse probabiliste de la protection d'un PWR contre les surpressions, EUR 5640.f, 1976.

[9] J. AMESZ, F. CESARI, R. RIGHINI, G. VOLTA, Probabilistic Analysis of Accidental Transients in a LMFBR, Int. Meeting on Fast Reactor Safety and Related Physics, Chicago, Ill., October 5-8, 1976.

[10] M. ASTOLFI, S. CONTINI, P. VAN DER MOYSENBERG, SALP-3 A Computer Program for the Analysis of Fault Tree with Dependencies between Primary Events and Gates, EUR Report (in press).

[11] M. ASTOLFI, The Program AWE-1, AWE-2, and BRUNA for Fault-Tree Analysis, EUR 4592.e, 1976.

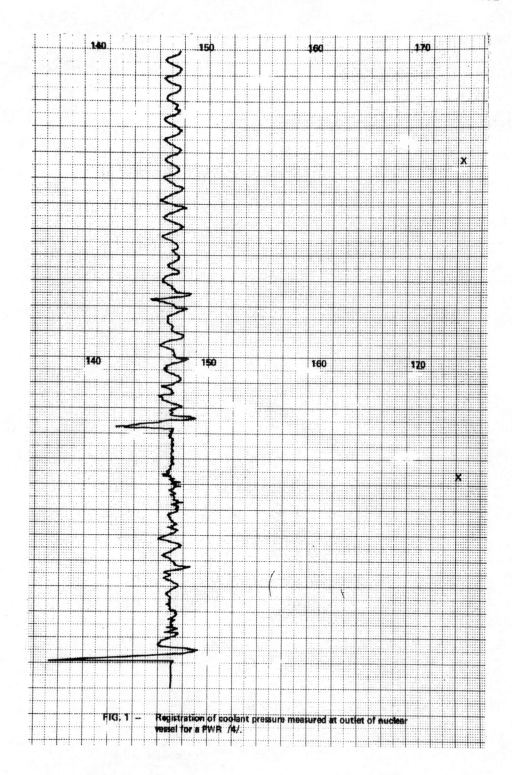

FIG. 1 — Registration of coolant pressure measured at outlet of nuclear vessel for a PWR /4/.

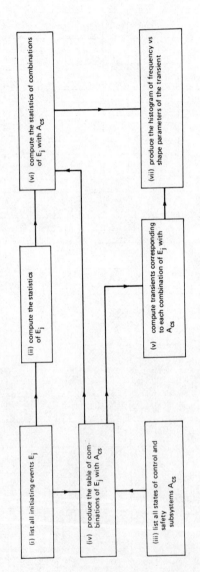

FIG. 2 - A block diagram displaying main steps taken towards a probabilistic analysis of operational and accidental transients in a nuclear reactor.

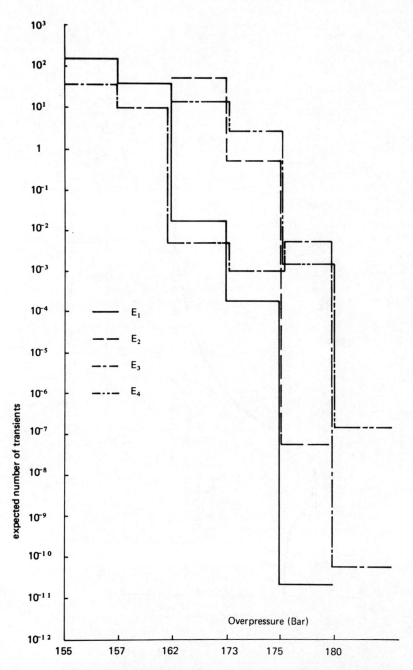

FIG. 3 — Histograms showing size and expected number of pressure transients in a PWR (outlet of pressure vessel). Data refer to 40 years of operation and consider four types of initiating events. (E_1 closure of throttle valves, E_2 closure of inlet regulation valves, E_3 shut down of condenser, E_4 closure of insulation valves /8/.

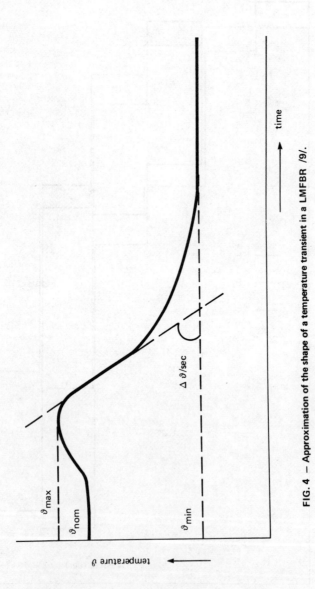

FIG. 4 — Approximation of the shape of a temperature transient in a LMFBR /9/.

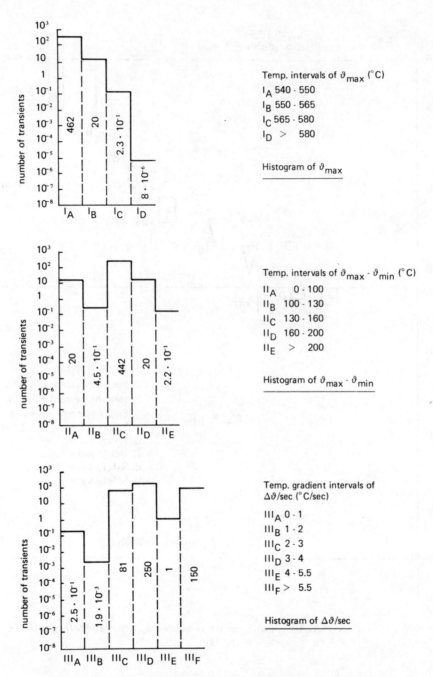

FIG. 5 — Histograms displaying shape parameters and expected number of accidental temperature transients for a LMFBR (outlet of reactor tank, 20 years of operation, transient shape approximated as shown in Fig. 4) /9/.

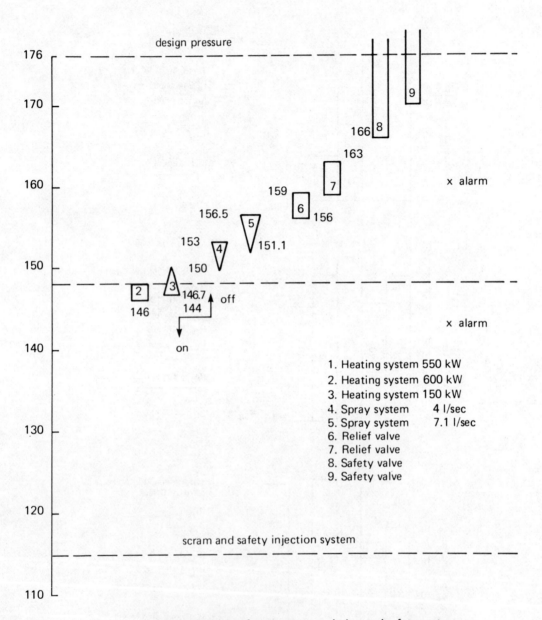

FIG. 6 — Provision for pressure regulation and safety systems
in a PWR /4/.

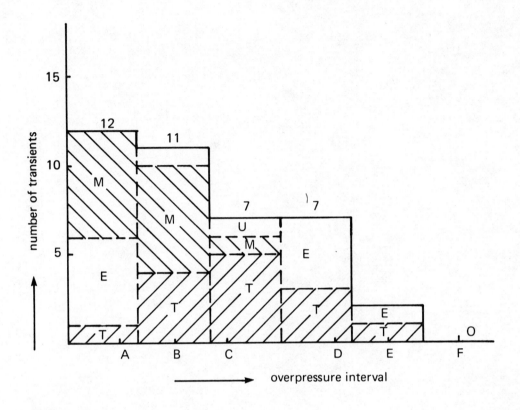

Interval A   148 – 150   atm
         B > 150 – 151,5 atm
         C > 151,1 – 156 atm
         D > 156 – 159   atm
         E > 159 – 166   atm
         F > 166   atm

FIG. 7 — Histogram of pressure transients as obtained from seven years of pressure records made at the outlet of nuclear vessel in a PWR /4/.

Topic 5
METHODOLOGY II

The papers in this section each address a particular problem area
in the field of reliability and safety assessment of nuclear power
plants. The first paper, by E. J. Henley and R. E. Polk, illustrates a
risk-cost technique which can be used to determine where administrative
time restrictions on downtimes for safety related systems and components
of operating plants could be relaxed at great savings in cost associated
with plant downtime and with no appreciable increase in risk to the
general public.

The second paper, by L. Caldarola and A. Wickenhäuser, reports on
"a very fast analytical computer program for fault tree evaluation". In
addition, the mathematical theory supporting a second computer program,
for multi-state analysis, under development at Karlsruhe, is also
presented. The program already operational locates minimal cut sets
deterministically using a "downward" algorithm like that developed by
Fussell [1]. Computational techniques are based on those derived by
Vesely [2]. Besides having the capability to rapidly determine a
variety of reliability characteristics for coherent systems, the program
can perform a limited (two-phase) phased mission analysis. System
unavailability is calculated for the first time period (phase) and
system unreliability is calculated for the second phase. Such

491

computations have application to safety systems which first must be available at the onset of their mission and then, given that they are available, must perform reliably through an additional period of time. The program under development is designed to deal with both coherent and noncoherent systems.

The third paper of this section, by R. E. Barlow, and B. Davis presents a technique for analyzing time-between-failure data to develop a graphical method for determining optimum replacement intervals for components. Although the example used to illustrate the technique uses data of tractors' ages and their engines' ages at failures, the method is general and is therefore applicable to the planning of maintenance actions for nuclear power plant systems.

The fourth paper, by J. D. Esary, extends the author's earlier work on phased mission analysis to an investigation of the effect of modeling depth on reliability predictions for systems required to perform a phased mission. A phased mission is a task, to be performed by a system, during which the system configuration or component basic event data changes at predetermined times. Some important examples of phased missions from the nuclear industry are shutdown heat removal for a liquid metal fast breeder reactor and emergency core cooling for a light water reactor. Each task requires various combinations of subsystems to perform for fairly well-determined periods of time ranging from a few hours to several days. An important point made in the paper is that significant overprediction of system reliability could result unless the phased mission analysis is carried out at a detailed component level. The author also develops a theory of degradable and nondegradable devices applicable to systems whose components encounter various service environments.

The fifth paper of this section, by R. A. Waller, M. M. Johnson, M. S. Waterman and H. F. Martz, Jr., deals with a very real problem encountered in reliability and risk assessments of advanced reactors: lack of data. The authors present two methods. One method is for obtaining estimates of component failure rates. The second method is for obtaining estimates of component reliability. Both methods require two percentiles (probability statements) specified by an engineer. The percentiles are used to select the parameters in a gamma density function, in the failure rate case, and negative-log gamma density function in the reliability case. In the latter case, the engineer must also specify a reference time with respect to which the percentile statements are made. Coupled with a Delphi approach to obtain the percentile estimates, the methods presented could produde quite accurate results.

REFERENCES

[1] J. B. Fussell et al, MOCUS - A Computer Program to Obtain
        Minimal Cut Sets, Aerojet Nuclear Company, ANCR-1156,
        1974.

[2] W. E. Vesely and R. E. Narum, PREP and KITT: Computer

Codes for the Automatic Evaluation of a Fault Tree,

IN-1349.

A RISK/COST ASSESSMENT OF ADMINISTRATIVE TIME RESTRICTIONS ON NUCLEAR POWER PLANT OPERATIONS

E. J. HENLEY* AND ROBERT E. POLK**

Abstract. This paper analyzes the effect of varying administrative time restraints on plant availability and the attending risk
and economics of continued plant operation. The approach involves a
Monte Carlo simulation to predict plant availability with variations
in operating restrictions due to degraded safeguard systems. The
model provides information on the cost of plant outages. The results
from an extensive study on the risks associated with the operation
of commercial power reactors are then coupled with the results of the
simulation to form the basis of a risk-cost comparison. From this
comparison, observations and recommendations can be made as to the
viability of relaxing such restrictions. A DC power system for a
pressurized water reactor plant is used for the simulation and risk
study.

1. Introduction. The licensing process for commercial nuclear

power plants involves many decisions on complex issues by members of

the regulatory agencies. As the industry matures and additional

questions arise as to safety and operational requirements of these

plants, it is evident that new tools for logical decision making are

required. In industries where reliability and safety of plant or

equipment operations is of primary concern, risk assessment and

reliability analyses have been applied to attempt to qualitatively and

quantitatively analyze plant behavior and consequences where subject

to critical system failures. However, not much work has been done to

*University of Houston, Houston, Texas   77004
**EG&G Idaho, Idaho Falls, Id.   83401

use similar techniques on administrative controls on the facets of
plant operation and maintenance. The objective of the analysis presented is to develop a procedure to obtain administrative controls on
the repair times for safeguard systems and is done in two parts. The
first involves a computer program to simulate the dependence of plant
availability on the administrative time limitation using a Monte Carlo
technique. This program provides the cost comparison between different
magnitudes of administrative downtime restrictions. The second portion
of the analysis is a risk study of the system on which the time limits
on repair are placed. This risk study requires a detailed examination
of system failure probabilities and consequences for the design basis
event using fault tree methodology. A risk equation is presented which
uses the results of the study to determine quantitative risk levels.

The analysis is performed on the safety related, standby DC power
system for a pressurized water reactor plant now in operation. The
system utilizes a one-out-of-two configuration and has a regulatory
limit for repair time of two hours. Beyond this limit a plant shutdown is required. Analysis of this system shows that a risk-cost
comparison indicates a potential savings of one million dollars per
year for the utility with negligible increase in risk to the public by
increasing the administrative time limit on repair to 8-10 hours.

2. Monte Carlo Simulation. The computer program uses a plant
model characterized by the failure and repair statistical distributions

of the critical system whose influence on plant availability is of

concern. The model considers the minimum degree of redundancy required

of this system for plant operation and also the effect of regulatory

or administrative time restrictions on plant operations with the

critical system in a degraded state. A degraded state is defined as

a condition in which a sufficient number of critical system parts are

available to allow the plant to operate but with inadequate redundancy

to meet acceptance criteria during accident or abnormal plant

conditions. A critical system part is one of the independent subsys-

tems that performs a safety function.

The program uses the following fundamental parameters to simulate

the effect on plant availability of a time restriction placed on

operation with a degraded critical system:

(1) Array of time restraints, TD, on plant operation with
     the critical system in a degraded state;

(2) Total number of critical system parts which perform a
     desired function, and minimum number necessary to perform
     this function, RED;

(3) Mean and standard deviation of a normal repair time
     distribution for each critical system part;

(4) Failure rate for each critical system part, λ;

(5) Time array, T, for which the simulation is to be performed.

Several important assumptions have been made in the simulation.

A normal repair distribution is used since it more accurately describes

a complex system than does a constant mean time to repair. However,

the program can be easily adapted to handle other distributions. A

constant hazard rate has been assumed in calculation of system
reliability. The mean and standard deviation for the repair distribution are determined by moment estimates of these parameters using
historical repair data.

The program contains four nested loops which control the introduction of the principle variables into the simulation (Figure 1). The
outermost loop varies the regulatory time restraint imposed on critical
system repair time. This variable, TD, is used in tests to determine
required plant shutdowns due to a repair event exceeding regulatory
limits. Following the TD loop is a loop which governs the number of
Monte Carlo trials used in the simulation. A sufficient number of
simulation operations are performed to allow a meaningful figure of
reliability to be determined. The next loop is a time loop which
provides the time points used in the calculation of the time of
failure, reliability, and elapsed time since failure for the critical
system parts. The innermost loop is a test loop of the critical system parts on which the plant operational status depends.

Three subroutines are used to perform several of the functions
shown in Figure 1.

The simulation process consists of the following principle
operations:

(1) A test for prior failure of a critical system part;

(2) A failure test to determine a current failure of a
     critical system part;

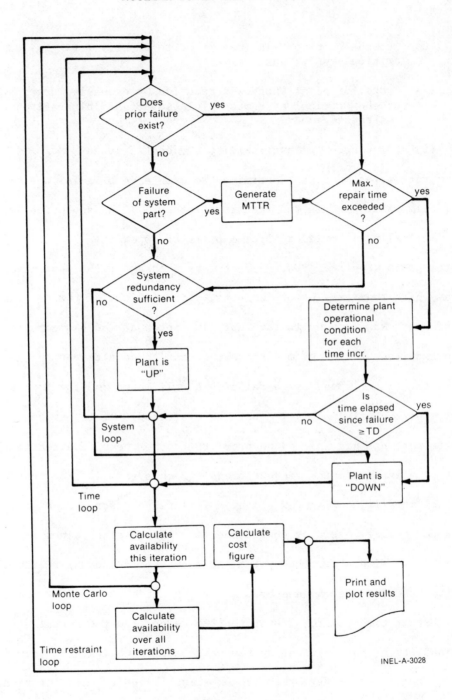

Figure 1

Simulation Flow Chart

(3)   Generation of a MTTR (mean time to repair) for a failed
        critical system part;

(4)   Tests for plant shutdowns requried by excessive degradation
        of the critical system or violation of administrative
        restraints; and

(5)   Calculation of availability and cost figures.

The first operation is to determine the existence of a failure of a

critical system part in the last time interval. An assumption is made

that initially all critical system parts are operational. Critical

system parts when failed are detected by a check of the content of an

array which registers prior failures. If a failure of the critical

system part was present in the last time interval and has not yet

been repaired, the program directs the test to an alternate path

where the time-elapsed-since-failure is calculated and the failure

status of the critical system part re-evaluated. If the critical

system part has not failed the previous test or been repaired in the

last time interval, the test proceeds. Each system is then tested for

operability (if no flags exist) by a failure test for each time point.

A random number generator subroutine is utilized to obtain a

reliability figure for the system under test. The actual reliability

figure is calculated by a subroutine using the system failure rate

and current time point. The subroutine calculates the actual

reliability of the critical system part using equation $R(t) = \exp(-\lambda t)$.

The failure test is performed by comparing the calculated and randomly

selected reliability figures. The system is recorded as "up" if the actual figure is greater or equal to the random figure. A system part is "down" if the reliability figure is less than the random value.

If a failure exists, a mean time to repair (MTTR) is selected at random from a normal distribution. The mean and standard deviation of the repair distribution for the failed system part are used as input to the program. Once a MTTR has been selected, a decision must be made whether or not the failure of this critical system part will result in a plant shutdown. If the system part can be repaired within the time period given by the time restraint, TD, then no shutdown is required. However, if the MTTR indicates that the repair time will exceed TD, a shutdown will be necessary when the elapsed time since failure reaches the value of TD. The shutdown must continue until repairs are completed and the critical system is no longer degraded. The time point at which this occurs is reached when the elapsed time since failure equals the MTTR. If MTTR > TD, a shutdown is eminent. If MTTR $\leq$ TD, no shutdown is anticipated. A time counter is established in each case to determine the status of a critical system part repair for each time point. The system part re-enters the failure test loop when repairs are completed, i.e., MTTR = time elapsed since failure. If a plant shutdown is necessary, the time counter will check at each time point to determine if the time elapsed since a critical system part failure has exceeded TD. When this occurs, a plant shutdown flag is raised.

If the system part has not failed, a counter is incremented to record the number of operable system parts for each time point. At the completion of the failure tests for all the system parts, this counter is compared to the minimum number of system parts necessary for plant protection. If the counter is greater than this minimum value required, no shutdown will occur due to insufficient protection capability. If less, a plant shutdown will result. Plant shutdowns are recorded by a downtime counter. Shutdowns may result from insufficient capability of the critical system or by a shutdown flag. A separate downtime counter is maintained for shutdowns due only to the condition where an insufficient number of subsystems are available for plant protection. This figure is used in calculating plant availability based on downtime from this source alone. A plant uptime counter is incremented if no plant shutdown occurs for that time interval.

An availability figure is calculated for each Monte Carlo iteration at the completion of the time loop. Two availability calculations are made. One considers all plant shutdowns and the other considers only shutdowns due to insufficient plant protection.

The availability figures are averaged over all Monte Carlo iterations to obtain the final values.

Revenue that is lost as a result of plant availability not reaching 100% is calculated by assessing a value per unit time on lost production, about $200,000/day for a 1000 MWe unit, and multiplying

this result by the total plant downtime over the time period covered.
The unavailability is determined by taking the complement of the
availability.

Data	Values
Protection system configuration	1 out of 2
System failure rate	5×10^{-6}/hr.
Mean of repair distribution	25 hours
Std. dev. of repair distribution	2 hours
Time restraint array, TDj	5, 10, 20, 25, 30, 50 hours
Initial time point, TI	20.0 hours
Number of time points, IK	5
Time point differential, DELTA	5.0 hours
Monte Carlo trial array, ij	5, 10, 25, 50, 100, 250

Table 1.   Information used in Example 1.

2.1. Example 1.   A simulation was performed for a plant with
the protection system and time restraints presented in Table 1.   Five
simulations were conducted in which the number of Monte Carlo trials
were varied in order to determine the minimum required to give
consistent results.   A plot of the availability vs. TDj is shown for
each run in Figure 2.   For runs with a low number of iterations, there
is poor consistency of results.   Above 250 iterations, no significant
improvement worth the increase in CPU time was expected.

In Figure 2, the availability is shown to be poor for time
restrictions less than the MTTR.   This result is expected since there

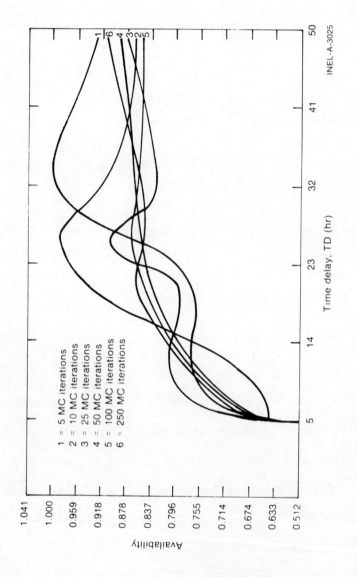

Figure 2

Simulation for Example 1

is a decreasing probability that repairs can be made as the time delay
becomes more restrictive. For values of TDj greater than the MTTR,
plant availability is limited by the availability of the protection
system and is independent of further increases in TDj. Past the
point where TD > MTTR, the major contribution to plant downtime is
the unavailability of a sufficient number of the protection systems
(critical system parts).

      2.2. Example 2. A 125 VDC system is used to supply power
to control and logic circuits of the Essential Safeguard Features (ESF)
systems of an existing nuclear plant. The plant, a pressurized water
reactor, received a detailed risk analysis in WASH 1400, thus making
available data on failure and event probabilities for this study.

      The ESF power system is described by the simplified one line
diagram shown in Figure 3. There are two redundant power systems with
five safety buses and three voltage levels. Each 4160 volt bus is
fed by an onsite power source and the offsite electrical grid. The
DC bus is fed by two inverters from the AC bus and a 125 volt battery
bus.

      In the WASH 1400 analysis, the fault tree of each bus was examined
to identify the dominant cut sets. The probabilities of bus failures
were determined in terms of these cut sets. The probability of
insufficient power to the ESFS during a LOCA event was calculated, and
computer codes were used to evaluate cut set unavailabilities and
failure rates for the DC system. Repair was not considered in the

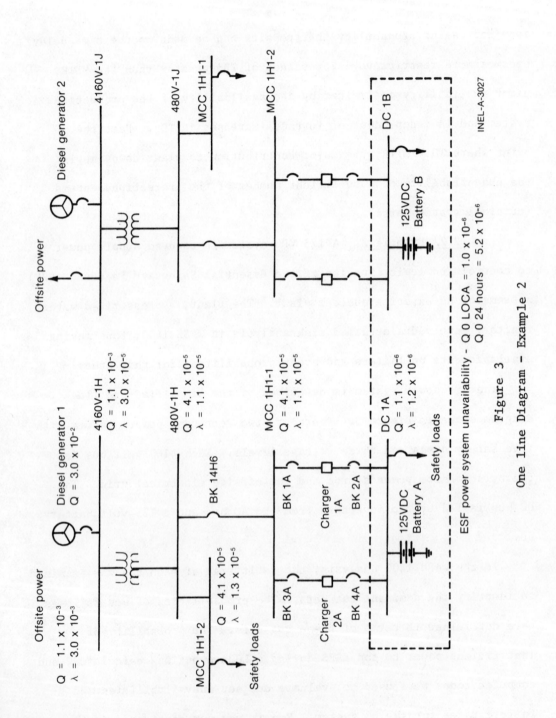

Figure 3

One line Diagram – Example 2

analysis except in the case of the diesel generators and the offsite
network. Failure probabilities of the ESF power systems are shown
in Figure 3 for the time period immediately after LOCA and 24 hours
after a LOCA event.

The 125 volt DC power system supplies logic and control power for
all the ESF systems. The system consists of two identical, redundant
buses which are electrically independent. Each bus has a battery bank
and inputs from the two battery chargers. Normally closed breakers
connect the buses to the chargers and to safety loads. During normal
operation the loads are fed through the charging system with the
batteries on floating charge. The boundary of the DC system is
designated by the dotted line. Only one source is required during
an accident to furnish power to its train of ESF systems in order to
mitigate the adverse effects of the event. Failure of both DC sources
would constitute a failure of the ESF power system since control power
for the ESF equipment is derived from the DC system.

An availability simulation was performed on the DC power system.
Initial time periods of 6 months and 50 weeks were chosen to "age" the
system to a period in time when a planned shutdown for refueling or
overhaul would take place. The "aging" is required so that failures
which could be expected to occur prior to a plant shutdown will be
accounted for in the reliability calculations performed in the
simulations. The period prior to a scheduled shutdown presents the
most desirable situation for extending the allowed repair time for

systems. If time restraints can be relaxed in this period, an
additional shutdown may be avoided by repairing such systems in the
scheduled shutdown period. For a 1:2 ESF DC power system, present
regulatory controls limit plant operation to two hours with the loss
of one DC source.[1] In the simulation, this restraint was varied
from 1.5 hours to 168 hours (approximately 1 week). Repair information
was obtained from the operating experience at Houston Lighting & Power
Co.'s fossil-fired generating stations. This data, listed in Table 2,
is the best estimate of the average repair time for the components.
The parameters which characterize a normal distribution, the mean and
standard deviation, were derived from this data by moment estimates.
Assumptions of a constant failure rate and independent component
failures were made.

Plots of the availability vs. time restraint for 50 week and
6 month initial time periods are shown in Figure 4 and 5, respectively.
The plots indicate that plant availability shows a marked improvement
for increasing values of the time restraint parameter TD until a
value of TD approximately equal to the MTTR of the DC system is
reached. Beyond this value of TD, the availability is independent
of this administrative restriction. The major difference between the
two plots is the absissa scale. The plot for the simulation for an
initial time = 6 months shows an overall availability greater than the
simulation in which the initial time was 50 weeks. The explanation
for the difference is that more failures can occur over a 50 week

period than a 6 month period. Table 3 presents the numerical output

of the simulations including lost revenue associated with plant

unavailability.

Repair Operation	MTTR
Replace cable	10.0
Fix open bus	6.5
Fix shorted bus	6.5
Replace battery	8.0
Replace breaker	8.0

Estimates of component repair times
(From Houston Lighting & Power Co.)

Mean estimate = $\hat{\mu}$ = 1/5 $\sum_{i=1}^{5}$ X_i = 7.8 hours

Variance estimate = $\hat{\sigma}^2$ = 1/5 $\sum_{i=1}^{5}$ $(X_i - \hat{\mu})^2$ = 1.6384 hours

Standard dev. estimate = $\hat{\sigma}$ = 1.28 hours

Table 2. Calculation of Moment Estimates for Normal Repair
Distribution Parameters.

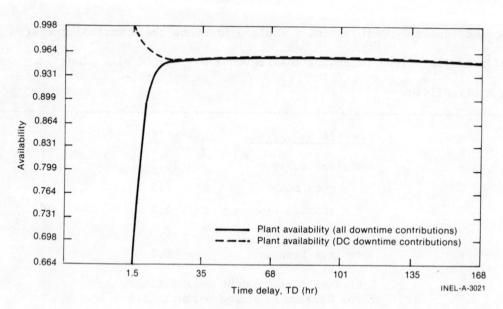

Figure 4

Simulation for 50 Week Period

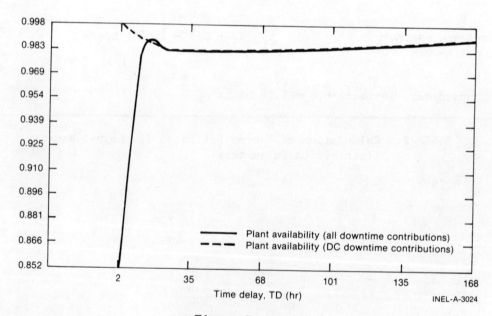

Figure 5

Case 1 - Initial time = 50 weeks

Time restraint TD (hours)	Plant availability (1)	Plant availability (2)	Lost revenue $ X 10^6
1.5	0.6642	0.9979	1.3090
2.0	0.7221	0.9916	1.8030
2.5	0.7227	0.9918	1.8010
3.0	0.7242	0.9919	1.0750
4.0	0.7840	0.9826	0.8424
8.0	0.9383	0.9525	0.2406
24.0	0.9498	0.9499	0.1957
168.0	0.9492	0.9492	0.1981

Case 2 - Initial time = 6 months

2.0	0.8519	0.9977	0.5778
4.0	0.8838	0.9952	0.4532
8.0	0.9783	0.9850	0.08463
10.0	0.9852	0.9859	0.05772
24.0	0.9844	0.9844	0.06084
168.0	0.9850	0.9850	0.05850

Table 3.   Numerical Results from Simulation of DC System.

3. Determination of Risk.   The risk associated with a degraded power system in a nuclear plant can be determined by applying equation 3.1. The initiating event is a loss of coolant accident (LOCA) event. Failure of the ESF power system will contribute to the severity of the consequences of the LOCA event.   The risk equation is

(3.1) $Risk = C_1 \times P_1 \times P_2 \times P_3$

where C_1 = damage resulting from LOCA event sequence with containment failure and no ESF power

P_1 = probability of initiating LOCA event

P_2 = probability of ESF power system failure

P_3 = probability of containment failure

The components of this equation have been determined in the draft

WASH 1400 report:

$$P_1 = 10^{-4} \text{ LOCA's/REACTOR YEAR}$$

$$C_1 = \$5.2 \text{ X } 10^9$$

$$P_3 = 10^{-1}/\text{REACTOR YEAR}$$

The failure probabilities, P_2, for several time periods after a LOCA event for the ESF power system were presented in Figure 3. The application of the risk equation will be used to determine reasonable plant operating restrictions with loss of an ESF DC power source. Thus, some discussion of the rationale of the decision to use the failure probabilities of the entire ESF Power System in the risk equation is required. The failure contributions to the ESF power system by the DC-system are interwoven with those of the higher level buses. The DC system performs control functions that allow the AC buses to provide power to critical equipment. Quantification of the failure probability of the ESF power system is necessary to isolate the contributions of the DC power system. The major contribution to ESF power system failure was identified as a loss of diesel generators coincident with a loss of the electrical network. The contribution by the DC system failure was lumped with 83 potential failures events. For the purposes of this study it was decided that a more conservative estimate of risk would result if the greater failure probability were used (that of the ESF power system). It could be argued (though no analysis was done to prove this) that the failure probability of the ESF power system would envelope that of the DC system in the early stages of the event (0-24 hours).

Increasing the time allowed for repair of the DC system with one failed channel would have some effect on the risk figure. With one channel failed, the probability is increasing with time that the second will also fail. However, for the short period of time considered for repair (in this case, 1.5-158 hours) no appreciable increase in risk would occur. Use of the failure probabilities of the ESF power system in the risk equation will produce conservative estimates and will be constant over the range of time restrictions considered.

The risk figure is calculated for two time periods, immediately after a LOCA and 24 hours following a LOCA event.

Applying equation 3.1, the risk figure for each period is:

$$\text{Risk (@ LOCA)} = CP_1\ P_2\ P_3$$

$$= (5.2 \times 10^9)\ (10^{-4})\ (10^{-6})\ (10^{-1})\ \$/\text{event-year}$$

$$= .052\ \$/\text{event-year}$$

$$\text{Risk (24 hours after) LOCA} = (5.2 \times 10^9)\ (10^{-4})\ (5.2 \times 10^{-6})\ (10^{-1})\ \$/\text{event-year}$$

$$= .27\ \$/\text{event-year}$$

4. Risk-Cost Comparison. The risk levels associated with the LOCA event and loss of power can be compared to the dollars lost from

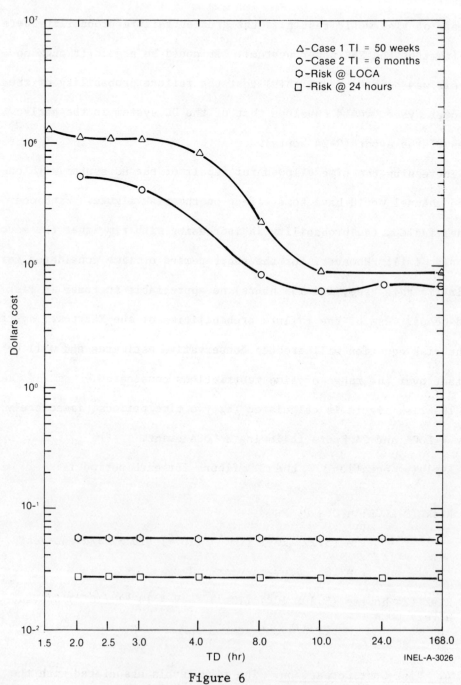

Figure 6

Risk/Cost Comparison

reductions in plant availability caused by the limitations on repair

time for the DC system. The object is to determine if the risk levels

are of a magnitude to justify the present regulations on repair time.

Figure 6 shows a plot of the "dollars lost" vs. the time restraints

considered in the simulation. Case 1 and Case 2 depict the simulation

runs with a 50 week and 6 month initial time point respectively.

Both risk figures are plotted to show the margin between the cost of

plant downtime and the risk.

It is evident that there is a large margin between the risk and

cost levels; approximately 4-5 decades. The margin approaches a

constant level for values of TD greater than the system MTTR. On the

basis of this result, relaxation of the time restraint from the current

limit of two hours to a value approximately equal to the MTTR of the

system (8-10 hours) would be acceptable and reduces the cost of plant

downtime as much as $1 million/year. A recent analysis performed by

Lambert suggests much the same thing.[2] While the level of risk will

increase for unlimited values of TD, the major influence on plant

availability of the time restraint is in the period prior to the MTTR

of the system. Thus, cost savings resulting from a relaxation of the

time restraint can only be realized in this time period if the risk

is shown to be constant or can be conservatively estimated. Otherwise,

an increase in risk during this period may justify time restraints

placed on system repair. Use of the final version of WASH 1400 rather

than the draft as a data source for this analysis would not have

changed the results significantly. Changes of several orders of
magnitude would not alter recommendation for relaxation of the time
restraint.

5. Conclusions. The results of the risk-cost comparison
indicate that a relaxation of present administrative controls on
plant operation with a degraded protection system may be warranted
without increasing the risk to the public a significant amount. A
detailed risk analysis is possible for complex systems such as a
nuclear power plant and should be a substantial input for any decisions
regarding restrictions on system or plant operations. There are
primarily two concerns regarding the use of quantitative analyses of
this type. The uncertainty of what constitutes an acceptable risk
level is one source of concern. Criteria for acceptable risk levels
have been proposed and are still being developed.[3] Another is the
lack of complete and accurate failure data on system components. Data
for use in reliability and risk analyses is now being gathered in data
banks established specifically for that purpose.[4]

It is hoped as risk methodology grows in acceptance and applications, that it will be used to a greater extent as a tool for
regulatory bodies to make realistic operating limitations.

REFERENCES

[1] Availability of Electric Power Sources, Regulatory Guide 1.93,
     USAEC, December 1974.

[2] H. Lambert, Fault Trees for Decision Making in System Analysis,
     UCRL 51829, October 1975.

[3] Joseph R. Penland, A Formulation for Risk Assessment and
     Allocation, Proceedings 1975 Annual Reliability and Maintainability
     Symposium, pp. 1-5.

[4] Nuclear Plant Reliability Data Reporting and Analysis System,
     DRAFT, Proposed American National Standard, ANSIN524P,
     May 3, 1973.

RECENT ADVANCEMENTS IN FAULT TREE METHODOLOGY AT KARLSRUHE

L. CALDAROLA AND A. WICKENHÄUSER*

Abstract. A very fast analytical computer program for fault tree evaluation has already been developed at the German nuclear research center of Karlsruhe. This program can evaluate coherent systems assuming binary component states. Four different classes of components can be handled by the program: (1) unrepairable components, (2) repairable components with revealed faults, (3) repairable components with faults remaining unrevealed until next demand occurs and (4) repairable components with faults which are detected upon inspection.

The program can perform also time dependent calculations. In particular the program can analyse systems characterized by two phases, one following the other in time (two time axis).

A new computer program is also being developed. This computer program will be capable to analyse non coherent systems with multistate components. The mathematical theory supporting the new program is described in the paper. In particular the algorithm for the identification of the prime implicants and the theory for the calculation of the occurrence probability and of the first occurrence probability of the TOP event are discussed.

Introduction. The evaluation of the occurrence and of the first occurrence probabilities of the top event of a fault tree can be carried out by means of simulation methods (Monte Carlo-type methods) or by means of deterministic methods. Numerical simulation allows reliability information to be obtained for systems of almost any degree of complexity. However, this method provides only estimates and no parametric relation can be obtained. In addition, since the failure probability of

* Institut für Reaktorentwicklung, Gesellschaft für Kernforschung mbH., Postfach 3640, 7500 Karlsruhe, Federal Republic of Germany.

518

a system is usually very low, precise results can be achieved only at the expense of very long computational times.

Deterministic methods give more insight and understanding because explicit relationships are obtainable. Results are also more precise because these methods usually give the exact solution of the problem. In 1970 Vesely [1] gave the foundations of the analytical method for fault tree analysis. The Karlsruhe computer program for the evaluation of the availability and reliability of complex repairable coherent systems with binary components is essentially based on Vesely's theory with some additional important and fundamental improvements. The description of this computer program is given in part 1 of this paper.

Part 2 deals with the mathematical theory of a second computer program which is now being developed at Karlsruhe. This program will be able to handle complex repairable systems (coherent and non coherent) with multistate components. The definition of coherent systems is given in [7]. In the following the components will be assumed to be statistically independent. This means that each component fails and is repaired independently of the other components. The case of statistically dependent failures has not been treated in this paper.

Part 1. The Karlsruhe computer program.

1.1. Generalities. Let us consider a repairable system S made of components which can be either in the unfailed or failed state. System S will, in general, have a large number of failed states. The set of all possible failed states is the top event (TOP) of the fault tree of the system. The fault tree shows in diagrammatic form the logic connections between the top event and the failed states of its components (primary events).

A minimal cut set is a minimal set of primary events which by occurring guarantee the occurrence of the top event.

The Karlsruhe computer program in its present form can handle coherent systems only and four different classes of components:

Class 1   Unrepairable components

Class 2   Repairable components with failures which are immediately
             detected (revealed faults)

Class 3   Repairable components with failures which are detected
             upon demand (faults remain unrevealed until next demand
             occurs)

Class 4   Repairable components with failures which are detected
             upon inspection (faults remain unrevealed until next in-
             spection is carried out).

The following quantities can be calculated by the computer program:

-    System point unavailability

- System average unavailability (unavailability averaged over the time)

- System asymptotic averaged unavailability $(t \to \infty)$

- System maximum unavailability

- System failure intensity (that is the limit (for $dt \to 0$) of the probability that the system is up at t and fails between t and t + dt divided by dt).

- System average failure intensity (failure intensity averaged over the time)

- System asymptotic averaged failure intensity $(t \to \infty)$

- System maximum failure intensity

- System integral of failure intensity. This quantity is an upper bound of the cumulative failure probability (unreliability) of the system. In the following we shall refer to this quantity as unreliability of the system.

The computer program has various options: each option allows the calculation of a selection of the quantities listed above. In particular one option can perform a time independent calculation (asymptotic and maximum values only) for a quick evaluation.

Particular features of the computer program are the following:

- Capability to identify the whole set of minimal cut sets (prime implicants) and to list them in order of importance (from the largest to the smallest unavailability).

- Capability to analyse systems characterized by two phases one
following the other in time (two time axes).

The computer program needs 480K in CPU (IBM 370/168). This allows
e.g. computing fault trees with a maximum of 256 elements and 200
points on each of the two time axis or with a maximum of 2000 elements
and a time independent calculation. Element means here either component or gate.

1.2. Basic mathematical methods used in the program. The calculation takes place in two steps:

Step No. 1   Identification of the minimal cut sets and of the higher
              order cut sets.

Step No. 2   Calculation of the probability of occurrence at time "t"
              of the TOP event (unavailability) and of the probability
              of the first occurrence within time "t" of the TOP event
              (unreliability).

1.2.1. Identification of the minimal cut sets and of the higher
order cut sets. The first step is that of ordering the fault tree in
the form of a list (table). The primary events are first listed. The
acceptance criterion of a gate in the list is the following: a gate is
accepted if and only if its predecessors have already been accepted.
If a gate satisfies the acceptance criterion it is written in the list.

The ordering process comes to an end when all gates have been accepted
in the list. The algorithm to identify the minimal cut sets is the so-
called "downward algorithm" already described in $\underline{/}\bar{6}\underline{/}$ by Fussell.
This algorithm begins with the TOP event and systematically
goes down through the tree from the highest to the lowest event, that
is from the bottom to the top of the ordered list of events. The fault
tree is developed in a matrix. The elements of the matrix are events.
Each row of the matrix is a cut set. The numbers of the elements contained in a row is called length of the row. Each time an OR gate is
encountered new rows will be produced. Each time an AND gate is
encountered the length of the rows (in which the gate appears) will be
increased. The idea is based on the fact that each time an AND gate
is reached in descending from the TOP, all the events beneath must
occur in order to produce the TOP. Thus all input events to the gate
will be written in the rows in which the gate appears. On the other
hand when an OR gate is found, any event beneath will cause the TOP,
thereby producing a row for each event beneath the gate. The process
comes to an end when all the elements of the matrix are primary events.
In addition the two following simplification rules are applied:

1. Repeated events in the same row are deleted

    $P \cap P = P$ (idempower law)

2. Any row which contains all elements of another row is deleted
    (absorption law).

After all simplifications have been carried out, each row of the matrix is a minimal cut set and the set of the rows is the complete set of the minimal cut sets.

In order to reduce the execution time and the storage region, some additional features have been incorporated in the program. The program is able e.g. to identify the so called "super events". If the domain of a gate is disjoint from those of the other gates which do not belong to the domain of the gate under consideration, the output event of that gate is called "super event". Once a super event has been identified, it replaces the whole branch under it and is treated as a primary event. A second feature is related to the way in which the matrix is developed. Based on some information associated to each gate, the program decides which gate among those belonging to the same row must be expanded first. Finally the program follows the criterion to expand a row down to its primary events before going on to expand the next row. In addition to the minimal cut sets (called also first order cut sets), the program identifies the cut sets of higher orders which are obtained by intersecting various minimal cut sets. The cut sets of order 2 are obtained by intersecting two minimal cut sets at a time. In general a cut set of order "k" is obtained by the intersection of "k" minimal cut sets. The symbol X_{j_k} indicates the cut set "j_k" of order "k". We shall indicate with L_k the total number of cut sets of order "k". By simple considerations of combinatorial analysis we can write the following equation:

$$L_k = \binom{L_1}{k}$$

(1)

where L_1 is the total number of minimal cut sets.

1.2.2. Calculation of the system unavailability and unreliability.

First the unavailability and the failure intensity of each component C_i as a function of time are calculated for both initial conditions: component intact (V_u) and component failed (V_d) at the initial time. The equations of V_u and V_d for each class of components are given in Table 1. For the meaning of the symbols see the nomenclature at the end of part 1 (section 1.4). With reference to Table 1, the equations of class 1 and 2 components are very well known equations obtainable from the current literature. The equations of class 3 and 4 components are however new $\underline{/}8\underline{/}$. The unavailability "V" of a component is given by the following equation:

$$V = (1 - P) V_u + P V_d$$

(2)

where P is the probability that the component is unavailable at the initial time. The failure intensity "h" of a component is given by the following equation:

$$h = \lambda (1 - V)$$

(3)

where λ is the component failure rate (assumed to be constant). The

Table **1**:   Component Unavailability

Class	Type of Component		Initial State		
			Intact (V_u)		Failed (V_d)
1	Unrepairable		$1-e^{-\lambda t}$		1
2	Repairable (revealed faults)		$\dfrac{\lambda}{\lambda+\mu}\left[1-e^{-(\lambda+\mu)t}\right]$		$\dfrac{\lambda}{\lambda+\mu}+\dfrac{\mu}{\lambda+\mu}e^{-(\lambda+\mu)t}$
3	Repairable (faults detected upon demand) $\varepsilon=\left(\dfrac{\lambda+\mu+\nu}{2}\right)^2-(\lambda\mu+\lambda\nu+\mu\nu)$	$\varepsilon<0$	$\dfrac{\lambda(\mu+\nu)}{\lambda\mu+\lambda\nu+\mu\nu}\left\{1-e^{-(\lambda+\mu+\nu)t/2}\left[\cos(t\sqrt{\lvert\varepsilon\rvert})+\dfrac{1}{2\sqrt{\lvert\varepsilon\rvert}}(\dfrac{\mu^2+\nu^2}{\mu+\nu}-\lambda)\sin(t\sqrt{\lvert\varepsilon\rvert})\right]\right\}$		$\dfrac{\lambda(\mu+\nu)}{\lambda\mu+\lambda\nu+\mu\nu}+\dfrac{\mu\nu}{\lambda\mu+\lambda\nu+\mu\nu}e^{-(\lambda+\mu+\nu)t/2}\left[\cos(t\sqrt{\lvert\varepsilon\rvert})+\dfrac{\lambda+\mu+\nu}{2\sqrt{\lvert\varepsilon\rvert}}\sin(t\sqrt{\lvert\varepsilon\rvert})\right]$
		$\varepsilon=0$	$\dfrac{\lambda(\mu+\nu)}{\lambda\mu+\lambda\nu+\mu\nu}\left\{1-e^{-(\lambda+\mu+\nu)t/2}\left[1+(\dfrac{t}{2}-(\dfrac{\mu^2+\nu^2}{\mu+\nu}-\lambda))\right]\right\}$		$\dfrac{\lambda(\mu+\nu)}{\lambda\mu+\lambda\nu+\mu\nu}+\dfrac{\mu\nu}{\lambda\mu+\lambda\nu+\mu\nu}e^{-(\lambda+\mu+\nu)t/2}\left[1+\dfrac{\lambda+\mu+\nu}{2}t\right]$
		$\varepsilon>0$	$\dfrac{\lambda(\mu+\nu)}{\lambda\mu+\lambda\nu+\mu\nu}\left\{1-e^{-(\lambda+\mu+\nu)t/2}\left[\cosh(t\sqrt{\lvert\varepsilon\rvert})+\dfrac{1}{2\sqrt{\lvert\varepsilon\rvert}}(\dfrac{\mu^2+\nu^2}{\mu+\nu}-\lambda)\sinh(t\sqrt{\lvert\varepsilon\rvert})\right]\right\}$		$\dfrac{\lambda(\mu+\nu)}{\lambda\mu+\lambda\nu+\mu\nu}+\dfrac{\mu\nu}{\lambda\mu+\lambda\nu+\mu\nu}e^{-(\lambda+\mu+\nu)t/2}\left[\cosh(t\sqrt{\lvert\varepsilon\rvert})+\dfrac{\lambda+\mu+\nu}{2\sqrt{\lvert\varepsilon\rvert}}\sinh(t\sqrt{\lvert\varepsilon\rvert})\right]$
4	Repairable (faults detected upon inspection) $q=\lg(2-\lg\theta\lambda)$ $\lambda_{eff}=2\sqrt{1-\Gamma(1/q)\sqrt{q}}\theta/n^2+\lambda\dfrac{n-\theta}{n}(\dfrac{n-\theta}{n}+\dfrac{2i}{n})$	$t\geq mn$	$1-e^{-(t-mn)\lambda_{eff}}\left[1-e^{-(\frac{t-mn}{\theta})q}\right]$	$0\leq t\leq n$	1
		$t=mn$ $(m>0)$	$1-e^{-n\lambda_{eff}}\left[1-e^{-(n/\theta)q}\right]$	$t\geq mn;\, m>0$	$1-e^{-(t-mn)\lambda_{eff}}\left[1-e^{-(\frac{t-mn}{\theta})q}\right]$
				$t=mn;\, m>1$	$1-e^{-n\lambda_{eff}}\left[1-e^{-(n/\theta)q}\right]$

unavailability U_{j_k} and the failure intensity w_{j_k} of a generical cut
set X_{j_k} are given respectively by the following equations:

$$U_{j_k} = \prod_{i=1}^{s} (V_i)^{\alpha_i^{j_k}} \tag{4}$$

and

$$w_{j_k} = \sum_{i=1}^{s} \alpha_i^{j_k} \lambda_i (1 - V_i) \prod_{\substack{q=1 \\ q \neq i}}^{s} (V_q)^{\alpha_q^{j_k}} \tag{5}$$

where

s = total number of components belonging to the fault tree

$$\alpha_i^{j_k} = 1 \text{ if } C_i \in |X_{j_k}| \; ; \; \alpha_i^{j_k} = 0 \text{ if } C_i \notin |X_{j_k}| \tag{6}$$

$$\alpha_q^{j_k} = 1 \text{ if } C_q \in |X_{j_k}| \; ; \; \alpha_q^{j_k} = 0 \text{ if } C_q \notin |X_{j_k}| \tag{6'}$$

and $|X_{j_k}|$ indicates the list of primary events from which cut set X_{j_k}
is generated.

The unavailability U and the failure intensity w of the system
are given respectively by the two following equations:

$$U = \sum_{k=1}^{L_1} \sum_{j_k=1}^{L_k} (-1)^{k-1} U_{j_k} \tag{7}$$

and

$$w = \sum_{k=1}^{L_1} \sum_{j_k=1}^{L_k} (-1)^{k-1} w_{j_k} \, . \tag{8}$$

Eq. 7 is derived from a very known equation of the probability theory

and its use has already been suggested by Vesely [1] . Eq. 8 instead, although being formally similar to Eq. 7, is a new equation
[9] . The equation for w suggested by Vesely [1] is different, more
complicated and more restrictive in its application.

The computer program considers the cut sets of order 1, 2, and
3 only.

$$U \approx \sum_{k=1}^{3} \sum_{j_k=1}^{L_k} (-1)^{k-1} U_{j_k} \tag{9}$$

$$w \approx \sum_{k=1}^{3} \sum_{j_k=1}^{L_k} (-1)^{k-1} w_{j_k} . \tag{10}$$

The system unreliability Q is approximated by the integral of the system failure intensity (expected number of failures)

$$Q(t) \approx \int_{o}^{t} w(t') \, dt' . \tag{11}$$

1.2.3. <u>Calculation with two time axis</u>. This type of calculation
is carried out when one wants to analyse a system characterized by two
phases one following to the other in time. Here one identifies two
phases, namely Phase A and B defined as follows:

Phase A - The system is ready for operation

Phase B - The system is in operation.

One is interested in calculating the system unavailability during phase A and the system unreliability during phase B. The first quanti-

ty is calculated according to the procedure of section 1.2.2. The cal-

culation of the system unreliability during phase B is instead more

complicated.

The program has two time axis, one for each phase: t_A and t_B. The

unavailability of each component during phase B depends upon the un-

availability of the same component during phase A. If we indicate with

$P(t_A)$ this unavailability during phase A, the component unavailability

during phase B is given by the following equation:

$$V(t_A;t_B) = \underline{/}\overline{1} - P(t_A)\underline{/} \, V_u(t_B) + P(t_A) \, V_d(t_B).$$ (12)

The failure intensity of a component during phase B is given by

$$h(t_A;t_B) = \lambda \, \underline{/}\overline{1} - V(t_A;t_B)\underline{/}.$$ (13)

The program calculates the system failure intensity $w(t_A;t_B)$ by using

the same procedure shown in section 1.2.2. The unreliability $Q(t_B)$ is

given by

$$Q(t_B) \simeq \frac{1}{T_A} \int_{t'_B=0}^{t_B} \int_{t_A=0}^{T_A} w(t_A;t'_B) \, dt_A \, dt'_B$$ (14)

where T_A is the system operating time during phase A.

1.3. <u>Numerical Results</u>. Fig. 1 shows one of the fault trees used

to test the program with the list of the components of the same fault

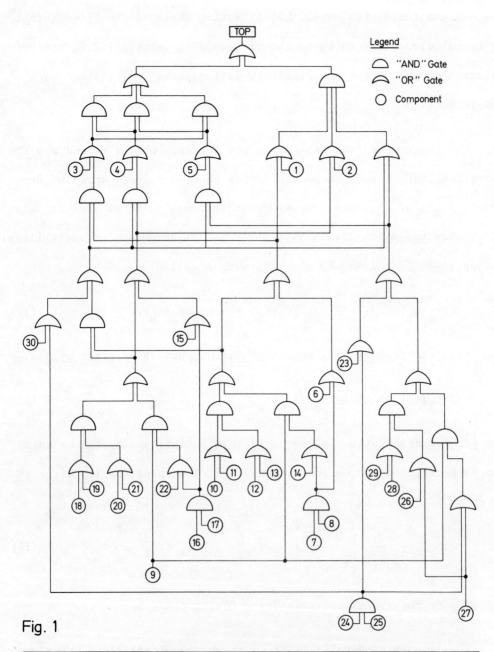

Fig. 1

Component	1	2	3	4	5	6	7	8	9	10	11	12	13	14	15	16	17	18	19	20	21	22	23	24	25	26	27	28	29	30
Failure Rate [$10^{-6} \cdot h^{-1}$]	1.5	1.5	0.2	0.2	0.2	10	50	50	10	50	50	50	50	10	10	50	50	50	50	50	50	50	10	10	10	50	50	50	50	10
Insp. Time Int. [10^3 h]	1.3	1.3	1.0	1.0	1.0	1.0	0.5	0.5	1.0	0.5	0.5	0.5	0.5	1.0	1.0	0.5	0.5	0.5	0.5	0.5	0.5	0.5	1.0	1.0	0.5	0.5	0.5	0.5	0.5	0.1

tree. They all belong to class 4. The fault tree of Fig. 1 contains

282801 paths to the top, but only 418 paths are minimal cut sets.

Fig. 2 shows both system average unavailability and unreliability as

functions of time with a summary of the results. The CPU time on an

IBM 370/168 computer was 8 secs. for the time independent calculation

and 72 secs. for the time dependent calculation with 130 points on the

time axis.

The Karlsruhe computer program is very fast and can be applied

to practical problems. This has also been demonstrated by the results

of the comparison among various German computer programs   [3] , where

the program was found to be the best among the German analytical pro-

grams together with the analytical program of the University of Berlin.

1.4. Nomenclature of the symbols used in Table 1

V_u =   component unavailability with component being intact at the
initial time.

V_d =   component unavailability with component being failed at the
initial time.

t =   time

λ =   component failure rate (constant)

μ =   component repair rate   (constant)

ν =   average demand frequency (constant)

η =   time interval between two successive inspections

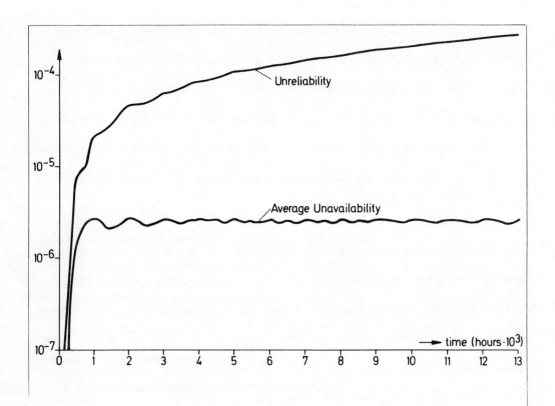

Table of Results

Case	Component Unavailability	System Asymp. Av. Unavailability	System Unreliability (at 8760 hours)
1 [x]	Average	$1.04 \cdot 10^{-6}$	$1.08 \cdot 10^{-4}$
2 [xx]	Point	$2.67 \cdot 10^{-6}$	$1.83 \cdot 10^{-4}$
3 [x]	Maximum	$1.31 \cdot 10^{-5}$	$7.32 \cdot 10^{-4}$

[x]   CPU Time = 8 secs. (both cases 1 and 3)

[xx] CPU Time = 72 secs. - Time dependent calculation performed
      until 13.000 hours (130 points on the time axis).

Fig. 2

Θ =   time needed to inspect an unfailed component (constant)

τ =   repair time of a failed component during inspection (constant)

$\Theta + \tau$ =   time needed to inspect a failed component

λ_{eff} =   component effective failure rate

q   =   numerical coefficient

$\Gamma(..)$ =   Gamma function.

Part 2. Fault tree analysis with multistate components

2.1. Boolean operations. Components will be indicated by the capital letter C followed by an integer possitive number as index (C_1,
C_2, C_3 etc.). This number identifies the component. Primary events are
states of components and they will be indicated by the capital letter
"E" with an index which on its turn has another index. The two indices
together identify the primary event and the second index alone identifies the component to which the primary event belongs. E.g. "E_{i_b}" means
primary event "i_b" which belongs to component C_b. We indicate with "n_b"
the total number of states that component C_b can occupy and with "s"
the total number of components present in the fault tree. We have

$$\sum_{b=1}^{s} n_b = N \tag{1}$$

where N = total number of primary events.

We assign a binary indicator to each event. The indicator is
given the value 1 if the event occurs and 0 if it does not occur. In
particular we shall have

$$\text{TOP} = \begin{cases} 1 \text{ if TOP occurs} \\ 0 \text{ if TOP does not occur} \end{cases} \tag{2}$$

$$E_{i_b} = \begin{cases} 1 \text{ if } E_{i_b} \text{ occurs} \\ 0 \text{ if } E_{i_b} \text{ does not occur.} \end{cases} \tag{3}$$

Since a component must occupy one of its states and can occupy only one state at a time, it follows that the primary events belonging to the same component constitute an universal set and are pairwise mutually exclusive. This is equivalent to writing

$$\bigcup_{i_b=1}^{n_b} E_{i_b} = 1 \tag{4}$$

and

$$E_{i_b} \bigcap E_{g_b} = 0 \qquad \text{for } i_b \neq g_b . \tag{5}$$

A fault tree is a graphic representation of the boolean function (Φ) which links its top event (TOP) to its primary events (E_{i_b}). This is equivalent to writing

$$\text{TOP} = \Phi (E_{1_1}, E_{2_1}, \ \ldots\ldots\ ; E_{n_s}). \tag{6}$$

The Boolean function is also called "structure function". The words "structure" and "system" are synonymous.

Given a fault tree one can apply to it the downward algorithm described in part 1 of this paper. The only difference here is that

the simplification rules are now three instead of two. The additional
rule is the following:

"Delete zero monomials, that is monomials (rows) which contain at
least a pair of mutually exclusive primary events. $E_{i_b} \cap E_{k_b} = 0$
(exclusion law)."

Once that the downward algorithm with the three above simplification rules has been applied to the fault tree, one obtains the
structure function Φ in a specific normal disjunctive form, that is
that form which is associated to the fault tree under consideration.
We shall call this form "associated normal disjunctive form" in order
to distinguish it from the other possible disjunctive forms of the
structure function Φ. The monomials of the associated normal disjunctive form do not coincide in general with the minimal cut sets (prime
implicants).

We shall call any sum of prime implicants, which is equivalent
to the function Φ, a "base of the function Φ". The sum of all prime
implicants has this property. We shall call it the "complete base".

We shall describe as an "irredundant base" a base which ceases
to be a base if one of the prime monomials occuring in it is removed
(deleted).

The associated normal disjunctive form does not coincide in
general with the smallest irredundant base.

The identification of the smallest irredundant base is carried
out in two steps:

Step No. 1   Identification of the complete base of Φ starting from
               its associated normal disjunctive form
Step No. 2   Extraction of the smallest irredundant base of Φ from
               its complete base.

Various algorithms for the identification of the complete base of
a Boolean function (step No. 1) are available from the literature
[2] . An algorithm due to Nelson [4] is particularly convenient.
This algorithm consists simply in complementing the boolean function
$\bar{\Phi}$. After each of the two complement operations the result is expanded
in a normal disjunctive form. Nelson's algorithm was improved by Hulme
and Worrell [5] to reduce the computing time. A modified Nelson's
algorithm has been developed at Karlsruhe. The execution times of the
three algorithms are compared in table 2 . The examples have been taken
from [5] .

Various algorithms for the extraction of the smallest irredundant base of a Boolean function from its complete base (step No. 2)
are available from the literature [2] . A new fast algorithm for
step No. 2 is being developed at Karlsruhe, but work has not yet been
completed. The theoretical foundations of the method are described in
[9] . Preliminary results show that the Karlsruhe algorithm in the

case of example 7 (table 2) is able to extract one of the smallest irredundant bases (19 prime implicants) in 2 secs. CPU time.

Table 2

Example	Number of prime impli-cants	CPU time (sec)		
		Nelson (CDC6600)	Sandia (CDC6600)	Karlsruhe (IBM 370/165)
1	4	0.158	0.156	0.11
2	3	0.367	0.182	not performed
3	15	221.418	0.391	
4	15	1413.580	0.388	
5	32	> 5300 [1]	3.868	0.26
6	61	> 4600 [1]	303.657	0.93
7	87	> 6000 [1]	417.371	1.09

[1]   Execution terminated without completing the algorithm.

2.2. Occurrence probability and transition intensity of primary
events. We shall start by defining these two quantities. We shall indicate with V_{i_b} the occurrence probability of the primary event E_{i_b}
and with h_{p_b,i_b} the transition intensity from state E_{p_b} to state E_{i_b}.
We shall assume that the process describing component behaviour is
Markovian.

$$V_{i_b} = V_{i_b}(t) = P\left\{E_{i_b} = 1 \text{ at } t\right\} = P\left\{E_{i_b}\right\} \tag{7}$$

and

$$h_{p_b,i_b}(t) = \lim_{dt \to 0} \frac{1}{dt} P\left\{E_{p_b} = 1 \text{ at } t \text{ and } E_{i_b} = 1 \text{ at } t+dt\right\}. \tag{8}$$

Note that the transition intensity between two primary events belonging to two different components is obviously zero.

$$h_{p_a, i_b} = 0 \quad \text{for } C_a \neq C_b. \tag{9}$$

The transition rates λ_{p_b, i_b} are defined as follows:

$$\lambda_{p_b, i_b} = h_{p_b, i_b}(t) / V_{p_b}(t) = \text{constant.} \tag{10}$$

The λ_{p_b, i_b} are assumed to be known. They can be used to calculate the V_{i_b} and the h_{p_b, i_b}. A system of differential equations can be written

$$\frac{dV_{i_b}}{dt} = \sum_{\substack{p_b=1 \\ p_b \neq i_b}}^{n_b} \lambda_{p_b, i_b} \cdot V_{p_b} - V_{i_b} \cdot \sum_{\substack{p_b=1 \\ p_b \neq i_b}}^{n_b} \lambda_{i_b, p_b} \quad (i_b = 1, 2 \ldots, n_b) \tag{11}$$

with the associated condition

$$\sum_{i_b=1}^{n_b} V_{i_b} = 1. \tag{12}$$

In addition the initial conditions $V_{i_b}^o$ at time 0 must be known. The system of differential equations with the associated conditions can be solved to get the $V_{i_b}(t)$. From Eq. (10) one can now obtain $h_{p_b, i_b}(t)$.

$$h_{p_b, i_b}(t) = \lambda_{p_b, i_b} \cdot V_{p_b}(t). \tag{13}$$

In the following we shall assume that the functions $V_{i_b}(t)$ and $h_{p_b, i_b}(t)$ are known because they have been previously calculated.

We shall call total transition intensity the quantity h_{i_b} defined as follows:

$$h_{i_b} = \sum_{p_b=1}^{n_b} h_{p_b,i_b} \qquad (p_b \neq i_b).$$

(14)

2.3. Occurrence probability and transition intensity of the TOP

event. We shall start by defining these two quantities. We shall indicate with "U" the occurrence probability of the TOP and with "w" the transition intensity to the TOP. We write now the two equations which define respectively U and w.

$$U(t) = P\left\{ TOP = 1 \text{ at } t \right\} = P\left\{ TOP \right\}$$

(15)

and

$$w(t) = \lim_{dt \to o} \frac{1}{dt} P\left\{ TOP = 0 \text{ at } t \text{ and } TOP = 1 \text{ at } t+dt \right\}.$$

(16)

The top event is given in one of its disjunctive forms (see section 2.1).

$$TOP = \bigcup_{j_1=1}^{L} X_{j_1}.$$

(17)

Eq. 17 can be the associated normal disjunctive form or the smallest irredundant base. One uses the smallest irredundant base in order to reduce the execution time of the calculations. However the quantities U and w can be calculated starting from the associated normal disjunctive form as well as from the smallest irredundant base. The monomials X_{j_1} will be called "cut sets of first order". If Eq.(17) is the smallest

irredundant base, then the monomials X_{j_1} are also minimal cut sets
(prime implicants). The higher order cut sets have been defined in
section 1.2.1. The occurrence probability "U_{j_k}" and the transition
intensity "w_{j_k}" of a generic cut set X_{j_k} are given respectively by
the expressions summarized in table 3. The results of table 3 have
been obtained in $\underline{/}^-9\_\underline{/}$.

The coefficients $\alpha_{i_b}^{j_k}$ and $\alpha_{g_a}^{j_k}$ of table 3 are determined according to the following equations:

$$
\alpha_{i_b}^{j_k} = \begin{cases} 1 & \text{if } E_{i_b} \in |X_{j_k}| \\ 0 & \text{if } E_{i_b} \notin |X_{j_k}| \end{cases} \tag{18}
$$

where $|X_{j_k}|$ indicates the list of the primary events from which cut
set X_{j_k} is generated.

The same equation holds also for $\alpha_{g_a}^{j_k}$.

The occurrence probability "U" and the transition intensity "w" of the
top event are given respectively by the two following equations:

$$
U = \sum_{k=1}^{L_1} \sum_{j_k=1}^{L_k} (-1)^{k-1} U_{j_k} \tag{19}
$$

and

$$
w = \sum_{k=1}^{L_1} \sum_{j_k=1}^{L_k} (-1)^{k-1} w_{j_k} . \tag{20}
$$

Eq. (19) is derived from the very well known equation of the occurrence

Table 3

Case	Cut set X_{j_k}	Occurrence probability U_{j_k}	Transition intensity w_{j_k}
1	$X_{j_k} \neq 0$	$\prod_{b=1}^{s} \prod_{i_b=1}^{n_b} (V_{i_b})^{\alpha_{i_b}^{j_k}}$	$\sum_{b=1}^{s} \sum_{i_b=1}^{n_b} \alpha_{i_b}^{j_k} h_{i_b} \prod_{a=1}^{s} \prod_{g_a=1}^{n_a} (V_{g_a})^{\alpha_{g_a}^{j_k}}$, $\quad g_a \neq i_b; \ a \neq b$
2	$X_{j_k} = 0$ because $\lvert X_{j_k} \rvert$ containes only one pair of mutually exclusive primary events, namely E_{i_b} and E_{P_b}	0	$(h_{P_b,i_b} + h_{i_b,P_b}) \prod_{a=1}^{s} \prod_{g_a=1}^{n_a} (V_{g_a})^{\alpha_{g_a}^{j_k}}$, $\quad g_a \neq i_b; \ g_a \neq P_b; \ a \neq b; \ P_b \neq i_b$
3	$X_{j_k} = 0$ because $\lvert X_{j_k} \rvert$ containes more than one pair of mutually exclusive primary events	0	0

probability of the union of different events. Eq. (20) instead, although being formally similar to Eq. (19), is a new equation [9]. The first occurrence probability "Q" of the TOP is approximated by the integral of the transition intensity (expected number of transitions).

REFERENCES

[1] W.E. VESELY, A time dependent methodology for fault tree evalua-
     tion, Nucl. Eng. Des., 13 (1970), 337-360.

[2] J. KUNTZMANN, Fundamental Boolean Algebra, Blackie and Sons Ltd.
     (1967).

[3] K. KOTTHOFF, W. OTTO, Vergleich von Rechenprogrammen zur Zuver-
     lässigkeitsanalyse von Kernkraftwerken, IRS-RS 172.

[4] R.J. NELSON, Simplest normal truth functions, The Journal of Sym-
     bolic Logic, vol. 20, No. 2, June 1954, 105-108.

[5] B.L. HUME and R.B. WORREL, A prime implicant algorithm with
     factoring, IEEE Transaction on computers, November 1975,
     vol. C-24, Nr. 11, 1129-1131.

[6] J.B. FUSSELL, Fault tree analyses: Concept and techniques, NATO
     Conference on Reliability, Liverpool, England (1973).

[7] R.E. BARLOW, F. PROSCHAN, Statistical Theory of Reliability and
     Life Testing - Probability Methods, Holt, Rinehart 1975.

[8] L. CALDAROLA, Unavailability and failure intensity of components,
     (being published in Nuclear Engineering and Design).

[9] L. CALDAROLA, Fault tree analysis with multistate components,
     (being published).

ANALYSIS OF TIME BETWEEN
FAILURES FOR REPAIRABLE COMPONENTS

RICHARD E. BARLOW AND BERNARD DAVIS*

Abstract. A method for analyzing time between failure data is developed. The method uses total time on test plots for a non-homogeneous Poisson process failure model. Engine failure data is used to illustrate the method. A graphical method for determining optimum replacement intervals is presented.

1. Introduction. Most of the statistical literature concerned with analyzing failure data assumes that observations are independent and identically distributed. Although engineering reports often purport to give Mean Time Between Failure (MTBF) estimates, the estimates are, in fact only valid in general if times between failures are exponentially distributed random variables. Since this assumption is often not valid, especially for mechanical components, more sophisticated techniques are required to analyze this type of data. To focus on the kind of problem we have in mind, we will first consider some failure data on caterpillar tractor engines. The actual data is given in the appendix. The data consists of the age of the tractor at engine failure, the age of the engine at failure, and the calendar date of the failure event.

---

*Operations Research Center, University of California, Berkeley, California 94720. This research was supported by the Air Force Office of Scientific Research under Grant AFOSR-77-3179.

Figure 1 shows engine failure removal times on each of 22 tractors as a function of tractor age. The large number of failures at
6000 hours was due to a piston failure problem and not a planned
maintenance action. In order to plan maintenance actions, we need a
mathematical model for predicting engine failures as a function of
tractor age as well as engine age. Sections 2 and 3 present a technique for solving this problem. In Section 4 a graphical method for
determining optimum component replacement based on component life
cycle costing is presented.

2. A Non-homogeneous Poisson Model for Times Between Failures.
In examining 59 engine failures on 22 D9G-66A caterpillar tractors
we found that engine age at failure depended on the operating age of
the tractor when the engine was last placed in the tractor. Except
for the original engines in new tractors, engines replacing failed
engines were often repaired engines. Figure 2 is a plot of engine
age at failure versus tractor age when the engine was last placed in
the tractor. Thus crosses on the y-axis corresponding to x = 0 are
ages at failure of the original engines. It is fairly clear from
Figure 2 that original engines tend to have a longer mean life than
repaired engines. The mean life of original engines is 6149 hours
versus 3241 hours for repaired engines. The standard deviation in
both cases is about 2000 hours.

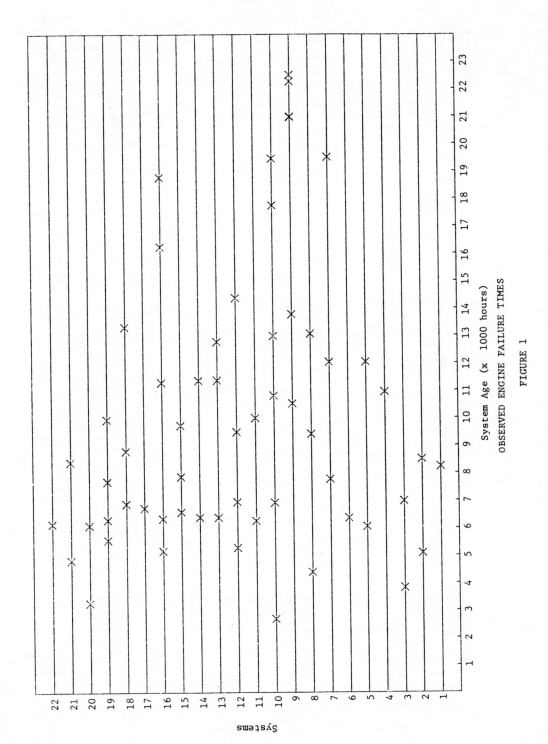

System Age (x 1000 hours)

OBSERVED ENGINE FAILURE TIMES

FIGURE 1

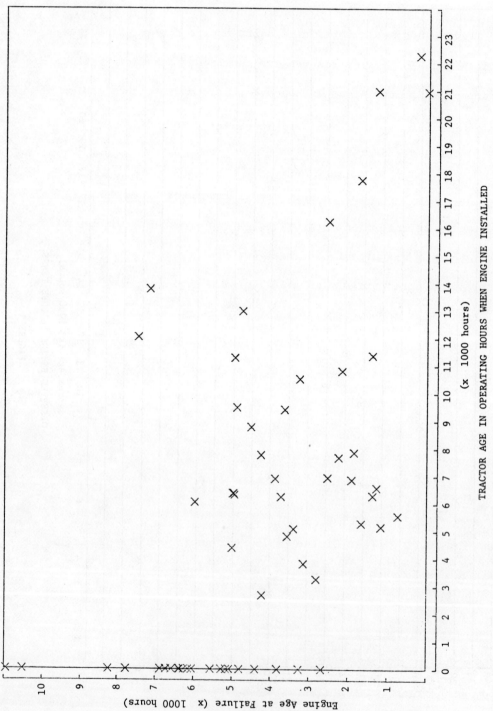

TRACTOR AGE IN OPERATING HOURS WHEN ENGINE INSTALLED

(x 1000 hours)

FIGURE 2

Figure 3 shows a total time on test plot for original (new) engines and also a plot for repaired engines. The fact that the new engine plot is more strongly concave indicates that the life of repaired engines is more nearly exponential. [See Barlow and Campo (1975) for a discussion of total time on test plots.]

Analysis of Failure Events from Independent Processes. Intuitively, engine failure processes will depend on both tractor age and engine age. However, Figures 2 and 3 suggest that tractor age may be the more significant variable in modelling the engine failure processes. We assume that the successive failure events of engines, say, in a given tractor can be described probabilistically by a non-homogeneous Poisson process. If $N(t)$ is the number of engine failures in $[0,t]$, for a particular tractor, then

$$P[N(t) = k] = \frac{[\Lambda(t)]^k}{k!} e^{-\Lambda(t)}$$

for $k = 0,1,2, \ldots$ where $\Lambda(t)$ is the mean number of engine failures in $[0,t]$. [See Çinlar (1975) for an introduction to non-homogeneous Poisson processes.] Since we do not know $\Lambda(t)$, it must be estimated from the data. Our approach is to use an appropriate total time on test plot to make a preliminary model identification. (The model is the analytic form of $\Lambda(t)$.) A final model identification will be made using a Bayesian approach.

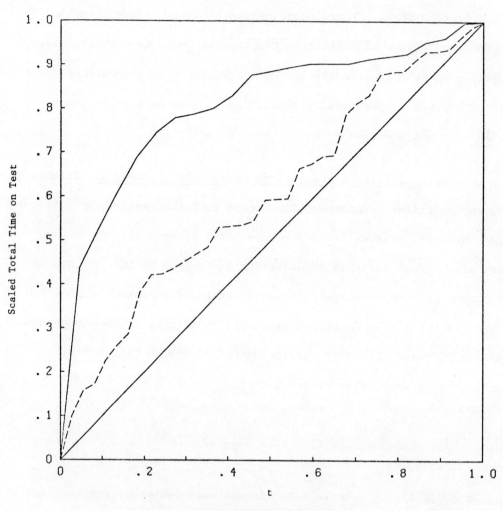

TOTAL TIME ON TEST PLOTS

D9G-66A CATERPILLAR TRACTOR ENGINES

$\bar{x}_{New}$ = 6149 hours $\bar{x}_{Old}$ = 3241 hours

$S_{x_{New}}$ = 2000 hours $S_{x_{Old}}$ = 1842 hours

n = 22   n = 37

FIGURE 3

The superposition of n   independent non-homogeneous Poisson

processes each with mean function, $\Lambda(t)$, will again be a non-

homogeneous Poisson process with mean function $n\Lambda(t)$. Now let each

process run for the same time interval [0,T] . Let

$$Z_{(1)} \leq Z_{(2)} \leq \cdots \leq Z_{(N(T))}$$

be the ordered superposed event times on a common age axis, where N(T)

is the total number of events in [0,T] . In Figure 1, if all points

were superposed on the x-axis, the ordered values would correspond

to $Z_{(i)}$'s . (In our case, however, we do not actually observe engine

failures over the same tractor age interval [0,T] for each tractor.

We make this assumption in order to derive our theoretical result.)

Let n(u) be the number of processes under observation at trac-

tor (or system) age u . In our example (see Figure 1), n(0) = 22

and n(u) = 22 up to about age 6000 hours at which age it drops to

21, etc. Finally, at age u = 22507 hours, n(u) = 0 . The scaled

total time on test plot for the non-homogeneous Poisson process model

is a plot of

$$\frac{\int_0^{Z_{(i)}} n(u)du}{\int_0^{Z_{(N(T))}} n(u)du} \quad \text{versus} \quad \frac{i}{N(T)} \ ,$$

for $i = 1,2, \ldots, N(T)$. Figure 4 is a Total Time on Test Plot of
the data in Appendix 1. We have used linear interpolation to produce
a smooth plot. For this data, $T = 22507$ hours.

The following theorem is the basis for our preliminary model
identification procedure.

Theorem 2.1. Assume that n independent non-homogeneous
Poisson processes are observed on $[0,T]$ where T is fixed. Then,
for $0 \leq p \leq 1$.

$$\lim_{n \to \infty} \frac{Z_{([pN(T)])}}{Z_{(N(T))}} = \frac{\Lambda^{-1}(p\Lambda(T))}{\Lambda^{-1}(\Lambda(T))}$$

$$= \frac{\Lambda^{-1}(p\Lambda(T))}{T}$$

almost surely. $[pN(T)]$ denotes the largest integer in $pN(T)$.
[Note that $n(u) = n$ for $0 \leq u \leq T$; i.e., all processes are completely observed.]

Proof. Let $Y_1, Y_2, \ldots, Y_k, \ldots$ be independent identically
distributed exponential random variables each with unit mean. Let
$S_k = \sum_{i=1}^{k} Y_i$ and $\Lambda_n(x) = n\Lambda(x)$. Then

$$Z_{([pN(T)])} \overset{d}{=} \Lambda_n^{-1}\left(S_{([pN(T)])}\right)$$

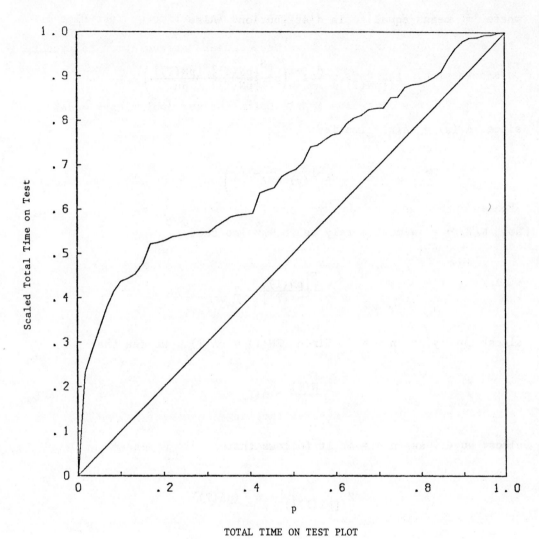

TOTAL TIME ON TEST PLOT

FOR NON-HOMOGENEOUS POISSON MODEL - ENGINE FAILURE DATA

FIGURE 4

where $\overset{d}{=}$ means equal to in distribution. Also

$$Z_{([pN(T)])} \overset{d}{=} \Lambda^{-1}\left(p \; \frac{S_{[pN(T)]}}{[pN(T)]} \; \frac{[pN(T)]}{pn}\right)$$

since $\Lambda_n(x) = n\Lambda(x)$ implies

$$\Lambda_n^{-1}(y) = \Lambda^{-1}\left(\frac{y}{n}\right) .$$

Now $N(T) \to \infty$ almost surely as $n \to \infty$ so that

$$\frac{S_{[pN(T)]}}{[pN(T)]} \to 1$$

almost surely as $n \to \infty$. Since $EN(T) = n\Lambda(T)$, we see that

$$\frac{N(T)}{n} \to \Lambda(T)$$

almost surely as $n \to \infty$. It follows that

$$Z_{([pN(T)])} \to \Lambda^{-1}(p\Lambda(T))$$

almost surely as $n \to \infty$.

Q.E.D.

If $\lambda(u)$, $u \geq 0$ is the intensity function for our failure process then $\Lambda(x) = \int_0^x \lambda(u)du$ is the expected number of failures in

$[0,x]$. If $\lambda(u) = \alpha\lambda^\alpha u^{\alpha-1}$, then $\dfrac{\Lambda^{-1}(p\Lambda(T))}{T} = p^{1/\alpha}$ for $0 \le p \le 1$.

In this case $\dfrac{\Lambda^{-1}(p\Lambda(T))}{T}$ is concave in $0 \le p \le 1$ for $\alpha > 1$ and

convex in $0 \le p \le 1$ for $0 < \alpha < 1$. This does not appear to be a

good model for the plot in Figure 4. If

$$\lambda(u) = \frac{u^{\alpha-1} e^{-\beta u}}{\displaystyle\int_u^\infty w^{\alpha-1} e^{-\beta w} dw}$$

then $\lambda(u)$ is a gamma intensity function and $\Lambda^{-1}(x)$ will be approxi-

mately linear for large values of x . For this reason, the gamma

intensity function with $\alpha > 1$ may be a better model for the plot in

Figure 4.

Since the plot in Figure 4 was based on incomplete data (i.e.,

not all tractors were observed for 22507 hours), Theorem 2.1 does not

strictly apply. However we can make a valid comparison to a homo-

geneous Poisson process using the following theorem.

Theorem 2.2. If $\dfrac{\Lambda(x)}{x}$ is nondecreasing in $x \ge 0$ and

$n(x) \le \dfrac{1}{x} \displaystyle\int_0^x n(u) du$ almost surely, then conditional on $N(T) = N$

$$\frac{\int_0^{Z_{(i)}} n(u)\,du}{\int_0^{Z_{(N(T))}} n(u)\,du} \underset{st}{\geq} U_{i:N-1}$$

where $U_{i:N-1}$ is the i-th order statistic from $N - 1$ independent
uniform $[0,1]$ random variables. $\left(\underset{st}{\geq}\right.$ means stochastically greater
or equal than. $\left.\vphantom{\underset{st}{\geq}}\right)$

Note that if $\lambda(u)$ is nondecreasing, then $\Lambda(x) = \int_0^x \lambda(u)\,du$

satisfies the condition of Theorem 2.2. It follows from Theorem 2.2.
that if the failure rate is nondecreasing, then the scaled total time
on test plot will tend to lie above the 45° line $\left(\text{since } EU_{i:N-1} = \frac{i}{N}\right)$.
Figure 4 indicates an increasing failure rate process. The distribu-
tion of crossings of the scaled total time on test plot in the case
of a homogeneous Poisson model has been derived by Bo Bergman (1976).
The proof of Theorem 2.2 is similar to that of Theorem 2 in Barlow and
Proschan (1969) and will not be given here.

3. A Bayesian Approach to Model Identification. In Section 2
we discussed a method for preliminary model identification. On the
basis of Figure 4 and the discussion in Section 2, we could choose a
gamma failure process model; i.e., $\Lambda(x) = \int_0^x \lambda(u)\,du$ where

(3.1)
$$\lambda(u) = \frac{\beta^{\alpha} u^{\alpha-1} e^{-\beta u}}{\int_{u}^{\infty} \beta^{\alpha} w^{\alpha-1} e^{-\beta w} dw} .$$

The parameters α and β are to be estimated. Given the model, the likelihood best summarizes the information in the data concerning the parameters [cf. Basu (1975)].

Let $n(u)$ be the number of systems (e.g., tractors) under observation at age u and $N(T) = N$, the total number of failures. The likelihood function, given $\underline{Z} = (Z_{(1)} \leq Z_{(2)} \leq \cdots \leq Z_{(N)})$ is easily found to be

$$L(\alpha,\beta \mid \underline{Z}) = \left[\prod_{i=1}^{N} n(Z_{(i)}^{-})\lambda(Z_{(i)}) \right] \exp\left[-\int_{0}^{Z_{(N)}} n(u)\lambda(u) du \right]$$

where $n(Z_{(i)}^{-})$ is the number of systems under observation just prior to the i-th observed failure and $\lambda(u)$ is given by (3.1). Given a prior density $\pi_{o}(\alpha,\beta)$, the posterior density on α, β is

$$\pi(\alpha,\beta \mid \underline{Z}) = \frac{L(\alpha,\beta \mid \underline{Z})\pi_{o}(\alpha,\beta)}{\int_{0}^{\infty}\int_{0}^{\infty} L(\alpha,\beta \mid \underline{Z})\pi_{o}(\alpha,\beta) d\alpha d\beta}$$

by Bayes Theorem. If a diffuse prior; i.e., $\pi_{o}(\alpha,\beta) \equiv c$ is chosen, the $\pi(\alpha,\beta \mid \underline{Z})$ is proportional to the likelihood function. For

illustrative purposes we could graph $L(\alpha,\beta \mid \underline{Z})$ for the data of
Appendix 1. The maximum likelihood estimates can be obtained from
contour plots of the likelihood function.

4. Graphical Determination of Optimum Replacement Policies.

For our non-homogeneous Poisson process model, $\Lambda(t)$ is the expected
number of component (engine) failures in $[0,t]$. Let c_1 be the
average cost of repairing the component and c_2 the cost of buying
a new replacement component. Then, if a new component is bought at
time t, the long run average cost of replacing components is

$$C(t) = \frac{c_1 \Lambda(t) + c_2}{t} .$$

The numerator is just the expected cost of a life cycle of length t.
Let t_o be the optimum replacement age, if it exists; i.e.,

$$C(t_o) = \underset{t>0}{\text{Minimum}} \frac{c_1 \Lambda(t) + c_2}{t} .$$

If c_1 is interpreted as the average time to repair the component and c_2 is the average time to replace with a new component,
then $C(t)$ is the long run average <u>unavailability</u> where an old component is replaced by a new component at age t.

To determine t_o graphically using the scaled total time on
test plot as in Figure 4, first plot $- c_2/c_1 \Lambda(T)$ on the x-axis [see
Figure 5]. ($\Lambda(T)$ is estimated from the data as in the previous

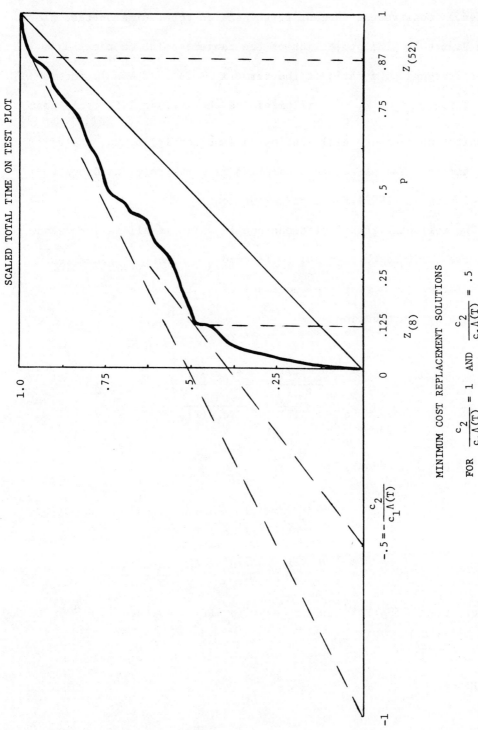

SCALED TOTAL TIME ON TEST PLOT

MINIMUM COST REPLACEMENT SOLUTIONS

FOR $\dfrac{c_2}{c_1 \Lambda(T)} = 1$ AND $\dfrac{c_2}{c_1 \Lambda(T)} = .5$

FIGURE 5

section.) Construct a tangent to the scaled total time on test plot

as in Figure 5. The projection of the tangent point on the x-axis

will correspond to a ratio of the form $i_o/N(T)$. The solution is

$Z_{(i_o)}$; i.e., $t_o = Z_{(i_o)}$ will minimize the average life cycle cost

per unit time against $\Lambda(t)$ estimated (implicitly) by the scaled

total time on test plot. This solution is completely analogous to

that of Bergman (1977) for a different model.

The following graphical technique is valid if all failure proc-

esses are observed throughout $[0,T]$ and the number of processes,

n , is large. To verify the above solution, let $p_o = i_o/N(T)$ and

note that by construction

$$\frac{\Lambda^{-1}(p_o\Lambda(T))}{p_o + \dfrac{c_2}{c_1\Lambda(T)}} \geq \frac{\Lambda^{-1}(p\Lambda(T))}{p + \dfrac{c_2}{c_1\Lambda(T)}}$$

for $0 \leq p \leq 1$. Hence

$$\frac{p + \dfrac{c_2}{c_1\Lambda(T)}}{\Lambda^{-1}(p\Lambda(T))} \geq \frac{p_o + \dfrac{c_2}{c_1\Lambda(T)}}{\Lambda^{-1}(p_o\Lambda(T))} .$$

Now let $t = \Lambda^{-1}(p\Lambda(T))$ and $t_o = \Lambda^{-1}(p_o\Lambda(T))$ so that $p = \dfrac{\Lambda(t)}{\Lambda(T)}$

and $p_o = \dfrac{\Lambda(t_o)}{\Lambda(T)}$.

Hence

$$\frac{c_1 \Lambda(t) + c_2}{t} \geq \frac{c_1 \Lambda(t_o) + c_2}{t_o}$$

which implies t_o is the optimum replacement interval. But

$t_o = \Lambda^{-1}(p_o \Lambda(T)) \sim Z_{([p_o N(T)])}$ by Theorem 2.1 Hence $Z_{(i_o)}$ will

be (approximately) the optimum replacement age.

Acknowledgement. We would like to thank William Vesely of Nuclear

Regulatory Commission for suggesting this problem and Paul Teicholz of

the Guy F. Atkinson Construction Company for supplying the data.

REFERENCES

[1]   R. E. BARLOW and R. CAMPO, Total time on test processes and
      applications to failure data analysis, in Reliability and
      Fault Tree Analysis, edited by R. E. Barlow, J. Fussell and
      N. Singpurwalla, Conference Volume, SIAM, Philadelphia, 1975.

[2]   R. E. BARLOW and F. PROSCHAN, A note on tests for monotone fail-
      ure rate based on incomplete data, Annals of Mathematical
      Statistics, 40(1969), No. 2, pp. 595-600.

[3]   D. BASU, Statistical information and likelihood, Sankhyā, 37,
      Series A, Part 1, pp. 1-71.

[4]   B. BERGMAN, Crossings in the total time on test plot, Technical
      Report, University of Lund, Department of Mathematical
      Statistics, LUNFD6/(NFMS-3043)/1-21/(1976).

[5]   B. BERGMAN, Some graphical methods for maintenance planning,
      Proceedings 1977 Annual Reliability and Maintainability
      Symposium, Philadelphia, 1977.

[6]   E. ÇINLAR, Introduction to Stochastic Processes, Prentice-Hall,
      Inc., 1975.

Appendix 1.   Hours on Tractor and Engine at the Time of Failure and
               the Date of Failure

Tractor	Engine	Hrs. on Tractor	Hrs. on Engine	Date of Failure
1	1	8230	8230	06-16-71
2	1	5085	5085	04-16-70
2	2	8501	3416	06-24-71
3	1	3826	3826	10-11-71
3	2	6983	3157	11-10-72
4	1	10950	10950	05-08-72
5	1	6052	6052	06-01-70
5	2	12040	5988	08-21-74
6	1	6367	6367	06-07-71
7	1	7774	7774	08-10-70
7	2	12035	4261	01-11-72
7	3	19520	7485	12-28-73
8	1	4394	4394	08-08-69
8	2	9415	5021	02-24-71
8	3	13069	3654	03-08-72
9	1	10517	10517	09-21-70
9	2	13783	3266	07-12-71
9	3	20970	7187	07-11-73
9	4	20988	18	08-14-73
9	5	22273	1285	03-12-74
9	6	22507	234	04-16-74
10	1	2690	2690	05-08-67
10	2	6922	4232	04-30-70
10	3	10815	3893	10-20-71
10	4	12988	2173	06-14-72
10	5	17751	4763	11-19-73
10	6	19458	1707	08-15-74
11	1	6259	6259	03-26-68
11	2	9994	3735	02-14-72
12	1	5278	5278	06-28-65
12	2	6949	1671	07-26-66
12	3	9484	2535	10-09-67
12	4	14383	4899	03-04-71
13	1	6378	6378	08-01-66
13	2	11374	4996	05-11-72
13	3	12771	1397	01-09-73
14	1	6385	6385	09-14-66
14	2	11359	4974	05-15-70

Tractor	Engine	Hrs. on Tractor	Hrs. on Engine	Date of Failure
15	1	6578	6578	08–03–66
15	2	7860	1282	03–22–67
15	3	9719	1859	05–22–68
16	1	5161	5161	04–15–65
16	2	6332	1171	11–12–65
16	3	11288	4956	11–04–69
16	4	16249	4961	07–11–72
16	5	18780	2531	06–27–73
17	1	6717	6717	10–26–66
18	1	6869	6869	11–01–67
18	2	8790	1921	12–10–68
18	3	13315	4525	05–08–71
19	1	5556	5556	04–03–67
19	2	6293	737	08–09–67
19	3	7679	1386	05–18–68
19	4	9931	2252	12–20–72
20	1	3268	3268	07–30–71
20	2	6091	2823	05–31–72
21	1	4815	4815	01–21–72
21	2	8388	3573	03–12–73
22	1	6150	6150	10–31–69

THE EFFECT OF MODELING DEPTH ON RELIABILITY PREDICTION FOR SYSTEMS SUBJECT TO A PHASED MISSION PROFILE

J. D. ESARY*

Abstract. The term "phased mission profile" describes a situation in which the factors that influence the longevity of a system change in the course of a sequence of distinct, successive periods of time which are the mission "phases." Phased mission profiles tend to be associated with more general phased missions, in which there can also be changes in the system configuration that is relevant to mission success, but many systems with a stable configuration are exposed to phased mission profiles.

Predictions of the probability of mission success for a system typically result from combining predicted probabilities of mission success for its components according to a logic model for the system's configuration. We investigate the effect that the depth to which the logic model is carried has on predictions, when the predictions at the component level are made using a "standard" methodology.

1. Introduction. Reliability predictions for complex systems typically begin with predictions of the probabilities of mission success for the components in a system. Then the component predictions are combined in accordance with a logic model which describes how the components interact in the system, e.g. a block diagram or a fault tree. The result is a predicted mission success probability for the system. Safety predictions follow a mathematically equivalent pattern which predicts the probability of occurrence for a catastrophic event by using a

---

*
Department of Operations Research, Naval Postgraduate School, Monterey, CA 93940. This research was supported by the Office of Naval Research (NR 042-300).

logic model to combine predicted occurrence probabilities for various
contributory events. In both cases it is reasonable to expect that the
validity of the prediction process can be affected by the *depth* of the
logic model, i.e. by the level of detail to which the block diagram or
fault tree is developed, and at which "component" predictions are introduced.

The purpose here is to investigate an optimistic bias which can
arise from using a logic model which is too shallow in conjunction with
the standard methodologies for making component level predictions from
historical experience, available test data, or similar sources. The
bias in question can be illustrated by a simple example.

Example 1.1. A device D (perhaps an actuator or a control) will
be required to complete two identical, brief cycles of operation during
the course of a mission. Previous experience with the device in a similar service environment is confined to a single operational cycle and
indicates a .99 probability that the device will function once. The
duration of the operational cycles is so short that hardware aging is
not expected to occur. Extrapolating that the probability that the device will function a second time is another .99 leads to a predicted
success probability of $(.99)^2 = .9801$ for two cycles of operation,
as is indicated in Figure 1.1.

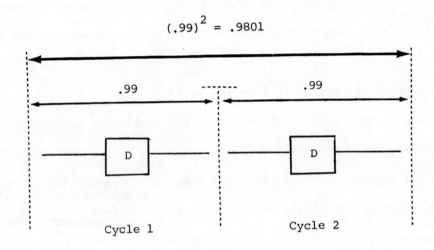

$$(.99)^2 = .9801$$

FIGURE 1.1

However, if viewed in greater detail, the device turns out to be a construct of two identical components, 1 and 2, that operate independently and in parallel. Its single cycle reliability of .99 results from a single cycle success probability of .9 for each component, i.e. $.9 \vee .9 = .99$, where $p_1 \vee p_2 = 1 - (1-p_1)(1-p_2) = p_1 + p_2 - p_1 p_2$ is a convenient notation for the reliability of a system with two independent components that function in parallel with reliabilities p_1 and p_2 (see Figure 1.2).

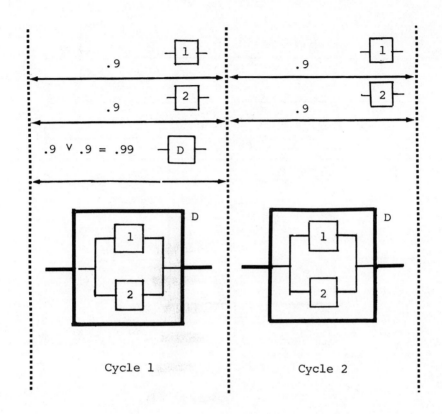

FIGURE 1.2

Then it can be recognized that the device will complete two operational cycles if either component 1 or component 2 does so. Extrapolating one cycle survival probabilities at the new level of component detail leads to a predicted probability $(.9)^2 = .81$ that component 1 will survive two cycles, the same probability that component
2 will survive two cycles, and to a predicted probability $.81 \vee .81$
$= .9639$ that the device will survive two cycles, as in Figure 1.3.

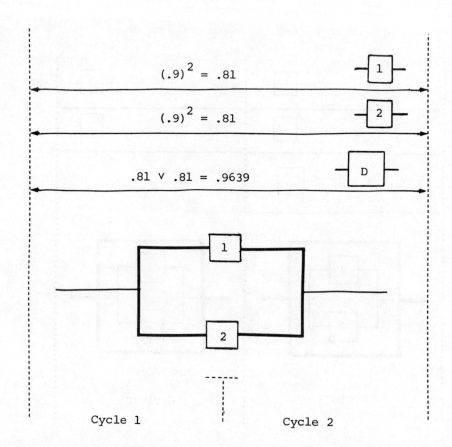

FIGURE 1.3

In this example the assumption that there will be no hardware aging over the course of two operational cycles has been incorporated with experience at two different modeling depths. The prediction based on the more detailed model is the more conservative. □

The scenario considered in Example 1.1 is an almost trivial example of a phased mission. It has successive periods of time in which environmental stresses are altered or repeated which can be regarded as mission phases, but the logic model for the system is the same in each period.   More general phased missions can involve successive epochs of time in which there are changes in the logic model that is relevant to system success as well as in the applied environmental stresses.   For such missions the depth of the logic models employed in making reliability predictions can have an effect similar to that noted in Example 1.1.

The pioneering work on reliability analysis for phased missions was motivated by the need to predict mission success and crew safety for manned spaceflights.   Rubin [6, 1964] and Schmidt and Weisberg [7, 1966] described an approximate, but conservative, method of making reliability predictions for phased missions.   Certain weapons systems are designed to perform phased missions.   Esary and Ziehms [4, 1975] studied a transformation technique that, at least in principle, reduces the prediction problem for a phased mission to that for a single-phase mission.   Ziehms [9, 1975] compared a variety of approximate methods for making phased mission reliability predictions, and identified those which are conservative and relatively the most accurate.   Bell [2, 1975] considered a class of multi-objective phased missions in which sub-missions diverge from a main mission, and described methods for predicting success probabilities for single objectives and composite figures of merit for combinations of objectives.   Bell also considered allowing for an "operational readiness" phase in making predictions.   This is a preliminary phase of indeterminate duration, prior to the inception of the active mission, during which components can be repaired if they fail in an effort to maintain the readiness of the system. Pilnick [5, 1977] emphasized the use of graphical techniques in conducting an expository analysis of a hypothetical mission proposed by Bell.

Recently Burdick, Fussell, Rasmuson, and Wilson [3, 1977] have discussed the analysis of phased missions from the safety perspective, using fault trees to represent the relevant logic models, and considering exact, and selected approximate, methods for making predictions. They suggest, accompanied by examples, possible applications in predicting the safety of nuclear reactors.

The papers just cited contain assorted examples of phased missions, and discuss some of the computational practicalities involved in their analysis. These papers are focused on a proper accounting for shifts in system configuration from phase to phase of a mission, under the assumption that the reliabilities of the components throughout the course of the mission have been correctly established.

Attention here is confined to a different aspect of the phased mission problem, the origins of the bias noted in Example 1.1 and the effect it has on predicted probabilities for mission success. We will seek to characterize those devices whose reliability over a phased environmental profile can be predicted by "standard" methods, and then to establish the modeling depth at which such predictions can be introduced into the analysis of a phased mission. For the present, only systems whose configuration is stable throughout the mission are considered.

2. Standard reliability predictions for phased mission profiles.

Any mission that is contemplated for a device will expose it to one or more service environments. From the physical and human factors point of view, a service environment for a device is an amalgam of the stresses (temperature, vibration) and other factors (corrosion, careless operation) that influence its longevity. From the stochastic point of view, the impact of a service environment on a device can be summarized by a probability distribution for the amount of time the device will survive if exposed in that environment.

We will assume that a fully up device introduced into a service environment e has a random, nonnegative time to failure T_e. For our purposes the probability distribution of T_e can conveniently be described by a *survival function*

(2.1) $\bar{F}_e(t) = P[T_e > t] , \quad t \geq 0 ,$

which gives the probability that the device will survive a mission of whatever duration t in environment e. Or, in some cases, the distribution of T_e can be described by a *failure* rate for the device in environment e, i.e. by a nonnegative function $r_e(t)$, $t \geq 0$, such that

(2.2) $\bar{F}_e(t) = e^{-\int_0^t r_e(s)\,ds} , \quad t \geq 0 .$

It is usually the case that there is a multiplicity of service en-vironments in which a device may be used. We will suppose that a de-vice can be exposed to a *range* E of possible service environments e, each characterized by a survival function $\bar{F}_e$ for the device in that environment, or perhaps by a failure rate r_e.

For many devices a typical mission requires exposure, for various periods of time, to a sequence of distinct service environments. For such a device, a *phased mission profile* will be a sequence $e_1, e_2, \ldots, e_m$ of environments to which it is successively exposed, accompanied by a sequence $d_1, d_2, \ldots, d_m$ of times which are the durations of the ex-posures in each environment.

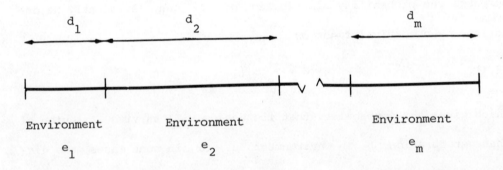

There is often a need to predict the probability that a device will operate successfully throughout a phased mission profile, using knowledge of its reliability in each of the service environments in-volved as a point of departure. A basic motivation for this paper is the presumption that there is a widely used (standard) methodology for doing this which is illustrated by the following example.

Example 2.1. A device (perhaps a generator) has two modes of operation, active and passive. Its failure rate in the passive mode is believed to be a constant λ_1 failures/hr. Its failure rate in the active mode is believed to be a constant λ_2 failures/hr (presumably
$\lambda_2 > \lambda_1$).

For a mission in which d_1 hours of passive operation are followed by d_2 hours of active operation, our standard methodology draws
the failure rate profile shown in Figure 2.1.

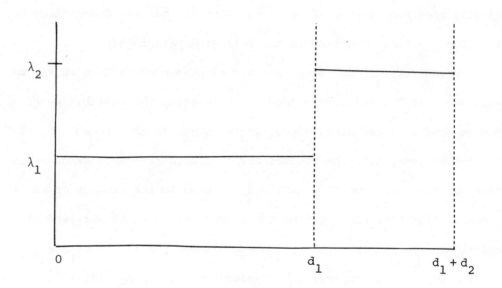

FIGURE 2.1

Then in keeping with equation (2.2), the area $\lambda_1 d_1 + \lambda_2 d_2$ under
the failure rate curve is found, and the probability of success for the
mission is predicted to be $e^{-(\lambda_1 d_1 + \lambda_2 d_2)}$.

Equivalently, the probability of mission success is predicted to be

$$\bar{F}_1(d_1)\,\bar{F}_2(d_2) = e^{-\lambda_1 d_1} \cdot e^{-\lambda_2 d_2} \quad .$$

where $\bar{F}_1(t) = e^{-\lambda_1 t}$ and $\bar{F}_2(t) = e^{-\lambda_2 t}$ are the survival functions for the device in the passive and active operating modes. $\square$

The reader can consider his own variations on the scenario of Example 2.1, involving shifts in stresses, repeated duty cycles, or similar features, to see if he agrees with the general description of feasible practice contained in the following paragraph.

In general, without requiring the existence of failure rates, we will say that the *standard method* for predicting the reliability of a device over a phased mission profile is to equate the probability of mission success, i.e. the probability that each period of exposure to each service environment is survived in turn, to the product of the probabilities that each environmental exposure would be survived if undertaken separately. For the phased mission profile e_1, d_1; e_2, d_2; $\ldots$; e_m, d_m we can express the standard prediction by writing

$$(2.3) \qquad \bar{F}(d_1, d_2, \ldots, d_m) = \bar{F}_1(d_1) \cdot \bar{F}_2(d_2) \cdot \ldots \cdot \bar{F}_m(d_m) \,,$$

where $\bar{F}(d_1, d_2, \ldots, d_m)$ is notation for the probability of surviving the sequence of exposures of durations $d_1, d_2, \ldots, d_m$, and $\bar{F}_j$ is a shortened notation for the survival function of the device in environment e_j, $j = 1, \ldots, m$.

For example, the Review Committee for this manuscript has indicated that the standard method is essentially that implemented by the KITT-2 computer code in treating phased mission profiles. See Veseley and Narum [8, 1970].

3. Degradable and nondegradable devices. The standard prediction method considered in Section 2 assumes that a device enters each new service environment with its survival potential unimpaired. Although failure is permitted in the course of a mission, deterioration is not.

More formally, we will say that a device is *nondegradable* if

(3.1) $$\bar{F}(d_1,d_2) = \bar{F}_1(d_1)\, \bar{F}_2(d_2)$$

for all periods of exposure d_1, d_2 to all service environments e_1, e_2 in E the range of possible environments to which the device may be exposed. As an alternative, a device is *degradable* if

(3.2) $$\bar{F}(d_1,d_2) \leq \bar{F}_1(d_1)\, \bar{F}_2(d_2)$$

for all exposures d_1, d_2 and environments e_1, e_2 in E. The inclusion of the class of nondegradable devices within the class of degradable devices as a boundary case reflects a convention that has proved convenient in treating similar notions.

Systems formed from nondegradable components can be either nondegradable or degradable, as is shown by the following example.

Example 3.1. A two component series system functions as long as both its components function. If the components fail independently, then $\bar{F}(d_1,d_2) = \bar{G}(d_1,d_2) \; \bar{H}(d_1,d_2)$ and $\bar{F}_j(d_j) = \bar{G}_j(d_j) \; \bar{H}_j(d_j)$,

$j = 1,2$, where $\bar{F}$ denotes a survival function pertaining to the system

and $\bar{G},\bar{H}$ denote survival functions pertaining to the components.

If the components in a two component series system are nondegradable, then

$$\bar{F}(d_1,d_2) = \bar{G}(d_1,d_2) \; \bar{H}(d_1,d_2)$$

$$= \bar{G}_1(d_1) \; \bar{G}_2(d_2) \cdot \bar{H}_1(d_1) \; \bar{H}_2(d_2)$$

$$= \bar{G}_1(d_1) \; \bar{H}_1(d_1) \cdot \bar{G}_2(d_2) \; \bar{H}_2(d_2)$$

$$= \bar{F}_1(d_1) \; \bar{F}_2(d_2)$$

for all d_1,d_2 and e_1,e_2 in E, the range of service environments

for the system. Thus the system is nondegradable. There is a tacit,

but reasonable assumption made that the range of service environments

for the system is contained in the range of service environments for

each of its components.

A two component parallel system functions as long as either of its

components functions. If the components fail independently, then

$\bar{F}(d_1,d_2) = \bar{G}(d_1,d_2) \lor \bar{H}(d_1,d_2)$ and $\bar{F}_j(d_j) = \bar{G}_j(d_j) \lor \bar{H}_j(d_j)$, $j = 1,2$.

If the components in a two component parallel system are nondegradable, then

$$\bar{F}(d_1,d_2) = \bar{G}(d_1,d_2) \lor \bar{H}(d_1,d_2)$$

$$= \bar{G}_1(d_1) \; \bar{G}_2(d_2) \lor \bar{H}_1(d_1) \; \bar{H}_2(d_2)$$

$$\leq \{\bar{G}_1(d_1) \lor \bar{H}_1(d_1)\}\{\bar{G}_2(d_2) \lor \bar{H}_2(d_2)\}$$

$$= \bar{F}_1(d_1) \; \bar{F}_2(d_2)$$

for all d_1, d_2 and e_1, e_2 in E (for the system). Thus the system is degradable. The crucial step in the argument depends on the inequality $p_1 p_2 \vee q_1 q_2 \leq (p_1 \vee q_1)(p_2 \vee q_2)$, where p_1, p_2, q_1, q_2 are probabilities. This inequality can be verified by inspection if block diagrams are compared for a system with reliability equal to the left side of the inequality, and a system with reliability equal to the right side of the inequality. $\square$

A trivial extension of the argument used in Example 3.1 for a two component series system justifies the following remark.

Remark 3.1. *If the components in a series system fail independently, and each component is nondegradable, then the system is nondegradable.*

A general class of systems that contains the two component systems considered in Example 3.1 is the class of *coherent* systems (see Barlow and Proschan [1, 1975], Chapters 1 and 2). These systems are characterized by the conditions:

(i) If all the components in the system function, then the system functions.

(ii) If all the components in the system fail, then the system fails.

(iii) Restoring a failed component will not cause a functioning system to fail.

Systems whose logic models can be represented by conventional block

diagrams, or by fault trees using only "and" and "or" gates are

coherent.

The *reliability function*

(3.3) $p = h(p_1, \ldots, p_n)$

of a system (coherent or not) relates the probability p that the

system will function to the probabilities $p_1, \ldots, p_n$ that its n

components will function when the components fail independently. The

reliability function of a coherent system satisfies the inequality

(3.4) $h(p_1 q_1, \ldots, p_n q_n) \leq h(p_1, \ldots, p_n) \, h(q_1, \ldots, q_n)$

for all probabilities $p_i, q_i, i = i, \ldots, n$ ([1], Theorem 1.3, page 23).

Equality holds when $0 < p_i < 1, 0 < q_i < 1, i = 1, \ldots, n$, only if the

system is a series system.

A system of independent components is degradable if

(3.5) $\bar{F}(d_1, d_2) = h\{\bar{G}^{(1)}(d_1, d_2), \ldots, \bar{G}^{(n)}(d_1, d_2)\}$

$$\leq h\{\bar{G}_1^{(1)}(d_1), \ldots, \bar{G}_1^{(n)}(d_1)\}$$

$$\cdot h\{\bar{G}_2^{(1)}(d_2), \ldots, \bar{G}_2^{(n)}(d_2)\}$$

$$= \bar{F}_1(d_1) \, \bar{F}_2(d_2),$$

where $\bar{F}$ denotes a survival function pertaining to the system, and

$\bar{G}^{(i)}, i = 1, \ldots, n$, denote survival functions pertaining to the compo-

nents. The system is nondegradable if equality holds in (3.5).

If the components in a system are nondegradable, then

(3.6) $h\{\bar{G}^{(1)}(d_1,d_2), \ldots, \bar{G}^{(n)}(d_1,d_2)\}$

$$= h\{\bar{G}_1^{(1)}(d_1)\ \bar{G}_2^{(1)}(d_2), \ldots, \bar{G}_1^{(n)}(d_1)\ \bar{G}_2^{(n)}(d_2)\}.$$

If the components in a coherent system are degradable, then

(3.7) $h\{\bar{G}^{(1)}(d_1,d_2), \ldots, \bar{G}^{(n)}(d_1,d_2)\}$

$$\leq h\{\bar{G}_1^{(1)}(d_1)\ \bar{G}_2^{(1)}(d_2), \ldots, \bar{G}_1^{(n)}(d_1)\ \bar{G}_2^{(n)}(d_2)\},$$

since the reliability function of a coherent system is increasing in each of its arguments ([1], Theorem 1.2, page 22).

In view of (3.5), augmented by (3.7), the following theorem is a direct consequence of inequality (3.4).

Theorem 3.2. *A coherent system of independent, degradable (including nondegradable) components is degradable.*

The following remark can be established from the condition for equality in inequality (3.4).

Remark 3.3. *If a coherent system of independent, nondegradable components is itself nondegradable, and if amongst the range of its possible service environments there is one environment in which, for some period of exposure, the survival of each component is neither impossible or certain, then the system must be a series system.*

The practical import of Remark 3.3 is that only series systems of nondegradable components can be treated as nondegradable, unless there are some atypical constraints on the range of service environments

embraced by a mission.

Remark 3.3 also serves to emphasize that the notions of a nonde-

gradable or a degradable device are defined by the relationships (3.1)

and (3.2) relative to some range of possible service environments E.

These definitions are strengthened, in a natural and appropriate way,

if the range of service environments to which the device may be exposed

is assumed to have the following closure property.

A range E of possible service environments is *complete* if, when-

ever e_1, e_2, ... are environments in E, then the environment e

which consists of an exposure of arbitrary duration d_1 to e_1,

followed by an exposure of arbitrary duration d_2 to e_2, and so on,

is also in E.

In essence, E is complete if every phased mission profile that

can be constructed from environments present in E is also to be found

in E.

If a device is nondegradable with respect to a complete range of

service environments E, then for each phased mission profile e_1, d_1;

e_2, d_2; ... , e_m, d_m constructed from environments in E,

$$(3.8) \qquad \bar{F}(d_1, \ldots, d_m) = \bar{F}_{1,\ldots,m-1}(d_1 + \cdots + d_{m-1}) \, \bar{F}_m(d_m) \, ,$$

where $\bar{F}_{1,\ldots,m-1}(d_1 + \cdots + d_{m-1})$ is notation for the probability

that the device will survive an exposure of duration $d_1 + \cdots + d_{m-1}$

to the composite environment $e_1,d_1; \dots ; e_{m-1},d_{m-1}$ which is now in

E.   Iterating the argument leads to

(3.9) $$\bar{F}(d_1, \dots , d_m) = \bar{F}_1(d_1) \, \bar{F}_2(d_2) \, \cdots \, \bar{F}_m(d_m) \, .$$

Similarly, if a device is degradable with respect to a complete range

of service environments   E,   then

(3.10) $$\bar{F}(d_1, \dots , d_m) \leq \bar{F}_1(d_1) \, \bar{F}_2(d_2) \, \cdots \, \bar{F}_m(d_m) \, .$$

Thus the standard method for predicting the reliability of a

device over a phased mission profile is precise if the device is non-

degradable with respect to the complete range of service environments

embraced by the mission, and is optimistic if the device is degradable.

4.   System reliability predictions for phased mission profiles.

There is an elaboration of the standard method for predicting the prob-

ability of mission success over a phased mission profile $e_1,d_1; \; e_2,d_2;$

$\dots ; e_m,d_m$ which is frequently used for complex systems.   This method

has two stages:

(a)   For each component $i = 1, \dots , n$ in the system, the

probability $\bar{G}^{(i)}(d_1, \dots , d_m)$ of mission success is

predicted by the standard method to be $\bar{G}_1^{(i)}(d_1) \cdots \bar{G}_m^{(i)}(d_m)$.

(b)   The system probability $\bar{F}(d_1, \dots , d_m)$ of mission

success is predicted by combining the component pre-

dictions using the system reliability function   h,

i.e. by

$$h\{\bar{G}_1^{(1)}(d_1) \cdots \bar{G}_m^{(1)}(d_m), \ldots, \bar{G}_1^{(n)}(d_1) \cdots \bar{G}_m^{(n)}(d_m)\}.$$

We will call this procedure the *refined standard prediction method*.

Assuming that the components in the system perform independently, the precise relationship which the refined standard prediction method approximates is

(4.1) $$\bar{F}(d_1, \ldots, d_m) = h\{\bar{G}^{(1)}(d_1, \ldots, d_m), \ldots, \bar{G}^{(n)}(d_1, \ldots, d_m)\}.$$

If the components are independent and are nondegradable with respect to the complete range of service environments embraced by the mission, then the refined standard prediction method is exact. This observation is confirmed by using (3.9), at the component level, in conjunction with (4.1).

However, if the system is coherent, its components are independent, and are degradable with respect to the complete range of service environments embraced by the mission, then the refined standard prediction method is optimistic, i.e. it over-predicts the probability of mission success. This observation is confirmed by using (3.10), at the component level, in conjunction with (4.1) and the fact that h is increasing.

It is interesting to compare the result of predicting the system mission success probability $\bar{F}(d_1, \ldots, d_m)$ by direct application of the standard method with the result of using the refined standard method. In the direct approach $\bar{F}(d_1, \ldots, d_m)$ is predicted according to (2.3) with

(4.2) $$\bar{F}_j(d_j) = h\{\bar{G}_j^{(1)}(d_j), \ldots, \bar{G}_j^{(n)}(d_j)\},$$

$j = 1, \ldots, m.$

With $\bar{F}(d_1, \ldots, d_m)$ defined by (4.1) and $\bar{F}_j(d_j)$, $i = 1, \ldots, m$, defined by (4.2), the inequality

(4.3) $$\bar{F}(d_1, \ldots, d_m) \leq h\{\bar{G}_1^{(1)}(d_1) \cdots \bar{G}_m^{(1)}(d_m), \ldots, \bar{G}_1^{(n)}(d_1) \cdots \bar{G}_m^{(n)}(d_m)\}$$

$$\leq \bar{F}_1(d_1) \cdot \cdots \cdot \bar{F}_m(d_m)$$

holds for a coherent system with independent components that are de-gradable with respect to the complete range of service environments embraced by the mission. As was the case in the arguments supporting Theorem 3.2, the first inequality in (4.3) holds because h is in-creasing, and the second inequality is a consequence of (3.4).

Thus the refined standard prediction method, while optimistic if applied using degradable components, is less optimistic than the direct application of the standard prediction method to the system itself.

In many cases the degradable components in a coherent system are themselves *modules* (coherent subsystems with nonoverlapping component subsets) of more basic degradable components, and these components may in turn be modules, and so on. Component independence at the most basic level is reflected as modular independence at the higher levels of amalgamation. In this situation it is easy to extend the preceding considerations to show that the refined standard prediction method be-comes less optimistic as the modeling depth at which standard component

predictions are introduced is increased. As previously noted, if the
modeling depth can be carried to a level at which the components are
nondegradable, then the refined standard method becomes exact.

Acknowledgment. The author would like to express his gratitude to
the Review Committee, G. R. Burdick, R. W. Butterworth, J. B. Fussell,
and K. T. Marshall for a number of helpful comments, and to Rosemarie
Stampfel who prepared the mats.

REFERENCES

[1]   R. E. BARLOW and F. PROSCHAN, *Statistical Theory of Reliability
      and Life Testing*, Holt, Rinehart, and Winston: New York, 1975.

[2]   M. G. BELL, *Multi-phase-mission reliability of maintained standby
      systems*, Naval Postgraduate School Research Report, NPS55Ey75121,
      Dec. 1975.

[3]   G. R. BURDICK, J. B. FUSSELL, D. M. RASMUSON, and J. R. WILSON,
      *Phased mission analysis: A review of new developments and an
      application*, IEEE TRANS. REL., R-26, #1 (April 1977), pp. 43-49.

[4]   J. D. ESARY and H. ZIEHMS, *Reliability analysis of phased missions,
      Reliability and Fault Tree Analysis*, SIAM, Philadelphia, 1975,
      pp. 213-236.

[5]   S. E. PILNICK, *An example of phased mission reliability analysis
      for a hypothetical naval weapons systems*, Naval Postgraduate
      School Research Report NPS55-77-28, June 1977.

[6]   J. C. RUBIN, *The reliability of complex networks, Proceedings of
      the Aerospace Reliability and Maintainability Conference*, June
      1964, p. 263.

[7]   J. H. SCHMIDT and S. A. WEISBERG, *Computer technique for estimating
      system reliability, Proceedings of the 1966 Annual Symposium on
      Reliability*, pp. 87-97.

[8]   W. E. VESELY and R. E. NARUM, *PREP and KITT: Computer Codes for the Automatic Evaluation of a Fault Tree*, IN-1349, Idaho Nuclear Corporation, August 1970.

[9]   H. ZIEHMS, *Approximations to the reliability of phased missions*, Naval Postgraduate School Research Report NPS55Ey75091, Sept. 1975.

GAMMA PRIOR DISTRIBUTION SELECTION FOR BAYESIAN ANALYSIS OF FAILURE RATE AND RELIABILITY

R. A. WALLER,* M. M. JOHNSON,* M. S. WATERMAN* AND H. F. MARTZ, JR.**

Abstract. We assume that the phenomenon under study is such that
the time-to-failure may be modeled by an exponential distribution with
failure rate λ. For Bayesian analyses of the assumed model, the family of gamma distributions provides conjugate prior models for λ.
Thus, an experimenter needs to select a particular gamma model to conduct a Bayesian reliability analysis. The purpose of this paper is to
present a methodology which can be used to translate engineering information, experience, and judgment into a choice of a gamma prior
distribution.

The proposed methodology assumes that the practicing engineer can
provide percentile data relating to either the failure rate or the
reliability of the phenomenon being investigated. For example, the
procedure will select the gamma prior distribution which conveys an
engineer's belief that the failure rate λ simultaneously satisfies the
probability statements, $P(\lambda<1.0 \times 10^{-3}) = 0.50$ and $P(\lambda<1.0 \times 10^{-5}) = 0.05$.
That is, we use two percentiles provided by an engineer to determine
a gamma prior model which agrees with the specified percentiles. For
those engineers who prefer to specify reliability percentiles rather
than the failure rate percentiles illustrated above, we can use the
induced negative-log gamma prior distribution which satisfies the
probability statements, $P(R(t_0)<0.99) = 0.50$ and $P(R(t_0)<0.99999)$
$= 0.95$ for some operating time t_0. Also, the paper includes graphs
for selected percentiles which assist an engineer in applying the
procedure.

1. Introduction. A widely used assumption in reliability analy-

ses is that the time-to-failure variable T is exponentially distributed

with failure rate λ. That is, the time-to-failure variable T has den-

sity function

*Los Alamos Scientific Laboratory, Los Alamos, New Mexico 87545. Work
performed under the auspices of the Energy Research and Development
Administration.
**Texas Tech University, Lubbock, Texas 79409.

$$f(t|\lambda) = \begin{cases} \lambda e^{-\lambda t} & , \quad t > 0, \\ \\ 0, & \text{otherwise}. \end{cases}$$

When conducting a Bayesian reliability analysis, conjugate prior models [1] for the failure rate λ are given by the family of gamma density functions

$$(1) \quad h(\lambda) = \begin{cases} \dfrac{\lambda^{\alpha-1} e^{-\lambda/\beta}}{\beta^{\alpha}\Gamma(\alpha)} & , \quad \lambda > 0 \\ \\ 0, & \text{otherwise}. \end{cases}$$

A particular Bayesian analysis requires that values be assigned to the prior parameters, α and β. The purpose of this paper is to present two methods by which engineering experiences, judgments, and beliefs can be used to assign values to α and β. One method relies on expertise and knowledge concerning the failure rate λ while the other technique assumes information about the reliability $R(t) = e^{-\lambda t}$. Both methods require an engineer to provide two percentile values which are used to solve a pair of simultaneous equations to determine values for α and β. Since the solutions do not exist in closed form, we present graphs which aid the engineer in applying the procedure.

The subsequent development is divided into three sections and three appendixes. Section 2 presents the development based on failure rate percentiles. The technique using reliability percentiles is given in Section 3. Section 4 provides discussion of the procedures which considers some of the consequences which follow from the assumption of

a gamma prior model. Appendix A provides a table and graphs for applying the results in Section 2. Tables and graphs for application of
the results in Section 3 are given in Appendix B. Justification for
mathematical results given in Section 4 is provided in Appendix C.

2. <u>Technique for Failure Rate Percentiles</u>. In this section, we
suppose that an engineer can best summarize his experiences, judgments,
and beliefs about the performance of an item by making statements about
the failure rate λ. The type of information desired is called a percentile. The pth percentile, say λ_p, is that value of λ such that the
probability that λ is less than λ_p is p. In symbols we use $h(\lambda)$ in
Eq. (1) to write

$$(2) \quad P(\lambda < \lambda_p) = \int_0^{\lambda_p} h(\lambda) d\lambda = p.$$

In practice there exists a set of values for (α, β) which satisfies
Eq. (2) with any given value of p. Therefore, the method we propose
requires that the engineer provide two distinct percentiles which generate a pair of simultaneous equations. Explicitly we ask that the en-
gineer provide two percentiles of λ, say λ_1 and λ_2, such that $\lambda_1 < \lambda_2$
and

$$P(\lambda < \lambda_1) = p_1 \; ,$$

(3)

$$P(\lambda < \lambda_2) = p_2 \; .$$

Clearly, the specifications of λ_1 and λ_2 are made with reference to the probabilities, p_1 and p_2, where $p_1 < p_2$. The simultaneous solution of Eq. (3) will select the pair of values for (α, β) which determine the gamma prior that summarizes the engineer's information.

A specific outline of the methodology is as follows:

Step 1: The engineer specifies the values for λ_1, λ_2, p_1, and p_2 which best represent the totality of his experiences, judgments and beliefs about the failure rate λ. These values provide Eq. (3).

Step 2: A search procedure is used to determine α and β for Eq. (1) which simultaneously satisfy the conditions of Step 1.

The table and graphs in Appendix A present values of α and β for selected choices of λ_0 and p_0 where

(4) $P(\lambda < \lambda_0) = p_0$.

By overlaying transparencies of the two graphs which present the specific choices of λ_1, λ_2, p_1, and p_2 of interest, we can determine graphically the desired values of α and β.

Example: Suppose we are studying the reliability of an item for which the available engineering information indicates that the failure rate λ is such that

$P(\lambda < 1.0 \times 10^{-5}) = 0.05$ and $P(\lambda < 1.0 \times 10^{-3}) = 0.50$.

By overlaying transparencies of the graphs in Figs. A1 and A2 we determine that $\alpha = 0.505$ and $\beta = 0.004$ in Eq. (1) provide a

gamma prior distribution which possesses the percentile properties given by the stated conditions. The selected prior is
described by

$$h(\lambda) = \frac{\lambda^{-0.495} \, e^{-\lambda/0.004}}{(0.004)^{0.505} \, \Gamma(0.505)} \, , \, \lambda > 0.$$

and is graphically represented in Fig. 1.

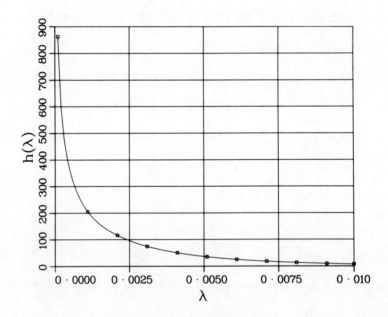

Fig. 1. A gamma prior distribution with $\alpha = 0.505$ and $\beta = 0.004$.

3.   Technique for Reliability Percentiles.   An alternative approach to the failure rate method in the preceding section is to use
percentiles of the reliability. This procedure would be preferable to
the previous method when the engineer providing the percentile

information is more knowledgeable about reliability properties than
failure rate characteristics of the item under study.

To begin the development we transform the gamma density for λ (Eq.
(1)) into a negative-log gamma distribution for the reliability,
$R = R(t) = e^{-\lambda t}$. The resulting density is

$$(5) \quad p(r) = \begin{cases} \dfrac{(-\ln r)^{\alpha-1} \, r^{\frac{1}{t\beta}-1}}{(t\beta)^{\alpha} \, \Gamma(\alpha)}, & 0 < r < 1, \\ \\ 0, & \text{otherwise.} \end{cases}$$

The density in Eq. (5) has been used by Springer and Thompson [2,3],
Mann [4], and Mastran and Singpurwalla [5]. Locks [6] provides a discussion of the negative-log gamma distribution. Thus, the prior distribution induced on the reliability by assuming a conjugate gamma
prior for the failure rate occurs frequently in the literature. Our
purpose here is to present a methodology that allows a reliability engineer to use available information as a tool for selecting values of
α and β to be used in a Bayesian reliability analysis.

From Eq. (5) it is clear that there is a different prior on R for
each choice of time t. Yet for analysis, it is convenient to re-
parameterize the density in Eq. (5) by using $\gamma = t\beta$. With that change,
the density of interest becomes

$$(6) \quad k(r) = \begin{cases} \dfrac{(-\ln r)^{\alpha-1} \, r^{\frac{1}{\gamma}-1}}{\gamma^{\alpha} \, \Gamma(\alpha)}, & 0 < r < 1, \\ \\ 0, & \text{otherwise .} \end{cases}$$

The pth percentile of the reliability at time t is $R_p = R_p(t)$. In symbols we write

$$P(R < R_p) = \int_0^{R_p} k(r) \, dr = p.$$

Once the engineer supplies two percentiles with respect to a reference time, say t_0, we can set up two simultaneous equations whose solution provides values for α and γ. The desired values of (α, β) are then given by α and $\beta = \gamma/t_0$. An outline of the method is as follows:

Step 1:   The engineer provides a reference time t_0.

Step 2:   With respect to t_0, the engineer specifies two percentiles R_1 and R_2 for p_1 and p_2, respectively, which best summarize his experiences, judgments, and beliefs about the reliability and are such that

$$P(R < R_1) = p_1 \text{ and } P(R < R_2) = p_2 .$$

Step 3:   A search procedure is used to determine α and γ which satisfy the probability statements in Step 2 for the density in Eq. (6).

Step 4:   Solve for $\beta = \gamma/t_0$. The values so determined for (α, β) are the selected parameters for either the gamma prior on the failure rate $\big(\text{Eq. (1)}\big)$ or the negative-log gamma prior on the reliability $\big(\text{Eq. (5)}\big)$.

Appendix B presents tables and graphs giving values of α and γ

which satisfy

(7) $P(R < R_0) = p_0$

for selected values of R_0 and p_0. By overlaying transparencies of the

two graphs which contain the values of $R_1(t_0)$, $R_2(t_0)$, p_1, and p_2 used

in Step 2 of the procedure, we determine the values of α and γ. The

value of β is given by γ/t_0 where t_0 is the reference time provided by

the engineer in Step 1.

    Example: Suppose a reliability engineer believes that the re-

    liability of a motor is such that for t_0 = 100 hrs:

            P(R(100) < 0.99) = 0.50 ,

            P(R(100) < 0.99999) = 0.95 .

    By overlaying transparencies of the graphs in Figs. B2 and B3,

    we determine that α = 0.35 and γ = 0.10. Therefore, α = 0.35 and

    β = 0.001 in Eq. (5) yield the selected negative-log gamma prior

$$
p(r) = \begin{cases} \dfrac{(-\ln r)^{-0.65}\, r^{\frac{1000}{t}-1}}{\left(\dfrac{t}{1000}\right)^{0.35}\,\Gamma(0.35)} , & 0 < r < 1 , \\[4mm] 0, & \text{otherwise.} \end{cases}
$$

Recall that we get a different prior for each choice of mission time t.

A graph of the selected prior is given for t = 50, 100, 500, 4000 in

Fig. 2.

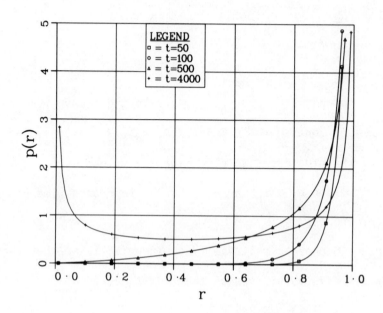

Fig. 2. A negative-log gamma prior for $\alpha = 0.35$,
$\beta = 0.001$, and $t = 50, 100, 500, 4000$.

4. Discussion. In this section we discuss some of the mathematical properties of the prior distributions selected by the percentile methods of Sections 2 and 3. Mathematical justification of the properties is provided in the appendixes.

4.1. The Conjugate Gamma Prior on λ. By differentiating $h(\lambda)$ in Eq. (2) with respect to λ [7], we find that $h(\lambda)$ is a unimodal function of λ with a mode at $\lambda = \beta(\alpha-1)$ when $\alpha > 1$. For $\alpha = 1$, we have an exponential distribution with parameter β. For $0 < \alpha < 1$, $h(\lambda)$ is either L-shaped or decreasing. These properties are summarized in Table 1.

β > 0	L-shaped or decreasing	Exponential	Unimodal with mode at $\lambda = \beta(\alpha-1)$
	$0 < \alpha < 1$	$\alpha = 1$	$\alpha > 1$

Table 1.   Shape properties of a conjugate gamma
prior on λ.

Since error bounds on engineering estimates of failure rates are
frequently large the useful gamma prior distributions are often those
with $0 < \alpha < 1$.   That is, the selected gamma prior models have density
functions which are heavily concentrated on values of λ near zero.
With $\alpha = 0.505$ and $\beta = 0.004$, the gamma prior in Fig. 1 exhibits a concentration of density near zero.

The precision on some incomplete gamma subroutines deteriorates as
α approaches zero.   Then precise calculation of the necessary gamma
percentiles for small values of α becomes difficult.   To address that
problem we can use the theorem of Appendix C.1.   Let λ_p denote the pth
percentile of a gamma prior with parameters α and β.   Then

$$p = \frac{1}{\beta^\alpha \, \Gamma(\alpha)} \int_0^{\lambda_p} \lambda^{\alpha-1} \, e^{-\lambda/\beta} \, d\lambda = \frac{1}{\Gamma(\alpha)} \int_0^{\frac{\lambda_p}{\beta}} x^{\alpha-1} \, e^{-x} \, dx \ .$$

From the theorem, $\lambda_p/\beta \sim p^{\frac{1}{\alpha}}$, as defined in Appendix C.1. Thus,

$\lambda_p \sim \beta p^{\frac{1}{\alpha}}$ for small values of α. This result can be used to extend the

tables in Appendix A to smaller values of α.

    4.2 The Negative-Log Gamma Prior on R. By differentiating

k(r) in Eq. (6) with respect to r (see Appendix C.2), we find that the

possible shapes of k(r) can be summarized as in Table 2.

	$0 < \alpha < 1$	$\alpha = 1$	$1 < \alpha$
$1 < \gamma$	U-shaped with antimode at $r = \exp\left[\frac{\gamma(\alpha-1)}{\gamma-1}\right]$	L-shaped or decreasing	L-shaped or decreasing
$\gamma = 1$	J-shaped or increasing	Uniform	L-shaped or decreasing
$0 < \gamma < 1$	J-shaped or increasing	J-shaped or increasing	Unimodal with mode at $r = \exp\left[\frac{\gamma(\alpha-1)}{\gamma-1}\right]$

Table 2. Shape properties of a negative-log
          gamma prior on R.

Our interest, as stated in Section 3, is to select a pair of

values (α,β) for the negative-log gamma prior in Eq. (5). However, since $\gamma = t\beta$, it follows that γ increases as t increases for a fixed value of β. Thus, we have a different prior on the reliability for each choice of t. That property was illustrated in Fig. 2 for $\alpha = 0.35$, $\beta = 0.001$ and t = 50,100,500,4000. The portion of Table 2 frequently encountered in reliability analyses is for $0 < \alpha < 1$. Now for $0 < \alpha < 1$ and t small enough so that $0 < \gamma = t\beta \leq 1$, the prior distribution p(r) is an increasing function of r. But, when t is such that $\gamma = t\beta > 1$ and $0 < \alpha < 1$, p(r) is U-shaped with its minimum (antimode) at $r = \exp\left[\frac{\gamma(\alpha-1)}{\gamma-1}\right]$. Thus, the negative-log gamma prior on the reliability is concentrated near one in early life (small t). As t increases to large values (late life), the distribution becomes U-shaped. That is, values of r near either zero or one are more likely to occur than other values. Thus, in "old-age" the prior predicts the item will be either quite reliable or quite unreliable. The density function in Fig. 2 for t = 4000 illustrates that point.

As for the gamma prior on λ, precision problems may be encountered when computing the percentiles of Eq. (6) for small α values. The theorem in Appendix C.1 again assists us by providing good approximations to the reliability percentiles as follows. Let R_p be the pth percentile in Eq. (6) for parameters α and γ. Then

$$p = \int_0^{R_p} k(r)\ dr = \frac{1}{\Gamma(\alpha)} \int_{\frac{-\ln R_p}{\gamma}}^{\infty} e^{-x}\ x^{\alpha-1}\ dx$$

or

$$1 - p = \frac{1}{\Gamma(\alpha)} \int_0^{\frac{-\ln R_p}{\gamma}} e^{-x} x^{\alpha-1} \, dx \ .$$

From the theorem in Appendix C.1

$$\frac{-\ln R_p}{\gamma} \sim (1-p)^{1/\alpha} \ .$$

To apply the result we act as if the "$\sim$" is an equality even though it is only an asymptotic result.

To examine the performance of the approximation we refer to Table B3 in Appendix B. Let $\alpha = 0.05$, $p = 0.95$, and $\gamma = 1.795 \times 10^{23}$. Then the approximation gives $R_0 = 0.9983$ in place of 0.9990 in the table. Similarly, for $R_0 = 0.999$, $\alpha = 0.05$, and $p = 0.95$, we find $\gamma = 1.049 \times 10^{23}$ in place of 1.795×10^{23} in the table. Thus, the approximation is of reasonable quality even for moderately large values of α.

5. Acknowledgments. The authors acknowledge the assistance of David Kahaner of Los Alamos Scientific Laboratory in developing the theorem in Appendix C.1. The necessary computation of gamma percentiles was facilitated by use of codes developed by Amos and Daniel [8] of Sandia Laboratories, Albuquerque.

Appendix A.

Table and Graphs of α and β for Gamma Prior Distributions. Table A1 gives β values which satisfy Eq. (4) for p = 0.05, 0.50, 0.95, a selected set of α values, and λ_0 = 1.0 x 10^{-6}. We use the fact that β is a scale parameter to obtain values of β which correspond to failure rate percentiles different from λ_0 = 1.0 x 10^{-6}. For given values of p_0 and α in Eq. (4), the ratio λ_0/β is constant. Therefore, in Table A1 for α and p_0, multiplication of the β value by $(\lambda_s/1.0$ x $10^{-6})$ yields the β value corresponding to a p_0th percentile of λ_s.

Example: Let p_0 = 0.05 and α = 0.25. Then

$$\beta = 2.3705 \text{ x } 10^{-1} \text{ for } \lambda_0 = 1.0 \text{ x } 10^{-6} .$$

Thus, for λ_s = 1.0 x 10^{-9}

$$\beta = (1.0 \text{ x } 10^{-9}/1.0 \text{ x } 10^{-6})(2.3705 \text{ x } 10^{-1}) = 2.3705 \text{ x } 10^{-4} .$$

That is, the 5th percentile of a gamma distribution with α = 0.25 and β = 2.3705 x 10^{-1} is 1.0 x 10^{-6} while the 5th percentile of a gamma distribution with α = 0.25 and β = 2.3705 x 10^{-4} is 1.0 x 10^{-9}.

In Figs. A1, A2, and A3, we have graphed α and β for selected values of λ_0. The results are derived from Table A1 as illustrated by the above example. To provide better resolution in the graphs, a logarithmic scale is used for β. The notation in Table A1 and Figs. A1, A2, and A3 is defined as follows: $nEm = n$ x 10^m.

$$\lambda_0 = 1.E-06$$

$\alpha \backslash p_0$	0.05	0.50	0.95
0.05	1.7941E+20	1.7941E+00	3.7604E-06
0.10	1.6861E+07	1.6852E-03	1.7228E-06
0.15	7.4851E+02	1.6039E-04	1.2113E-06
0.20	4.9039E+00	4.8201E-05	9.7038E-07
0.25	2.3705E-01	2.2897E-05	8.2637E-07
0.30	3.1143E-02	1.3674E-05	7.2868E-07
0.35	7.2474E-03	9.3136E-06	6.5705E-07
0.40	2.4117E-03	6.8928E-06	6.0170E-07
0.45	1.0187E-03	5.3958E-06	5.5729E-07
0.50	5.0863E-04	4.3962E-06	5.2064E-07
0.55	2.8678E-04	3.6895E-06	4.8972E-07
0.60	1.7715E-04	3.1675E-06	4.6317E-07
0.65	1.1740E-04	2.7684E-06	4.4007E-07
0.70	8.2215E-05	2.4544E-06	4.1971E-07
0.75	6.0182E-05	2.2018E-06	4.0159E-07
0.80	4.5667E-05	1.9946E-06	3.8534E-07
0.85	3.5698E-05	1.8218E-06	3.7064E-07
0.90	2.8605E-05	1.6758E-06	3.5728E-07
0.95	2.3405E-05	1.5508E-06	3.4505E-07
1.00	1.9496E-05	1.4427E-06	3.3381E-07
1.10	1.4134E-05	1.2654E-06	3.1381E-07
1.20	1.0736E-05	1.1262E-06	2.9650E-07
1.30	8.4552E-06	1.0142E-06	2.8134E-07
1.40	6.8528E-06	9.2219E-07	2.6792E-07
1.50	5.6843E-06	8.4532E-07	2.5593E-07
1.60	4.8058E-06	7.8016E-07	2.4514E-07
1.70	4.1280E-06	7.2424E-07	2.3536E-07
1.80	3.5937E-06	6.7574E-07	2.2646E-07
1.90	3.1644E-06	6.3329E-07	2.1830E-07
2.00	2.8140E-06	5.9582E-07	2.1080E-07

Table A1. Values of β for p_0 = 0.05, 0.50, 0.95, and a selected set of α which solve Eq. (4) when λ_0 = 1.0 x 10^{-6}.

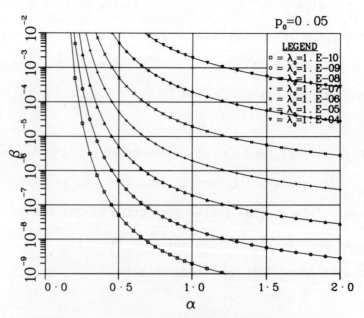

Figure A1. A graph of Table A1 for p_0 = 0.05 and a selected set of λ_0 values.

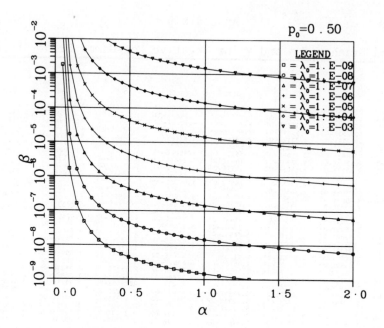

Figure A2. A graph of Table A1 for p_0 = 0.50 and a selected set of λ_0 values.

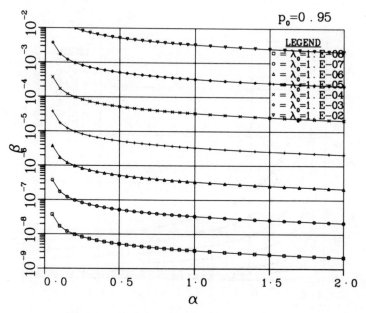

Figure A3. A graph of Table A1 for p_0 = 0.95 and a selected set of λ_0 values.

WALLER, JOHNSON, WATERMAN AND MARTZ

Appendix B.

Tables and Graphs of α and γ for Negative Log-Gamma Prior Distribution.

$p_0 = 0.05$

α\R₀	0.50	0.60	0.70	0.80	0.90	0.95	0.99
0.05	2.6065E+00	1.9209E+00	1.3412E+00	8.3910E-01	3.9619E-01	1.9288E-01	3.7793E-02
0.10	1.1942E+00	8.8007E-01	6.1450E-01	3.8444E-01	1.8152E-01	8.8370E-02	1.7315E-02
0.15	8.3958E-01	6.1874E-01	4.3203E-01	2.7028E-01	1.2762E-01	6.2129E-02	1.2174E-02
0.20	6.7261E-01	4.9569E-01	3.4611E-01	2.1653E-01	1.0224E-01	4.9774E-02	9.7526E-03
0.25	5.7279E-01	4.2213E-01	2.9474E-01	1.8440E-01	8.7066E-02	4.2387E-02	8.3053E-03
0.30	5.0508E-01	3.7223E-01	2.5990E-01	1.6260E-01	7.6774E-02	3.7376E-02	7.3235E-03
0.35	4.5543E-01	3.3564E-01	2.3435E-01	1.4662E-01	6.9227E-02	3.3702E-02	6.6036E-03
0.40	4.1707E-01	3.0736E-01	2.1461E-01	1.3427E-01	6.3395E-02	3.0863E-02	6.0473E-03
0.45	3.8628E-01	2.8468E-01	1.9877E-01	1.2436E-01	5.8716E-02	2.8585E-02	5.6009E-03
0.50	3.6088E-01	2.6595E-01	1.8570E-01	1.1618E-01	5.4854E-02	2.6705E-02	5.2326E-03
0.55	3.3945E-01	2.5016E-01	1.7467E-01	1.0928E-01	5.1597E-02	2.5119E-02	4.9218E-03
0.60	3.2105E-01	2.3660E-01	1.6520E-01	1.0335E-01	4.8800E-02	2.3758E-02	4.6551E-03
0.65	3.0503E-01	2.2480E-01	1.5696E-01	9.8198E-02	4.6366E-02	2.2572E-02	4.4228E-03
0.70	2.9092E-01	2.1440E-01	1.4970E-01	9.3655E-02	4.4220E-02	2.1528E-02	4.2182E-03
0.75	2.7836E-01	2.0514E-01	1.4324E-01	8.9612E-02	4.2312E-02	2.0599E-02	4.0361E-03
0.80	2.6709E-01	1.9684E-01	1.3744E-01	8.5985E-02	4.0599E-02	1.9765E-02	3.8727E-03
0.85	2.5691E-01	1.8933E-01	1.3220E-01	8.2706E-02	3.9051E-02	1.9011E-02	3.7251E-03
0.90	2.4764E-01	1.8251E-01	1.2743E-01	7.9724E-02	3.7643E-02	1.8326E-02	3.5907E-03
0.95	2.3917E-01	1.7626E-01	1.2307E-01	7.6995E-02	3.6354E-02	1.7699E-02	3.4679E-03
1.00	2.3138E-01	1.7052E-01	1.1906E-01	7.4487E-02	3.5170E-02	1.7122E-02	3.3549E-03
1.10	2.1751E-01	1.6030E-01	1.1193E-01	7.0024E-02	3.3063E-02	1.6096E-02	3.1539E-03
1.20	2.0552E-01	1.5146E-01	1.0575E-01	6.6162E-02	3.1240E-02	1.5209E-02	2.9799E-03
1.30	1.9501E-01	1.4372E-01	1.0035E-01	6.2779E-02	2.9642E-02	1.4431E-02	2.8276E-03
1.40	1.8571E-01	1.3686E-01	9.5559E-02	5.9784E-02	2.8228E-02	1.3742E-02	2.6926E-03
1.50	1.7740E-01	1.3073E-01	9.1283E-02	5.7108E-02	2.6965E-02	1.3127E-02	2.5722E-03
1.60	1.6992E-01	1.2522E-01	8.7434E-02	5.4701E-02	2.5828E-02	1.2574E-02	2.4637E-03
1.70	1.6314E-01	1.2023E-01	8.3948E-02	5.2520E-02	2.4798E-02	1.2073E-02	2.3655E-03
1.80	1.5697E-01	1.1568E-01	8.0772E-02	5.0533E-02	2.3860E-02	1.1616E-02	2.2760E-03
1.90	1.5132E-01	1.1151E-01	7.7863E-02	4.8713E-02	2.3000E-02	1.1197E-02	2.1940E-03
2.00	1.4611E-01	1.0768E-01	7.5187E-02	4.7038E-02	2.2210E-02	1.0813E-02	2.1186E-03

Table B1.   Table of γ values for p_0 = 0.05 and a selected set of values for α and R_0 in Eq. (7).

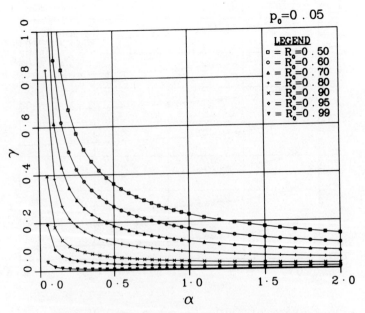

Figure B1.   A graph of Table B1.

$P_0 = 0.50$

$\alpha\backslash R_0$	0.7000	0.8000	0.9000	0.9500	0.9900	0.9990	0.9999
0.05	6.3990E+05	4.0034E+05	1.8903E+05	9.2024E+04	1.8031E+04	1.7950E+03	1.7942E+02
0.10	6.0108E+02	3.7605E+02	1.7756E+02	8.6441E+01	1.6937E+01	1.6861E+00	1.6853E-01
0.15	5.7207E+01	3.5790E+01	1.6899E+01	8.2269E+00	1.6120E+00	1.6047E-01	1.6040E-02
0.20	1.7192E+01	1.0756E+01	5.0785E+00	2.4724E+00	4.8444E-01	4.8225E-02	4.8204E-03
0.25	8.1668E+00	5.1093E+00	2.4124E+00	1.1745E+00	2.3012E-01	2.2908E-02	2.2898E-03
0.30	4.8772E+00	3.0513E+00	1.4407E+00	7.0139E-01	1.3743E-01	1.3681E-02	1.3675E-03
0.35	3.3219E+00	2.0783E+00	9.8129E-01	4.7773E-01	9.3605E-02	9.3183E-03	9.3141E-04
0.40	2.4585E+00	1.5381E+00	7.2623E-01	3.5356E-01	6.9275E-02	6.8963E-03	6.8932E-04
0.45	1.9246E+00	1.2040E+00	5.6851E-01	2.7677E-01	5.4230E-02	5.3985E-03	5.3961E-04
0.50	1.5680E+00	9.8099E-01	4.6319E-01	2.2550E-01	4.4183E-02	4.3984E-03	4.3964E-04
0.55	1.3160E+00	8.2330E-01	3.8873E-01	1.8925E-01	3.7081E-02	3.6914E-03	3.6897E-04
0.60	1.1298E+00	7.0682E-01	3.3373E-01	1.6247E-01	3.1835E-02	3.1691E-03	3.1677E-04
0.65	9.8741E-01	6.1774E-01	2.9168E-01	1.4200E-01	2.7823E-02	2.7698E-03	2.7685E-04
0.70	8.7544E-01	5.4769E-01	2.5860E-01	1.2590E-01	2.4668E-02	2.4557E-03	2.4546E-04
0.75	7.8534E-01	4.9132E-01	2.3199E-01	1.1294E-01	2.2129E-02	2.2029E-03	2.2019E-04
0.80	7.1143E-01	4.4508E-01	2.1015E-01	1.0231E-01	2.0046E-02	1.9956E-03	1.9947E-04
0.85	6.4980E-01	4.0653E-01	1.9195E-01	9.3448E-02	1.8310E-02	1.8227E-03	1.8219E-04
0.90	5.9770E-01	3.7394E-01	1.7656E-01	8.5955E-02	1.6842E-02	1.6766E-03	1.6758E-04
0.95	5.5312E-01	3.4605E-01	1.6339E-01	7.9544E-02	1.5586E-02	1.5515E-03	1.5509E-04
1.00	5.1457E-01	3.2193E-01	1.5200E-01	7.4001E-02	1.4500E-02	1.4434E-03	1.4428E-04
1.10	4.5133E-01	2.8236E-01	1.3332E-01	6.4906E-02	1.2718E-02	1.2660E-03	1.2654E-04
1.20	4.0169E-01	2.5131E-01	1.1866E-01	5.7767E-02	1.1319E-02	1.1268E-03	1.1263E-04
1.30	3.6174E-01	2.2631E-01	1.0686E-01	5.2022E-02	1.0193E-02	1.0147E-03	1.0143E-04
1.40	3.2892E-01	2.0578E-01	9.7163E-02	4.7302E-02	9.2684E-03	9.2265E-04	9.2224E-05
1.50	3.0150E-01	1.8863E-01	8.9063E-02	4.3359E-02	8.4957E-03	8.4574E-04	8.4536E-05
1.60	2.7826E-01	1.7409E-01	8.2198E-02	4.0017E-02	7.8408E-03	7.8055E-04	7.8019E-05
1.70	2.5832E-01	1.6161E-01	7.6306E-02	3.7149E-02	7.2788E-03	7.2460E-04	7.2427E-05
1.80	2.4102E-01	1.5079E-01	7.1196E-02	3.4661E-02	6.7914E-03	6.7608E-04	6.7577E-05
1.90	2.2588E-01	1.4131E-01	6.6724E-02	3.2483E-02	6.3648E-03	6.3361E-04	6.3332E-05
2.00	2.1252E-01	1.3295E-01	6.2776E-02	3.0562E-02	5.9882E-03	5.9612E-04	5.9585E-05

Table B2. Table of γ values for $P_0 = 0.50$ and a selected set of values α and R_0 in Eq. (7).

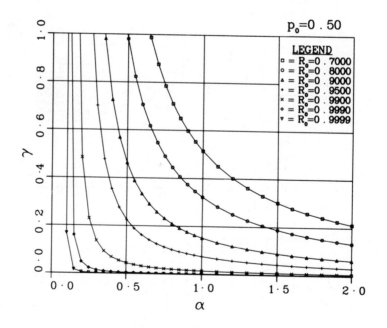

Figure B2. A graph of Table B2.

$P_0 = 0.95$

$\alpha \backslash R_0$	0.900000	0.950000	0.990000	0.999000	0.999900	0.999990	0.999999
0.05	1.8903E+25	9.2024E+24	1.8031E+24	1.7950E+23	1.7942E+22	1.7941E+21	1.7941E+20
0.10	1.7765E+12	8.6488E+11	1.6946E+11	1.6870E+10	1.6862E+09	1.6861E+08	1.6861E+07
0.15	7.8863E+07	3.8393E+07	7.5227E+06	7.4888E+05	4.9041E+02	4.9039E+01	4.9039E+00
0.20	5.1667E+05	2.5153E+05	2.3824E+03	2.3717E+02	2.3706E+01	2.3705E+00	2.3705E-01
0.25	2.4975E+04	1.2159E+04	2.3824E+03	2.3717E+02	2.3706E+01	2.3705E+00	2.3705E-01
0.30	3.2812E+03	1.5974E+03	3.1299E+02	3.1158E+01	3.1144E+00	3.1143E-01	3.1143E-02
0.35	7.6359E+02	3.7174E+02	7.2838E+01	7.2510E+00	7.2477E-01	7.2474E-02	7.2474E-03
0.40	2.5409E+02	1.2370E+02	2.4238E+01	2.4129E+00	2.4118E-01	2.4117E-02	2.4117E-03
0.45	1.0734E+02	5.2255E+01	1.0239E+01	1.0193E+00	1.0188E-01	1.0188E-02	1.0187E-03
0.50	5.3589E+01	2.6089E+01	5.1119E+00	5.0888E-01	5.0865E-02	5.0863E-03	5.0863E-04
0.55	3.0216E+01	1.4710E+01	2.8823E+00	2.8693E-01	2.8680E-02	2.8679E-03	2.8678E-04
0.60	1.8665E+01	9.0868E+00	1.7804E+00	1.7724E-01	1.7716E-02	1.7715E-03	1.7715E-04
0.65	1.2369E+01	6.0216E+00	1.1799E+00	1.1746E-01	1.1740E-02	1.1740E-03	1.1740E-04
0.70	8.6622E+00	4.2171E+00	8.2629E-01	8.2256E-02	8.2219E-03	8.2216E-04	8.2215E-05
0.75	6.3408E+00	3.0869E+00	6.0484E-01	6.0212E-02	6.0185E-03	6.0182E-04	6.0182E-05
0.80	4.8115E+00	2.3424E+00	4.5897E-01	4.5690E-02	4.5670E-03	4.5667E-04	4.5667E-05
0.85	3.7611E+00	1.8310E+00	3.5877E-01	3.5715E-02	3.5699E-03	3.5698E-04	3.5698E-05
0.90	3.0138E+00	1.4672E+00	2.8749E-01	2.8619E-02	2.8606E-03	2.8605E-04	2.8605E-05
0.95	2.4660E+00	1.2005E+00	2.3523E-01	2.3417E-02	2.3407E-03	2.3406E-04	2.3405E-05
1.00	2.0541E+00	1.0000E+00	1.9594E-01	1.9505E-02	1.9497E-03	1.9496E-04	1.9496E-05
1.10	1.4892E+00	7.2498E-01	1.4205E-01	1.4141E-02	1.4135E-03	1.4134E-04	1.4134E-05
1.20	1.1311E+00	5.5068E-01	1.0790E-01	1.0741E-02	1.0736E-03	1.0736E-04	1.0736E-05
1.30	8.9084E-01	4.3370E-01	8.4978E-02	8.4594E-03	8.4556E-04	8.4552E-05	8.4552E-06
1.40	7.2201E-01	3.5150E-01	6.8873E-02	6.8562E-03	6.8531E-04	6.8528E-05	6.8528E-06
1.50	5.9890E-01	2.9157E-01	5.7129E-02	5.6871E-03	5.6846E-04	5.6843E-05	5.6843E-06
1.60	5.0634E-01	2.4650E-01	4.8299E-02	4.8082E-03	4.8060E-04	4.8058E-05	4.8058E-06
1.70	4.3493E-01	2.1174E-01	4.1488E-02	4.1301E-03	4.1282E-04	4.1280E-05	4.1280E-06
1.80	3.7863E-01	1.8433E-01	3.6117E-02	3.5955E-03	3.5938E-04	3.5937E-05	3.5937E-06
1.90	3.3341E-01	1.6231E-01	3.1804E-02	3.1660E-03	3.1646E-04	3.1644E-05	3.1644E-06
2.00	2.9649E-01	1.4434E-01	2.8282E-02	2.8154E-03	2.8142E-04	2.8140E-05	2.8140E-06

Table B3.   Table of γ values for p_0 = 0.95 and a selected set of values for α and R_0 in Eq. (7).

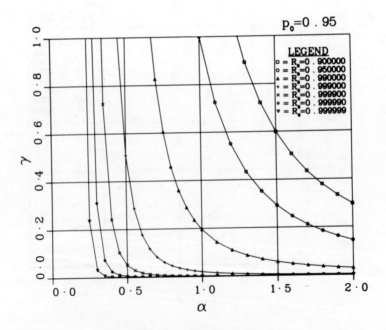

Figure B3.   A graph of Table B3.

Appendix C.

Mathematical Results.

C.1. A theorem on the asymptotic behavior of gamma percentiles.

Theorem: If, as $\alpha \to 0^+$, $K(\alpha)$ is defined by

$$\frac{1}{\Gamma(\alpha)} \int_0^{K(\alpha)} e^{-x} x^{\alpha-1} \, dx = p$$

and if $0 < p < 1$, then

$$K(\alpha) \sim p^{\frac{1}{\alpha}},$$

where $f(\alpha) \sim g(\alpha)$ means $\lim\limits_{\alpha \to 0} \dfrac{f(\alpha)}{g(\alpha)} = 1$.

Proof: Consider

$$F(\alpha) = \frac{1}{\Gamma(\alpha)} \int_0^K e^{-x} x^{\alpha-1} \, dx$$

for small α and fixed $0 < K < \infty$. Now

$$F(\alpha) = \frac{\displaystyle\int_0^K e^{-x} x^{\alpha-1} \, dx}{\displaystyle\int_0^K e^{-x} x^{\alpha-1} \, dx + \int_K^\infty e^{-x} x^{\alpha-1} \, dx}$$

and

(i) $\lim\limits_{\alpha \to 0^+} \int_0^K e^{-x} x^{\alpha-1} \, dx = +\infty$

(ii)   there is an M such that, for $0 < \alpha < \alpha_1$, $\int_K^\infty e^{-x} x^{\alpha-1} \, dx < M.$

Therefore, $F(\alpha) \to 1$ as $\alpha \to 0^+$. This implies, if $K(\alpha)$ satisfies

$$\frac{1}{\Gamma(\alpha)} \int_0^{K(\alpha)} e^{-x} x^{\alpha-1} \, dx = p \, ,$$

then $\lim\limits_{\alpha \to 0^+} K(\alpha) = 0$. Then, as $\alpha \to 0^+$,

$$p = \frac{1}{\Gamma(\alpha)} \int_0^{K(\alpha)} e^{-x} x^{\alpha-1} \, dx \sim \frac{1}{\Gamma(\alpha)} \int_0^{K(\alpha)} x^{\alpha-1} \, dx$$

$$= [\alpha \, \Gamma(\alpha)]^{-1} \, [K(\alpha)]^\alpha = [\Gamma(\alpha+1)]^{-1} \, [K(\alpha)]^\alpha \sim [K(\alpha)]^\alpha.$$

Therefore, the conclusion, $K(\alpha) \sim p^{1/\alpha}$, holds.

C.2. Differentiation of a negative-log gamma density. Let

$$k(r) = \frac{(-\ln r)^{\alpha-1} \, r^{\frac{1}{\gamma}-1}}{\gamma^\alpha \, \Gamma(\alpha)} \, , \qquad 0 < r < 1 \, .$$

Differentiation gives

$$\frac{\partial k(r)}{\partial r} = \frac{(-\ln r)^{\alpha-2} \; r^{\frac{1}{\gamma}-2}}{\Gamma(\alpha)\gamma^{\alpha+1}} \; [(\gamma-1) \ln r - (\alpha-1)\gamma] \; .$$

Now for $0 < r < 1$,

$$\frac{\partial k(r)}{\partial r} \begin{cases} > 0 \Leftrightarrow (\gamma-1) \ln r > (\alpha-1)\gamma \\ = 0 \Leftrightarrow (\gamma-1) \ln r = (\alpha-1)\gamma \\ < 0 \Leftrightarrow (\gamma-1) \ln r < (\alpha-1)\gamma \; . \end{cases}$$

The nine possible cases for different combinations of α and γ values are summarized in Table 2.

REFERENCES

[1] N. R. MANN, R. E. SCHAFER and N. D. SINGPURWALLA, Methods for Statistical Analysis of Reliability and Life Data, John Wiley & Sons, New York, London, 1974.

[2] M. D. SPRINGER and W. E. THOMPSON, Bayesian Confidence Limits for Reliability of Redundant Systems when Tests are Terminated at First Failure, Technometrics, 10(1965), pp. 29-36.

[3] M. D. SPRINGER and W. E. THOMPSON, Bayesian Confidence Limits for the Reliability of Cascade Exponential Subsystems, IEEE Transactions on Reliability, R-16(1967), pp. 86-89.

[4] N. R. MANN, Computer-Aided Selection of Prior Distributions for Generating Monte Carlo Confidence Bounds for System Reliability, Naval Research Logistics Quarterly, 17(1970), pp. 41-54.

[5] D. V. MASTRAN and N. D. SINGPURWALLA, A Bayesian Assessment of Coherent Structures, The George Washington University, School of Engineering and Applied Science Institute for Management Science and Engineering, Social T-293, 14 January 1974.

[6] M. O. LOCKS, Reliability, Maintainability and Availability Assessment, Hayden Books, Rochelle Park, NJ, 1973.

[7] N. L. JOHNSON and S. KOTZ, Continuous Univariate Distribution - 1, Houghton Mifflin Company, Boston, 1970.

[8] D. E. AMOS and S. L. DANIEL, Significant Digit Incomplete Gamma Ratios, SC-DR-72 0303, April 1972.

Topic 6
RELIABILITY APPROACHES AND PROGRAMS

DISCUSSION BY THE EDITORS

Topic 6
Reliability Approaches and Programs

In this section the papers present specific programs and approaches that are directed toward improvement of system reliability. In the first paper, G. L. Crellin, I. M. Jacobs and A. M. Smith, present an approach for utilizing experience data to improve nuclear power plant system performance by using the data to help determine how components acquire defects. Managing these defects requires the development of procedures for data evaluations as are presented in the paper.

In the second paper J. W. Bartlett, H. C. Burkholder and W. K. Winegardner present aspects of the program at Battelle Pacific Northwest Laboratories for investigating the safety of geologic repositories for nuclear wastes. The problems involved in this type assessment are unique because of the long time periods during which containment must be maintained. However, present risk assessment methods are recognized as being valuable although expected uncertainty in the calculated results is expected to be large.

In the next paper A. E. Green gives a study of the interaction of plant availability and safety system reliability. The problem of data is addressed. Results of analyses are given. Also, the role of target probability assignment for subsystems is discussed.

J. Graham and P. P. Zemanick present their view of the role of reliability and risk assessment in the design of liquid metal fast breeder reactors in the next paper. This paper emphasizes the importance of quantitative evaluations while pointing out existing limitations of these types of analyses. The paper is optimistic toward the possibility of using these assessments as a part on the licensing process.

The fifth paper in this section, by J. R. Penland, A. M. Smith and D. K. Goeser, addresses the use of reliability technology in the liquid metal fast breeder reactor industry. The paper emphasized the idea that the most effective use of reliability engineering techniques is for improving the design and operation of systems. The paper outlines the requirements for designing a reliability program.

The final paper is by R. C. Erdmann, J. E. Kelly, H. R. Kirch, F. L. Leverenz and E. T. Rumble. This paper described the "WAM" computer program for evaluating system reliability characteristics.

A DEFECT ANALYSIS PROGRAM APPLIED TO NUCLEAR PLANT EXPERIENCE DATA

G. L. CRELLIN, I. M. JACOBS AND A. M. SMITH*

Abstract. This paper deals with the initial phase of a development program for Defect Management in the design, construction and operation of nuclear power plants. The program objective is to capitalize on existing operating experience to form a basis for "lessons learned" applications to future nuclear power plant systems.

1. Introduction. The safety record of U.S. operating nuclear power plants is without peer in the world of complex industrial products; their record of the past twenty plus years is a clear testimonial to the success achieved in preventing even a remote hazard to the public. This is not to say that there have been no equipment failures but rather that these plants are designed to be tolerant of failure. The industry is proud of this impeccable safety record and continues to strive for improvement.

In spite of this record there is room for improvement. Failures do in fact occur, and when they do, the depth of defense against serious consequences is eroded. Broad industrial experience as well as our nuclear experience leads us to the fact that no complex product is totally defect** free. Thus, a basic issue of how to assure the

---

*Fast Breeder Reactor Department of the General Electric Company, P.O. Box 5020, Sunnyvale, California 94086.
**A defect is here defined to be the underlying cause, problem or error that can potentially lead to some form of product malfunction. A failure (or unusual occurrence or unplanned event) is simply the manifestation of the defect.

elimination of defects <u>before</u> a product reaches its operational phase
is still today a major consideration and factor to technician and
management alike in the structure of an efficient and comprehensive
product implementation plan.

This paper represents the early phase of work undertaken to
define a development program for Defect Management. Defect Management, in its simplest terms, is the development of a systematic
management approach to direct and utilize resources in an optimal
fashion for the control and elimination of defects. This, in turn,
implies that the nature of defect populations must be thoroughly
understood; that reasons for their origin must be clear; that the
way a system works to find them (or fails to find them) must be
ascertained.

As an initial step in this direction, emphasis is being placed
on "lessons learned" from existing operational experience; for
example, where do the generic weaknesses really lie, what are the
characteristics or dimensions of a defect population which are
important to understand, how do defects really originate, why do they
remain in the system so long, what can be done to prevent them or,
failing that, to detect their presence sooner and effectively treat
them? A basic understanding of the phenomena behind these questions
is central to the workings of Defect Management.

This paper discusses an approach being used to derive these "lessons learned" from the Light Water Reactor industry, and to note from such experience those factors which may be pertinent in the future design and operation of both Light Water Reactors and Breeder Reactors.

2. The Defect Management Concept.

A Mission Success Strategy. Central to the concept of Defect Management is the clear delineation of its objective and the formulation of the basic strategy by which this objective may be satisfied. Clearly, the overall objective of Defect Management is to have a product in use which fulfills its intended mission or function without compromise. Often, this mission is not easy to define to the total satisfaction of all concerned even though the product may perform to specification. Certain failures may be tolerable if the basic product function is not impaired prior to some predetermined time or end of mission. Failure in one of two redundant paths may be such a condition. Mission success definition, while not the subject of this paper, is important to safety and economic considerations, and it is appropriate to devote proper attention to its definition.

Given the mission success definition, the heart of the strategy now becomes the question of how to achieve this mission success. There are, in reality, only two ways to do this:

(1) Have a defect free product, or

(2) Have a product with the capability to mitigate the mission
     effects of defects that might ultimately lead to failure.

Defect free products just do not exist. Trade-off studies may be
employed to gauge the degree of resources that should be committed to
obtain a near defect-free product vs. the deliberate system design to
accommodate unplanned events or failures.

Actions to preclude product defects are twofold; first, there are
prevention actions that can be taken during the product definition
phase to preclude the introduction of defects in the specifications,
drawings, processes, manufacturing instructions, and operational
procedures; and second, there are removal actions that can be taken to
detect and screen out defects that may have been inadvertently introduced. These prevention and removal actions are thus taken before the
product goes into use, since such actions are aimed at providing a
defect free product to the customer.

Once the product is in service with the customer, certain preplanned actions are possible to mitigate the effects of unplanned
events or failures - and these again are basically two-fold in nature.
First, during the product definition phase, actions can be taken for
protection of the mission by including various types of redundancy,
diversity, fail to safety, and alternate mode features which actually
make the product "forgiving" if failures occur. This of course is a
very common practice in complex products such as nuclear power plants.

Second, the mission can be structured to allow for maintenance

actions - either preventive or corrective in nature. Both protection

and maintenance are preplanned actions and are built into the product

as a means of coping with defects that are passed on to the ultimate

user.

Thus, our overall Defect Management strategy has identified four

ways to cope with defects:

o   Prevent defects in the product outright.

o   Remove defects when they are detected before customer usage.

o   Protect against defects in service where prevention and

     removal have not been previously accomplished.

o   Maintain the product in service to preclude or correct for

     defects.

This strategy is shown on Figure 1, and represents the only four

options available to us.

Defect Flow. A study of product defects from historical data and

experience indicates that an understanding and control of defects is a

rather complex matter. This results from the fact that the intro-

duction of a defect into the product is usually a rather subtle

occurrence. True, certain defects (e.g. broken solder joints, gross

mismatch of tolerances preventing mechanical mating) become glaringly

obvious immediately - and here detection and correction is a simple

matter. But those defects which escape attention and go into the product's operational phase are the ones of real interest - and these are the essence of the problem which must be addressed.

In general, a defect is introduced into the flow of a product cycle at any one of a thousand places. Once there, it will stay in that flow until detected and corrected. Thus, defect flow is a real phenomena which occurs whenever a defect is not instantly caught at its point of introduction. For this situation, the flow must then contain various check points or screens whose job is to deliberately subject the product (or its definition documentation) to various rigors that will detect the presence of the defect. These screens, and the resources applied to both their number and quality, are the subject of management choice. Thus, we can begin to see the development of the options that confront us in the Defect Management strategy.

As the strategy develops further, two fundamental principles evolve for management attention. The first we shall call "Basic Quality Level" (BQL). This is simply a measure of the number of defects that are introduced during the product development/manufacture cycle. The objective is to have a high BQL which is achieved by having a diminishingly small number of defects introduced. Generally, as BQL increases, so does the cost of the product. For example, one way to reduce design defects is to utilize only senior engineers with extensive experience in the field. This approach of course is unpractical and uneconomic, but skill mix and methods of organization and management have nonetheless an important bearing on BQL. The same can be said for other functional organizations such as manufacturing, quality, test, and field operations. Whatever approach is used, a high BQL is desirable since the number of defects to be screened out is small, and the resulting possibility of defects escaping into operational use is diminished.

The second we shall call "Screen Effectiveness Level" (SEL). This is a measure of the effectiveness or ability of an organization to systematically search for, detect and eliminate defects that have been inadvertently introduced. The SEL is a parameter that can be applied to individual screens such as design review, specification/ drawing review, test, and inspection. A high SEL can also become costly; for example, comprehensive environmental testing. Generally, a high SEL is also desirable to assure a minimal escape of defects into operational use. Of course, the ultimate screen is operational use, and usually this screen is very effective at finding residual defects, but this is the wrong place to achieve our objective.

The Defect Management strategy is thus a trade-off between BQL and SEL. If a high BQL can be achieved, the necessity for costly screens diminishes. Conversely, a low BQL necessitates a very effective screening system. The proper mix is difficult to rigidly define and will differ from product to product and organization to organization. The only generalization that can be stated at this point is the necessity for management to recognize the importance of BQL and SEL, and above all to avoid the deadly combination of both a low BQL and a low SEL.

Defect Flow Analysis. One way of understanding the proper mix of BQL and SEL is to analyze past experience and to determine the nature of defect populations that historically occur on similar products. This process is called Defect Flow Analysis. In Defect Flow Analysis, the analyst basically takes a file of recorded product failures (hence known defects) and systematically evaluates them to determine their dimensional properties and the manner in which the screen system functioned or failed. Thus, we have Dimensional Analysis and Screen Analysis.

In Dimensional Analysis, the tacit assumption is made that similar products exhibit similar defect trends. Thus, if one can understand the dimensions or properties of previous defect populations, it is reasonable to assume that a similar product under development will exhibit like characteristics unless steps are taken to preclude a repeat of history. In Screen Analysis, we look to understand what the various screens did or did not provide in the way of defect detection, and to emulate or correct past history.

Let's illustrate these points with one example each from real life. Several firms have designed and built spacecraft systems. We are about to build a new and improved system. As an aid to

structuring the development and acceptance test program that is plan-

ned, the following questions are posed:

    (1)   What is the minimum level of assembly required to detect

           defects? (Dimensional Analysis)

    (2)   How strongly can traditional qualification tests be relied

           upon to qualify the design? (Screen Analysis)

For the first question, we select a group of operational (flight)

failures from existing spacecraft experience,and from our engineering

knowledge each failure point is categorized as shown on Figure 2.

The resulting message is very significant in that 50 percent of the

failures experienced _required_ a higher level of assembly (subsystem or

system) before the defect was even present and thus available for

detection. From a test screen point of view we should, based on this

observation, anticipate that component level testing is not sufficient;

that a significant portion of the defect population may well lie in

the assembled product only! While this is not welcome news due to the

more costly implications of higher level testing, it is nonetheless

very important to know that this dimensional property is character-

istic of this particular product. To neglect it could seriously

degrade the product. Since test resources are limited, it clearly

signals the necessity to do some system level testing, perhaps even at

the expense of component test deletion if such is necessary. (It

should be noted that not all products necessarily exhibit this
particular characteristic and that primary emphasis on component tests
alone may be fine if you have the knowledge to so guide that decision.)

For the second question, we select a group of system acceptance
test failures from an array of various spacecraft systems; each
failure is evaluated to determine if it could have been detected in
the previously conducted qualification test program. The results on
Figure 3 indicate a very disturbing pattern in that sizeable percentages of the acceptance test failures are shown to originate with
defects that could/should have been detected in qualification test.
The reasons for this are not clear from Figure 3 alone, but further
investigations of this data and their program history revealed that
small sample sizes (one or two at most) in the qualification test
program were a major factor. Why this? Essentially because various
problems of variability (in dimensions, material properties, process
control, test conditions, etc.) were not permitted to show themselves
in a small sample population. When the larger sample population
(i.e. 100%) was encountered in acceptance test, these problems revealed their presence. In summary, this example suggests that conventional qualification tests do not necessarily qualify the design;
subsequent testing continues to play a major role in product qualification and should be so treated. If this is indeed true but not
consciously recognized, the acceptance test design and the discipline

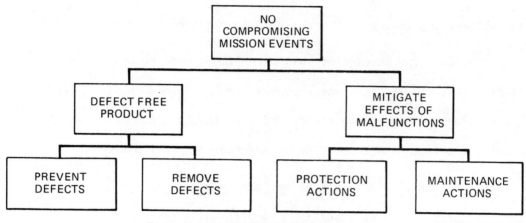

Figure 1. MISSION SUCCESS STRATEGY

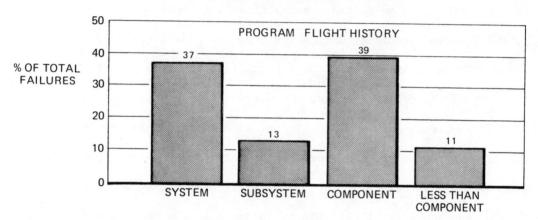

Figure 2. DEFECT OCCURENCE VERSUS LEVEL OF ASSEMBLY

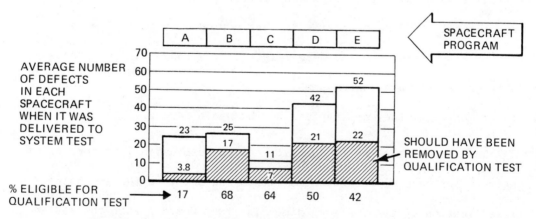

Figure 3. QUALIFICATION VERSUS ACCEPTANCE TESTING

in implementing it may not be all that is required.

The remaining portion of this paper is devoted to some preliminary Defect Flow Analysis that has been performed on Light Water

Reactor data. This is one element of work that has been initiated to

develop a Defect Management Program for nuclear power systems.

3. Defect Flow Analysis of Nuclear Experience. The first step

in this analysis was to identify a comprehensive source of failure

reports. The data base used was the Licensing Event Reports compiled

by the Nuclear Regulatory Commission from the Reportable Occurrence

reports filed by licensees. The reports list the name of the plant,

the date of the reported event, the date of the report, a description

of what occurred, and the disposition. The descriptions are brief and

generally informative. However, the brevity usually means that facts

which would have helped in the analysis are missing and other basic

information can only be obtained by inference. The data came from a

period between January 1, 1975 and April 29, 1976. There are

approximately 2200 data entries in that period of time. From that

base, a group of 63 entries were selected by a random number process

in order to obtain a random sample. The expectation is that the

results of this study are typical of the whole base of data.

The second step was to identify various dimensional properties of

interest for each of the reported events and set up a forced choice

classification system for each dimension that would permit the assign-

ment of each failure (defect) to a category that would be descriptive
for that dimension. For example, the systematic (repeatable)
tendencies of each event is a dimension of considerable interest.
Hence, each event was categorized as either a systematic or nonsystematic defect. If judgment indicated that the defect had $\geq$ 25
percent chance of recurring, it was classified as systematic. This
particular dimension (systematic tendency) is of some importance
since it identifies defects which are, in the main, part of the system
definition process (specification, drawing, procedures, etc.) and thus
are defects which will remain until the system definition itself is
corrected, as well as all affected hardware and software.

The third step was to identify a typical screen system that could
or should have been in place to detect the defect. These include
items such as design reviews, audits, acceptance tests, quality
control tests, preoperational tests, periodic tests, and self-
annunciation. Each screen was further analyzed as to the reason the
defect may have escaped the screen. For the data base used for this
paper, the screen analysis is highly subjective in that it presupposes the screens that were used. For example, a defect actually
discovered in operation should have, in the opinion of the analyst,
been discovered in a design review; and as a further refinement the
analyst may also have decided that design review was the "most
eligible screen", i.e. the one screen with the best opportunity for

detection. The analyst must then subjectively determine and code the most likely reason the defect escaped the (most eligible) design review screen, and, if possible, why it escaped other screens as well.

The fourth and final step is to accumulate the data in various groupings for evaluation purposes and to ascertain the messages that may be embodied therein. This is a discovery process aided by postulating a given correlation, then examining the data to see if the evidence supports the postulation.

In this analysis, some 20 dimensional characteristics were identified for evaluation. These include items such as equipment hazard classification, personnel hazard classification, radiological effect, source of incident, dominant cause, environmental sensitivity, failure mode, method of detection, alternate paths to success, prior history of failure, corrective action, reason for escaping detection screen, most eligible detection screen, skill level required to detect the defect, defect diversion status, reason for escaping a diversion screen, operational mode of plant, and years of service at time of defect discovery. There are carefully choosen force choice responses to each item. For example, the force choice responses to the method of detection item are meant to indicate how the defect was actually detected and include periodic test, self-annunciating, maintenance, walk around-visual, audit, accidental discovery, other, and no information. There follows a few of the observations that are typical of those that can be made.

Failure Mode, Personnel Hazard, and Systematic Tendency. Among
the many dimensions conveying information about defect population
characteristics are failure mode, personnel hazard, and systematic
tendency.

The failure mode dimension is assessed with respect to the
potential effect of the defect if the assumption is made that no
protective or mitigating systems are present. For purposes of this
analysis, defects were classified as revealing themselves either in
the safe or degraded modes of failure. Also, when the failure mode is
coupled with the actual personnel hazard associated with the defect,
a measure of the success of the various means to mitigate and/or
avoid hazards to personnel is revealed. In this analysis, the levels
of personnel hazard were taken as off-site radionuclide release,
on-site radionuclide release, no release, and other. It is significant to point out that while releases were observed, they were all
within allowable limits.

The systematic tendency of defects, as previously mentioned,
deals with the likelihood of such defects remaining in the operating
plants. Means of detection and actions to divert (correct) the
defects are influenced greatly by this systematic characteristic.
Thus, a knowledge of this particular dimension is imperative if
appropriate (permanent) corrective action is to result. Past

experience indicates that all too often defects are treated unknow-
ingly as nonsystematic; one-of-a-kind disposition results, and the
problem remains only to recur again and again.

Figure 4 combines the significant make-up of the population of
63 analyzed defects with respect to the three aforementioned dimensions. The majority of the population failure modes (54/63 = 86%)
were classified as degraded. This indicates that the LER's are properly attuned to those events which are potentially critical. The
characteristic of personnel hazard is shown by the double cross-
hatched bar which represents those defects which had some release
associated with them. Note that for the degraded mode, less than 10
percent (5/54) resulted in any release. Not shown is the fact that
none of these releases exceeded allowable limits. Thus, we might
conclude that attention is being placed in the correct areas and
that the various levels of protection are effective in avoiding
hazards.

Also shown in Figure 4 is the division of the population into
systematic and nonsystematic defects. About 62 percent of the total
defect population is systematic; about the same proportion of
systematic defects is seen in the degraded category. This is quite
revealing and indicates that a significant proportion of the degraded
defects would continue to surface unless the corrective actions taken
to resolve these defects are permanent in nature. Of even more

significance is that all five of those defects having release hazard
are also systematic.

Thus, while protection seems effective, and the industry should
be proud of its record, there is a subtle implication that unless the
proper actions are taken to remove systematic defects, a repeat
occurrence would be expected at some future time. The general nature
of systematic defects is such that simple redundancy does not
universally provide protection. We thus see emerging one element of
the problem commonly referred to as the Common Cause Failure problem.
Thus, some 60 percent of the degraded problems may have some relation
to Common Cause Failure; the success in mitigating the hazard
indicates significant success at providing diversity and/or functional
(not-similar) redundancy.

Defect Distribution vs Plant Age. Figure 5 shows the data sample
as a function of the age of the plant at its last birthday. The date
of birth is the date of first criticality*. The general trend is
something of an exponential decay with reported events even in the
14-15 year old bracket. However, this direct result presents a
biased picture because there are relatively more young plants than
old.

This young-old bias is normalized in Figure 6 by adjusting for
the total number of plants of a given age at the time of a given event.

*Commercial Nuclear Power Plants, NUS Corporation, Edition No. 8,
January, 1976

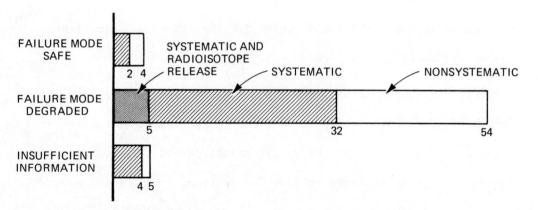

Figure 4. NUMBER OF EVENTS AS A FUNCTION OF FAILURE MODE

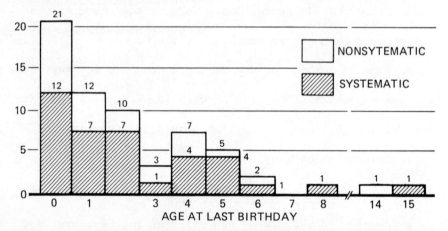

Figure 5. NUMBER OF EVENTS IN SAMPLE AS A FUNCTION OF AGE OF PLANT

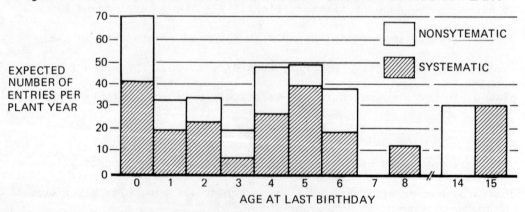

Figure 6. EXPECTED TOTAL NUMBER OF EVENTS/YEAR/ PLANT AS A FUNCTION OF AGE OF PLANT

The data is also adjusted to reflect the total expected entries per year in the data base instead of dealing only with the 63 in the study sample. This latter adjustment assumes that the data sample represents identically the distribution of defects vs. plant age for the entire data base; this assumption remains to be verified but is felt to be reasonably valid. The result is surprisingly much more uniform. The missing years at age 7, 9, 10, 11, 12, and 13 represent the several years when no new plants were being started.

The remarkable finding from Figures 5 and 6 is that systematic failures consistently constitute a major portion of the population and are being discovered throughout the whole span of ages, not predominantly in the younger plants as might be expected. This tends to also support a supposition that design type defects (not workmanship type defects) are a major cause of operational difficulties, an observation that has been made in other nonnuclear product lines.

Skill Level Required to Detect Defect in Most Eligible Screen.
A subjective judgment was made as to the skill level that would be required to detect the defect in the screen judged to be most eligible (i.e. the screen with the best opportunity for detection).
Three skill levels are defined. The reference or median level is that of the cognizant engineer, a level associated with the nominal skills of a component designer. At a lower skill level is the person who understands how the component functions, but could not be ex-

pected to design it. This lower skill level is designated simply as

less than cognizant engineer. The higher level is a senior engineer

who has had a depth of experience that gives him the insight into

subtleties of the design not readily recognized by the cognizant

engineer.

The results of this analysis are shown in Figure 7. The lowest

skill level is adequate in over 70 percent of the events and a senior

engineer is required for about 6 percent of the cases. The ratio of

less than cognizant engineer to senior engineer is greater than 10 to

1, an indication that manpower resources can be deployed efficiently

according to skill level.

Another observation from Figure 7 is that the systematic failures

become a higher proportion of the total in each skill level category

as the skill level increases. This observation tends to reinforce

an intuitive feeling that systematic defects are generally more

difficult to detect.

Timeliness of Defect Detection. Each of the 63 reports analyzed

represents a defect that was detected by one means or another. For

each reported event, the actual method of detection was noted. In

addition, for this analysis, a subjective judgment was made as to the

most eligible method of detection. This most eligible method of

detection (or screen) was conceptualized as a screen that would be ideal for detecting that defect, regardless of whether the screen was actually in place or not. Finally, for each event, the actual method of detection as reported was compared to the most eligible method and a judgment made as to whether it was detected early, on time, or late as referenced to the aforementioned most eligible screen.

The results of this investigation are shown in Figure 8. In eight reported cases, the actual detection was earlier than the most eligible screen. Typically, this might be a component whose failure was self-annunciated through normal operational processes rather than waiting on a periodic test. In most cases, 31 in all, the method of detection and the most eligible screen agree and detection was considered to be on time. The most frequent entry was a defect caught in a periodic test where the periodic test was also listed as the most eligible screen. A total of 20 defects were not detected in their most eligible screen but were found at a later time. As an example, a defect was caught in periodic test which ideally should have been caught in an acceptance test. Another example; a defect caught in an operational audit that ideally should have been detected and diverted in a design review. There are four cases in which the available information is insufficient to make a meaningful determination.

The lessons that may be learned from Figure 8 are these. In about two-thirds of the events reported, the method of detection was timely,

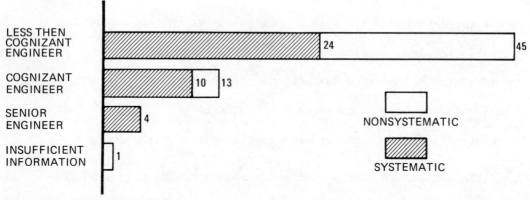

Figure 7. NUMBER OF EVENTS AS A FUNCTION OF SKILL LEVEL REQUIRED TO DETECT DEFECTS

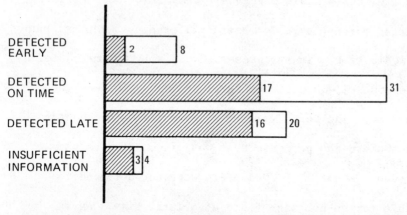

Figure 8. NUMBER OF EVENTS AS A FUNCTION OF TIMELINESS OF DETECTION

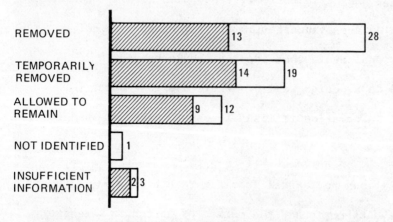

Figure 9. NUMBER OF EVENTS AS A FUNCTION OF DEFECT DIVERSION

either on time or early. But by the same token, it can be observed that the most eligible screen for these events occurred <u>after</u> the plant was operational; ideally it is more practical to have the most eligible screen situated in the design, development or pre-operational phase.

<u>Effectiveness of Defect Diversion</u>. An important element of a defect management system is that of detection. However, an industry cannot bring about a maturity of design unless action is also taken to permanently divert (remove) those defects which have been detected. This is especially true of defects which exhibit the dimension of systematic behavior.

Figure 9 displays the "diversion status" of the 63 analyzed defects. Somewhat less than half (44%) of the defects were actually diverted (removed) from the plant/design, while for 51 percent, the defects were only temporarily removed, allowed to remain, or were not identified. The remainder had insufficient information to determine the diversion status.

The breakdown into systematic/nonsystematic reveals some interesting facets. For the 25 nonsystematic (i.e., random) defects most (15/25) were removed. It is normally expected that corrective actions for nonsystematic defects attack the failed item directly and result in defect removal even though it might randomly occur again. Of interest, here, is that about 40 percent of the nonsystematic

defects were allowed to remain or were only temporarily removed and
thus the basic defect apparently still exists in the plant.

Finally, it is seen that for the 38 systematic defects, only 13
(34%) were actually removed and 23 (61%) were only temporarily
removed or were allowed to remain. The nature of systematic defects
is such that corrective actions required are usually more time consuming to implement (requiring design and/or procedural changes) and
thus it may be that corrective actions were in progress. However,
in order to take appropriate action requires recognition of the basic
systematic nature of the defect. This recognition did not appear to
visibly exist and one may thus conjecture whether appropriate action
was in fact implemented.

4. Conclusions. Defect Management is a systematic approach to
the measurement, control and elimination of defects in a product. In
the nuclear industry, the available experience data, mostly from Light
Water Reactors, are abundant, but often scattered and unorganized.
Defect Management provides a way to organize this information, draw
conclusions as to the effectiveness of past practices, and plan for
the effective allocation of resources and the development of meaningful program tasks and procedures.

The techniques of Defect Management are particularly important
in the nuclear industry where there is no data on ultimate system
failure and we must make our observations from discovered defects

that are scattered throughout a wide variety of equipments. The method uses subjective judgment where appropriate to classify existing information, but the subjective judgment is exercised principally to fill in missing details that can be inferred from the circumstantial evidence.

Although Defect Management has found application on aerospace projects, it is a relatively new concept for the nuclear industry. The specific results of this study must still be viewed as preliminary. However, the overview is that Defect Management is a concept that has much to offer.

SAFETY ASSESSMENT OF GEOLOGIC REPOSITORIES FOR NUCLEAR WASTE

J. W. BARTLETT, H. C. BURKHOLDER AND W. K. WINEGARDNER*

Abstract. Consideration of geologic isolation for final disposition of radioactive wastes has led to the need for evaluation of the
safety of the concept. Such evaluations require consideration of
factors not encountered in conventional risk analysis: consequences
at times and places far removed from the repository site; indirect,
complex, and alternative pathways between the waste and the point of
potential consequences; a highly limited data base; and limited
opportunity for experimental verification of results.

R&D programs to provide technical safety evaluations are under
way. Three methods are being considered for the probabilistic aspects
of the evaluations: fault tree analysis, repository simulation
analysis, and system stability analysis. Nuclide transport models,
currently in a relatively advanced state of development, are used to
evaluate consequences of postulated loss of geologic isolation.

This paper outlines the safety assessment methods, unique features of the assessment problem that affect selection of methods and
reliability of results, and available results. It also discusses
potential directions for future work.

1. Introduction. Western nations with significant commitments

to nuclear power currently anticipate that long-lived radioactive

wastes will be managed by isolation in geologic formations. This

management method acknowledges current lack of practical methods to

eliminate the wastes from existence on earth. It aims at eliminating

---

*Battelle Pacific Northwest Laboratories, Richland, WA 99352. This
research was supported by the U.S. Energy Research and Development
Administration under Contract EY-76-C-06-1830.

636

or minimizing potential for adverse consequences in the biosphere by separating the wastes from man and his environment until that potential is negligible.

All waste management operations preceding final commitment of wastes to geologic isolation can be accomplished with known technology. The safety of these operations can be estimated using techniques such as those used to estimate risk for reactor operations. Engineered designs to satisfy safety requirements can be developed, although perhaps at high cost.

In comparison with safety evaluations for design and operation of conventional engineered systems, evaluations for geologic isolation must incorporate some unusual considerations. They include: an analysis time frame related to half-lives of long-lived ($T_{1/2} \approx 30$, 25,000, and 400,000 yr) nuclides; potential for radiological consequences at locations remote from the repository and at times long after emplacement of wastes; and reliance on geologic rather than engineered structures for containment. In addition, the scope and content of the evaluations are potentially influenced by nontechnical issues such as temporal and geographic export of risk; longevity of human institutions; and the choice between concentration and dispersion of nuclide inventories. These issues are related to what may be conveniently summarized as "perceived risk." Much has been and will be written on these issues. They are beyond the scope of this paper but cannot be neglected.

2. <u>Technical Basis for Safety Estimates for Geologic Isolation</u>. Radioactive wastes are produced in a variety of forms in the nuclear fuel cycle. Actual quantities and types will depend on the way the fuel cycle is operated (e.g., once-through or U + Pu recycle). Wastes requiring isolation can be subjected to treatment, interim storage, and transportation operations prior to isolation. Conventional inductive and deductive safety analyses and safety assurance actions apply to these operations.

Criteria for geologic isolation are not yet firmly established, but lack of firm criteria is relatively unimportant to this discussion. What is important is the intent to isolate those nuclides which have long half-lives and high toxicity. The usual focus of public attention is plutonium. Other radioactive materials, including plutonium decay products such as ^{226}Ra, the fission products ^{129}I and ^{99}Tc, and the activation product ^{14}C may also be important.

To date, attention has been directed primarily at so-called "high-level" waste, the fission product waste resulting from Purex reprocessing of spent fuel to recover uranium and plutonium. This waste contains more than 99% of the radioactivity in fuel cycle wastes as measured a year or so after fuel discharge from the reactor.

It is expected to contain about 0.5% of the uranium and plutonium in spent fuel. In the long term, after decay of fission products cesium and strontium (ca 800 yr), radioactivity levels in the high-level waste are greatly reduced and associated primarily with the long-lived transuranics.

The high-level waste will contain on the order of 50% of all transuranic nuclides expected to be consigned to repositories for a fuel cycle operated to recover uranium and plutonium. Other waste forms such as cladding hulls and consolidated trash from MOX and reprocessing plants would contain the remainder of the transuranics. In a once-through fuel cycle, the uranium and plutonium levels in the spent fuel, which would be the waste form, would be about 200 times higher than that in high-level waste.

High-level and other transuranic wastes can be expected to be placed into repositories as solids. Specific final form requirements have not yet been established. They may depend somewhat on results of safety analyses such as those discussed here.

In broadest terms, technical safety assessment for geologic repositories of nuclear wastes requires analysis of two factors: loss of isolation, and its consequences in the biosphere. Loss of isolation is dependent on disruptive events (natural phenomena and human intrusion). Consequences in the biosphere can depend on

several factors: the mode of loss of isolation, transport in the
geologic medium, and transport and interactions in the biosphere.
For some potential loss-of-isolation scenarios such as extreme
faulting and meteorite impact, the repository can be postulated to be
directly exposed to the biosphere. For most, however, nuclides must
move from the repository through the geosphere and subsequently the
biosphere.

Excepting scenarios based on human intrusion of the repository,
the most likely scheme for transporting wastes to the biosphere with
adverse consequences is expected to be leaching of the waste by
groundwater in convective motion. Under this scenario, potential
consequences in the biosphere will depend on leach rates of the waste
form, age of the waste when release occurs, transport in the geosphere,
and pathways in the biosphere.

In this paper, past and anticipated analyses based on the various
scenarios are reviewed and discussed. Risk concepts, probability
estimation, and consequence estimation are emphasized.

3. Concepts of Risk for Geologic Repositories. The basic,
conventional concepts of technical risk estimation (i.e., evaluation
of likelihoods and consequences) apply to long-term safety assessments
for geologic repositories. Modified use of the concepts is necessary, however, because consequences of disruptions potentially

producing adverse consequences may not be immediate or occur at the location of the risk-producing event. Furthermore, the "event" potentially producing risk may be a slow, extended process rather than a discrete occurrence.

The technical estimations of potential long-term radiological risk from geologic repositories must include consideration of:

- delays between risk-initiating "events" and occurrence of consequences (e.g., as a result of slow nuclide transport in the geosphere)

- geographic translocation (consequences can occur at locations distant from the repository because of nuclide migration)

- risk dispersion (atmospheric resuspension, surface water transport, etc., can produce biosphere transport of the radioactive nuclides)

- nuclide accumulation in the biosphere (quantities of nuclides at a given location may increase with time because of continued input from the repository, slow decay, and lack of biosphere dispersion phenomena)

- populations and lifestyles (total human risk will depend on how many people are present and how they interact with radioactivity in the biosphere).

R&D programs aimed at developing risk estimation methods that account for these considerations are under way in the U.S. and other countries.

4. Probability Estimation Methods. Identified phenomena that have been postulated to potentially have an effect on the stability of a geologic repository are of four types:

1. rapid man-caused events (e.g., resource exploration or acts of war)

2. rapid natural events uninfluenced by humans (e.g., meteorite impact or seismically induced fracturing)

3. slow natural processes independent of the existence of the repository (e.g., tectonic activity or erosion)

4. slow processes caused by the existence of the repository (e.g., faulting caused by radiation, thermal and/or mechanical force gradients).

Three methods have been proposed to assist in identifying and evaluating potential loss-of-isolation scenarios involving the above phenomena:

1. fault tree analysis[1]

2. repository simulation analysis[2]

3. system stability analysis.[2]

Fault tree analysis is a technique by which the component failures leading to system failure can be logically deduced. Application of the technique yields combinations of basic events whose occurrence causes the undesired failure events. These event combinations can then be evaluated by various screening techniques to determine the high risk scenarios and their probability of occurrence. A risk-based fault tree analysis method for the identification, preliminary evaluation, and screening of potential accidental release sequences is presented in Reference 3. For its application the fault tree method requires probability information about all of the individual component failures. The fault tree technique is well suited to analyzing the rapid events (which have discrete probabilities). It is not well-suited to analyzing the slow processes for which event ordering, interdependencies, and time-phasing are important.

Repository simulation analysis begins by listing all phenomena in the four types, estimates probabilities for the rapid events and rates for the slow processes, and defines the generic failures which are possible.[2] The analysis then steps through time assuming random occurrence of the rapid events at the estimated probabilities and continuous occurrence of the slow processes at the estimated rates. This procedure simulates repository behavior and if applied until one of the postulated failures occurs produces an estimate of the time and conditions of repository failure. Performance of this repository simulation a large number of times produces a distribution

of times and conditions of repository failure. From this distribu-

tion the probabilities of failure in any given year can be calculated.

   System stability analysis begins by mathematically modeling the

natural geologic processes and the processes induced by the presence

of the waste (i.e., processes caused by introducing radiation, ther-

mal, and/or mechanical force gradients into the previously undisturbed

geology).[2] This modeling activity produces a set of coupled dif-

ferential equations (along with mathematical expressions for the

appropriate boundary conditions) which describe not only the indi-

vidual processes but also their interaction. Linear and nonlinear

stability analyses are then performed on the model system. The

linear analyses determine the systems response to small arbitrary

disturbances which are ever present in all systems. The nonlinear

analyses determine the systems response to large arbitrary distur-

bances caused by the occurrences of rapid events (e.g., earthquakes).

If the modeled system damps the disturbances, it is said to be stable;

if the system allows the disturbances to grow in amplitude, it is

said to be unstable and the analysis proceeds to determining the rate

of approach to nuclide release. The results of such instability

could be manifest in occurrences such as local fracturing, metamorpho-

sis of the local geology with subsequent repository collapse, movement

of the respository as a whole, or movement of individual canisters

with subsequent meltdown. The output of the analysis is either a conclusion that the system is stable or a conclusion that the system is unstable with release expected at some given time. The analysis is performed repetitively over the expected range of the important input parameters and produces a distribution of predicted failure times (an ever-stable system will have a null distribution) from which failure probabilities can be estimated. The system stability analysis technique may be adequate for analyzing the slow processes and their interactions, but it is not well suited for analyzing the rapid events whose processes are difficult to model by continuum mathematics.

None of the three analysis techniques taken alone seems ideal for analyzing the failure of geologic repositories, but together they provide a reasonably comprehensive analysis method. Thus a prudent plan for this portion of the repository safety analysis problem might include investigation of all three techniques simultaneously. The multiple technique approach is particularly justified because the repository simulation and system stability analysis techniques have not advanced beyond the conceptual stage and may not be ultimately successful. All techniques (but especially stability analysis) may require a detail of knowledge about the natural geologic processes which does not presently exist.

5. Probability Estimations for Geologic Repositories. All four types of loss-of-isolation phenomena have been investigated. Schneider

and Platt (1974) report a detailed fault tree approach to the problem.[1] These fault trees were not evaluated because of time constraints and a lack of probability information about the component
events. A simplified fault tree approach with enough detail to
permit interaction of all four types of phenomena but not so much
detail that the trees are too sophisticated for the quality of the
data has not been documented. Claiborne and Gera (1974) estimated
the individual probabilities of some Type 1, 2, and 3 phenomena for
the proposed Carlsbad Repository in southeastern New Mexico.[4] The
probability of a catastrophic meteorite impact was estimated to be
1.6×10^{-13}/yr, and the probability of faulting (which would not necessarily lead to containment failure) was estimated to be 4×10^{-11}/yr.
The interaction of these various events was not investigated, and
hence the significance of such interaction is not clear at this time.
Additional work on the first three types of phenomena can be found in
References 1 and 5. A number of investigations of some Type 4 processes have been completed.[1,6-10]

6. Consequence Estimation Methods. Estimation of radiological
consequences requires modeling of radionuclide transport from the
point where the nuclides become mobile to the point(s) of interaction
with humans. Radiation doses are then calculated. Radionuclide
transport modeling requires an understanding of the important physico-
chemical, biophysical, and biochemical processes which govern the

concentration and rate of nuclide movement and their interaction with humans. These processes must be known with sufficient accuracy to be described by abstract mathematical relationships and to be characterized by experimentally measurable parameters. Humans are affected by radionuclides either by external exposure from nuclides in the environment or by internal exposure from nuclides which have been ingested or inhaled. Thus water and atmospheric transport models as well as radiation dose models are needed. Models of both types exist, and comprehensive compilations of each type are under way at the present time. Examples of integrated water-dose and air-dose models can be found in References 11 and 12. These integrated consequence estimation models require input information concerning the time and conditions of nuclide release and produce output information about projected doses to individuals and populations at various times after failure. The models also require such input data as: radionuclide inventories; waste form volatilities and leach rates; groundwater and surface water directions, velocities, and volumetric flow rates; wind directions, durations, and velocities; sorption properties of geologic media; biosphere transfer and dose conversion factors; demography; and population living habits.

These requirements are addressed by geosphere transport models which have been specially developed for geologic isolation studies. Lester, Jansen, and Burkholder (1975) developed an analytical model

to describe the transport of radionuclides by flowing groundwater following a leach incident at an underground nuclear waste disposal
site.[13]   This model is applicable to particulate and faulted monolithic media and can be applied to heterogeneous media if a weighted
averaging technique can be applied to the relevant input parameters.[14]   A computer code named GETOUT has been developed to implement this model.

Burkholder and DeFigh-Price (1977) developed a companion model
to describe the diffusion of radionuclides in either the liquid or
gas phase from a geologic or seabed nuclear waste disposal site.[15]
All of the assumptions of the previous model were made except that
convective transport was neglected and diffusion was modeled instead
of dispersion. The generalizations of the migration model to faulted
monolithic and/or heterogeneous media applied to the diffusion model
as well. A computer code named DIFFUS has been developed.

A model to describe geosphere transport from a salt formation is
under development.[16,17]   This model describes both the transport of
radionuclides and dissolved salt following groundwater intrusion into
a geologic isolation site in a salt formation.

Together these models provide the framework for modeling the
important geosphere transport situations of geologic isolation. When
combined with already existing models for surface water transport,

atmospheric transport, and dose calculation, they provide a complete
set of models needed to estimate the consequences of disruptive
events at geologic repositories.

7. Estimation of Radiological Consequences. A number of investigations have described methods for making consequence estimations for
geologic isolation[1,4,12,18,19] but only three have thus far produced
specific dose estimates.[4,12,19] Claiborne and Gera (1974) estimated
the local contamination consequences of meteorite impact on a hypothetical nuclear waste repository located in a bedded salt formation
in southeastern New Mexico. The meteor was assumed to be of sufficient size to exhume the entire repository, and impacts were assumed
to occur 1,000 yr and 100,000 yr after disposal. Doses were calculated using a generalized model to predict quantitative radionuclide
movement through terrestrial food pathways (TERMOD) and dose codes
EXREM and INREM.[20,21,22] The radionuclide source inventory was not
specified. Fifty-yr bone (the critical organ) doses of 2.3×10^6
(event at 1,000 yr) and 9.8×10^6 (event at 100,000 yr) mrem were
predicted for individuals who moved into the local area after the
meteorite impact and reestablished a human society similar to our
present society. The dose controlling nuclide was ^{241}Am for the
1000-yr event and ^{226}Ra for the 100,000-yr event. Claiborne and Gera
noted that a cataclysmic meteorite impact would cause great devas-
tation, regardless of the simultaneous presence of radioactive wastes.

Carrying this line of reasoning further, it seems logical that any disruptive event for geologic isolation can be neglected if the consequences of the initiating event itself are much larger than the long-term integrated consequences of the released radionuclides. Claiborne and Gera also estimated the local consequences from ground-water intrusion into the repository at 1,000 and 100,000 yr after disposal. They assumed that the waste consisted of 2-mm-dia particles of borosilicate glass with a leach-rate of 10^{-7} g/cm^2 day and that 1% of the waste was contacted by groundwater. The resulting saturated salt brine with its contained radioactivity was assumed to be pumped directly to the surface and used to irrigate crops with no subsequent dilution. The crops were assumed to grow with the salt saturated water, and the surrounding populace was assumed to eat only contaminated crops. Fifty-yr bone doses of 5×10^5 (event at 1,000 yr) and 6×10^5 mrem (event at 100,000 yr) were predicted for these local individuals.

Burkholder, Cloninger, Baker, and Jansen (1975, 1976) used the geosphere transport code GETOUT and a currently unnamed biosphere transport-dose code containing the programs ARRRG and FOOD to predict the consequences of groundwater intrusion into a geologic isolation repository.[12,23-26] The radionuclide inventory in the repository was assumed to be accumulated high-level waste from the U.S. nuclear power economy through the Year 2000 including all tritium, carbon, and iodine from the spent fuel and all activation products from the

cladding. The surrounding geology was assumed to have the nuclide

retention properties of western U.S. desert subsoils as estimated in

Reference 1 and discussed in Reference 14. Following the intrusion

event, the groundwater and its dissolved nuclides were assumed to

flow at a velocity of 1 ft/day (10 times the velocity of "typical"

desert aquifers) back to the biosphere.

A "typical" biosphere scenario was assumed with nuclides entering

the biosphere where a surface river and underground aquifer inter-

sect. The nuclides were assumed to be diluted by a 10,000-cfs river

($\sim$1/10 the flowrate of the Columbia River) before exposure to humans

occurred. The nuclides were assumed to affect humans by external

exposure from shoreline activities and recreation on and in the river

and internal exposure from drinking the water and eating aquatic

foods or foods grown or raised using the river water for irrigation.

Doses were calculated to "maximum" individuals who were assumed to

live within the region of influence of the isolation site at various

times after disposal. In calculation of these doses, the nuclide

inflow rates to the biosphere at any time of interest were assumed to

have accumulated on the irrigated land at those rates for 50 yr. The

"maximum" individual was then assumed to be exposed to the results of

that 50-yr accumulation for another 50 yr. At the end of this 100-yr

period the nuclides were assumed to disperse throughout the regional,

national and international biosphere with no subsequent dose.

Thus the dynamic biosphere accumulation and dispersion phenomena were approximated by a steady-state accumulation. The 50-yr accumulation assumption seems conservative, but investigation is needed to provide supporting information.

In the Burkholder et al. studies, release consequences were calculated for combinations of three parameters. These parameters (called waste-management-control variables) are measurement scales for the effectiveness of the radioactivity release barriers (containment, leach resistance, and nuclide retention) provided by geologic isolation systems. The time that water first contacts the waste (the time of initial release) is a measure of canister integrity and site stability from water penetration. The time for complete dissolution of the waste (the leach time) is a measure of the resistance of the waste form to leaching. The distance that the dissolved nuclides move through the geologic medium in migrating from the isolation site to the biosphere (the path length) is a measure of geosphere isolation.

Burkholder (1976) extended the consequence estimations for high-level wastes to other nuclear fuel cycle waste such as spent fuel, TRU-contaminated cladding hulls, tritium solids, carbon solids, iodine solids, miscellaneous α, β, γ solids, miscellaneous β,γ solids, and ore tailings.[19] The results showed that for the situation investigated:

1. Spent fuel disposal requires greater assurance of nuclide isolation than high-level waste disposal because of the larger amounts of uranium and plutonium in the spent fuel.

2. Poor selection of isolation conditions can sometimes increase and good selection can greatly decrease the projected maximum dose consequences to humans compared to leaving the waste in the biosphere. The former effect is a result of the reconcentration phenomenon of radionuclide chain migration (discussed in Reference 27).

3. A strongly nonlinear relationship exists between the amount of radioactivity at a given site and the waste form leach resistance requirements necessary to meet specific site acceptability criteria. Thus multisite isolation of nuclear wastes can greatly reduce waste form leach resistance requirements for some wastes.

8. Constraints on This Use of Risk Analysis. Assessment of geologic isolation safety is unique relative to assessments for other engineered systems because some elements of the analysis are not amenable to uncertainty reduction by additional R&D. The limitations result from the long time periods of interest to the assessments, the hypothetical nature of the nuclide release scenarios, and limits on the ability to characterize the system without destroying its function. Examples are listed below for the groundwater intrusion scenario:

1. The circumstances surrounding the release initiation will be purely hypothetical because disposal sites will not be located where flowing water is initially present. Thus the origin, direction, and velocity of future groundwater flow cannot be accurately predicted.

2. The future dissolution or leach rate of the waste form will depend on the long-term stability of the waste form and the chemistry of interaction with intruding ground water.

3. Nuclide retention data for the geology in the immediate vicinity of the site will be constrained by need to avoid destroying the integrity of the site.

4. Capability for laboratory simulation of actual nuclide transport conditions will be limited.

5. Accumulation and dispersion processes for radionuclides in the biosphere over long time periods will always be uncertain.

6. The demography and living habits of future civilizations will always be conjecture.

Because of these limitations on input information, there will always be limitations on the accuracy and precision of the results of consequence estimations for geologic isolation. Consequently, appropriate sophistication levels for the models should be determined. Highly sophisticated models should be used with caution or perhaps not be used at all. They could create unwarranted "illusions of certainty."

Results of these safety assessments will be subjected to judgment in accordance with two types of acceptability criteria: 1) validity of the technical safety evaluations, and 2) social perceptions of what levels and types of risks are acceptable.

The above factors provide guidance on the direction and depth of development of safety assessments of this type. Experience to date suggests the following guidelines for future development:

- Sufficiency of scope of scenarios considered is more important than detail.

- Levels (i.e., magnitudes) of projected risks are more important than the precision of the estimates.

- Probabilities and consequences are important independently in addition to their multiplicative role as components of risk. Perceived risk tends to be highest at the extremes of the spectrum, i.e., for high P, low C scenarios and for low P, high C scenarios. These inequalities in perceived risk are the reason for interest in, and a role for, assessment methods such as consequence estimation.

- A primary target for the analyses should be estimation of the
  upper boundary of projected risks.

- Scenarios should include consideration of corrective action
  as a factor mitigating risk. For example, consequences of
  accidental intrusion of a repository during drilling can be
  minimized by stopping the activity. Consequences of broad
  dispersion in the environment can be reduced by control
  procedures and regulations. In general, time constants
  associated with repository releases are so large that there
  is ample time for corrective action.

- Potential consequences of loss of isolation should be placed
  in perspective. In a drilling intrusion, which presumably
  could result from searching for resources, accompanying
  analytic procedures should reveal the presence of transuranics; adverse consequences could then be limited by direct
  action. If released radioactivity is broadly dispersed in
  the environment, a release or its much-delayed entry to biosphere would be significant only if a significant perturbation
  in background resulted, could be detected, and produced
  adverse effects that could not be mitigated.

- A role should be anticipated for perceived risk in the acceptance of risk assessment results and, if possible, the performance of the assessments. At present, the level of perceived

risk associated with geologic isolation is unknown. The

issue is clouded by concerns associated with the fact

that isolation has not been put into practice.

In summary, present risk assessment methods provide a basis for

evaluation of geologic repositories, and there are guidelines for

future development of methods. The evaluations are essential as an

aid to system design and site selection because of the insights they

provide concerning factors that influence or control safety. The

evaluations also provide a framework from which to communicate those

insights to decision makers and the general public.

REFERENCES

[1] K. J. SCHNEIDER and A. M. PLATT, eds. High-Level Radioactive
     Waste Management Alternatives, BNWL-1900, Battelle Pacific
     Northwest Laboratories, Richland, WA, 1974.

[2] H. C. BURKHOLDER, J. A. STOTTLEMYRE, and J. R. RAYMOND, Safety
     Assessment and Geosphere Transport Methodology for the
     Isolation of Nuclear Waste Materials, BNWL-SA-6310, Battelle
     Pacific Northwest Laboratories, Richland, WA, 1977.

[3] T. H. SMITH, P. J. PELTO, D. L. STEVENS, G. D. SEYBOLD, W. L.
     PERCELL, and L. V. KIMMEL, A Risk Based Fault Tree Analysis
     Method for Identification, Preliminary Evaluation, and
     Screening of Potential Release Sequences in Nuclear Fuel
     Cycle Operations, BNWL-1959, Battelle Pacific Northwest
     Laboratories, Richland, WA, 1976.

[4] H. C. CLAIBORNE and F. GERA, Potential Containment Failure
     Mechanisms and Their Consequences at a Radioactive Waste
     Repository in Bedded Salt in New Mexico, ORNL-TM-4639,
     Oak Ridge National Laboratory, Oak Ridge, TN, 1974.

[5] F. GERA and D. G. JACOBS, Considerations in the Long-Term Manage-
     ment of High-Level Radioactive Wastes, ORNL-4762, Oak Ridge
     National Laboratory, Oak Ridge, TN, 1972.

[6] G. H. JENKS, Radiolysis and Hydrolysis in Salt-Mine Brines,
     ORNL-TM-3717, Oak Ridge National Laboratory, Oak Ridge, TN,
     1972.

[7] T. R. ANTHONY and H. E. CLINE, "Thermomigration of Liquid Drop-
     lets in Salt," in Northern Ohio Geological Society, Case
     Western Reserve University, Fourth Symposium on Salt, 1: 313,
     1974.

[8] K. A. HOLDOWAY, "Behavior of Fluid Inclusions in Salt During
     Heating and Irradiation," in Northern Ohio Geological
     Society, Case Western Reserve University, Fourth Symposium
     on Salt, 1: 303, 1974.

[9] G. H. JENKS, Gamma-Radiation Effects in Geologic Formations of
     Interest in Waste Disposal: A Review and Analysis of
     Available Information and Suggestions for Additional
     Experimentation, ORNL-TM-4827, Oak Ridge National Laboratory,
     Oak Ridge, TN, 1975.

[10] F. M. EMPSON, R. L. BRADSHAW, W. C. MCCLAIN, and B. L. HOUSER,
     "Results of the Operation of Project Salt Vault: A Demon-
     stration of Disposal of High Level Radioactive Solids in
     Salt" in Northern Ohio Geologic Society, Case Western Reserve
     University, Third Symposium on Salt, 1:455 1969.

[11] J. R. HOUSTON, D. L. STRENGE, and E. C. WATSON, DACRIN - A Com-
     puter Program for Calculating Organ Dose from Acute or
     Chronic Radionuclide Inhalation, BNWL-B-389, Battelle,
     Pacific Northwest Laboratories, Richland, WA, 1974.

[12] H. C. BURKHOLDER, M. O. CLONINGER, D. A. BAKER, and G. JANSEN,
     Incentives for Partitioning High-Level Waste, BNWL-1927,
     Battelle Pacific Northwest Laboratories, Richland, WA,
     1974.

[13] D. H. LESTER, G. JANSEN, and H. C. BURKHOLDER, "Migration of
     radionuclide chains through an adsorbing medium," AIChE
     Symposium Series No. 152, Adsorption and Ion Exchange,
     71:202, 1975.

[14] H. C. BURKHOLDER, "Methods and Data for Predicting Nuclide
     Migration in Geologic Media," Proceedings of the Inter-
     national Symposium on the Management of Wastes from the

LWR Fuel Cycle, CONF-76-0701, U.S. Energy Research and
Development Administration, p.658, 1976.

[15] H. C. BURKHOLDER and C. DEFIGH-PRICE, "Diffusion of Radionuclide
Chains Through an Adsorbing Medium," BNWL-SA-5787, Battelle
Pacific Northwest Laboratories, Richland, WA, 1977.

[16] H. C. BURKHOLDER and G. E. KOESTER, "Radionuclide Migration from
Salt Formations," in Nuclear Waste Management and Transportation Quarterly Progress Report January through March
1975, BNWL-1913, Battelle Pacific Northwest Laboratories,
Richland, WA, 1975.

[17] A. G. GIBBS and H. C. BURKHOLDER, "Radionuclide Migration from
Salt Formations," in Nuclear Waste Management and Trans-
portation Quarterly Progress Report April through June 1975,
BNWL-1936, Battelle Pacific Northwest Laboratories, Richland,
WA, 1975.

[18] S. F. LOGAN, A Technology Assessment Methodology Applied to
High-Level Radioactive Waste Management, PhD Dissertation,
University of New Mexico, Albuquerque, NM, 1974.

[19] H. C. BURKHOLDER, "Management Perspectives for Nuclear Fuel
Cycle Wastes," in Nuclear Waste Management and Transporta-
tion Quarterly Progress Report January through March 1976,
BNWL-2029, Battelle Pacific Northwest Laboratories, Richland,
WA, 1976.

[20] W. D. TURNER, S. V. KAYE, and P. S. ROHWER, EXREM and INREM
Computer Codes for Estimating Radiation Doses to Popu-
lations from Construction of a Sea-Level Canal with Nuclear
Explosives, K-1752, Oak Ridge National Laboratory, Oak
Ridge, TN, 1968.

[21] S. K. TRUBEY and S. V. KAYE, The EXREM III Computer Code for
Estimating External Radiation Doses to Populations from
Environmental Releases, ORNL-TM-4322, Oak Ridge National
Laboratory, Oak Ridge, TN, 1973.

[22] R. S. BOOTH, S. V. KAYE, and P. S. ROHWER, "A Systems Analysis
Methodology for Predicting Dose to Man from a Radioactively
Contaminated Terrestrial Environment," in Proceedings of the
Third National Symposium on Radioecology, CONF-710501, U.S.
Atomic Energy Commission, 1:877, 1973.

[23] H. C. BURKHOLDER, M. O. CLONINGER, D. A. BAKER, and G. JANSEN,
     Incentives for partitioning high-level Waste, Nuclear
     Technology, 31:202, 1976.

[24] D. H. DENHAM, D. A. BAKER, J. K. SOLDAT, and J. P. CORLEY,
     Radiological Evaluations for Advanced Waste Management
     Studies, BNWL-1764, Battelle Pacific Northwest Laboratories,
     Richland, WA, 1973.

[25] J. K. SOLDAT, N. M. ROBINSON, and D. A. BAKER, Models and Computer
     Codes for Evaluating Environmental Radiation Doses, BNWL-
     1754, Battelle Pacific Northwest Laboratories, Richland,
     WA, 1974.

[26] D. A. BAKER, G. R. HOENES, and J. K. SOLDAT, FOOD - An Inter-
     active Code to Calculate Internal Radiation Doses from
     Contaminated Food Products, BNWL-SA-5523, Battelle Pacific
     Northwest Laboratories, Richland, WA, 1975.

[27] H. C. BURKHOLDER and M. O. CLONINGER, The Reconcentration Phenome-
     non of Radionuclide Chain Migration, BNWL-SA-5786, Battelle
     Pacific Northwest Laboratories, Richland, WA, 1976.

RELIABILITY ASSESSMENT OF NUCLEAR SYSTEMS WITH REFERENCE TO SAFETY AND AVAILABILITY

A. E. GREEN*

ABSTRACT.   This paper concerns the evaluation of the reliability
of nuclear plant systems which usually depends upon some form of
prediction. This applies at the early design stage when the overall
system is being formulated. Some of the systems are dominated in
their design and concept by safety considerations. However, not only
must the reactor plant be shown to be adequately safe but also it must
be capable of performing its operational role when required to do so.
The paper discusses some of the experience and results of carrying out
reliability assessments in connection with the evaluation of safety and
the availability of these systems.

INTRODUCTION.   When considering the safety or availability of a

nuclear system, a reliability assessment is an important part of the

process. If the system is relatively new or is still in the design

stage then those aspects which are related to high risk usually

involve sparse data. As the events being considered become more rare

then it becomes likely that the appropriate data will not be available

at the complete system or sometimes the equipment level.

This leads to some method being used where estimates and actual

data applicable to different levels in the system may be combined to

give a reliability estimate for the overall system. Hence, by the use

of field experience, sample testing and theoretical prediction such

estimates are evolved. However, the reliability assessment undertaken

will have different characteristics according to whether safety or

---

*UKAEA, Wigshaw Lane, Culcheth, Warrington WA3 4NE, England.

purely availability is being considered.

In the safety case the analysis tends to be deeper and more searching than in the availability case which tends to have greater complexity. Of course these attributes arise where the appropriate data from the overall system are not available to answer directly the questions being posed. Another important distinction is that usually in the safety case a boundary or envelope approach is made and the actual quantification is not always as important as when considering the availability of a system and its output is directly earning money. This paper outlines some of the properties which arise in the reliability assessments involved.

SAFETY ASSESSMENT.

Allocation of Target Probabilities.   The allocation of target probabilities may be illustrated by considering the automatic protective system designed for the prototype fast reactor PFR which is a liquid-metal-cooled fast breeder reactor (LMFBR) at Dounreay. Taking one specific fault condition i.e. loss of electric power as the initiating event this had a frequency of once per year. From the Farmer type of criterion[1] a target was indicated of the probability of failure to shut-down on demand against loss of coolant flow due to loss of electric supply as 10^{-6}. From this information it was possible to allocate target probabilities for assessment as shown in Figure 1.

In the safety assessment the response and accuracy requirements

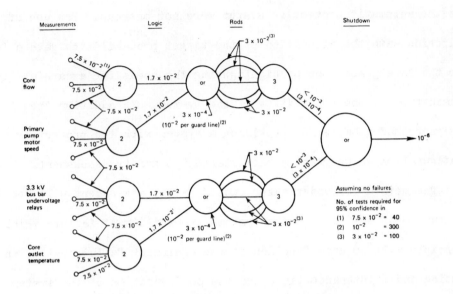

Figure 1   Target Probabilities for Assessment

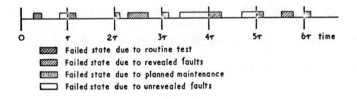

Figure 2   Typical Patterns of 2-state Behaviour of a System

for the automatic protective system were not onerous. The use of the
criterion with the allocation of the target probabilities gave a focus
for the safety assessor by highlighting the needs for a careful
examination of the reliability of the system. In addition the
designers had a basis for providing a system with the properties and
configuration which greatly facilitated the safety assessment.

The method for undertaking such a safety assessment will require
to lead to an estimate of the systems probability of failure which in
principle will be some function of the equipment failure rates, the
testing and maintenance times and the configuration of the system. It
is normal practice to use a form of synthesis in order to estimate the
overall system reliability characteristics.

<u>Failure rate estimation by consensus.</u> In the first analysis it
may be that data is not readily available and a consensus method of
estimating the equipment failure rate is used. An experiment was
conducted in the United Kingdom in 1966 and is described in Reference
2. A total of 73 people participated representing a cross-section of
both electronic and mechanical engineers, physicists, chemists,
statisticians and plant operating staff. Each participant was asked
to give his best estimate of the failure rate for each of 16 equipments used in nuclear reactor applications. The results showed that
typically one-third of the estimates for each equipment were within a
factor of two and the overall average was pessimistic.

Although such a method has been tried on a number of occasions

informally, it was not repeated until the 1977 Reliability and
Maintainability Symposium held in Philadelphia, USA. In this experiment different aircraft navigation equipments were exhibited and
estimates of failure rates obtained from 62 participants. The results
are shown in Table 1, for Digiprox equipment and inertial navigation
equipment. It will be seen that for each equipment at least one-third
of the estimates lie within a factor of 3 of the actual value. A
normal distribution was indicated for the logarithms of the individual
estimated failure rates. Therefore the estimated failure rate used for
Table 1 was derived from the mean of this normal distribution which is
equivalent to taking the geometric mean of the individual estimated
failure rates.

This type of technique extending to modified Delphi techniques
have been applied to basic data and equipment[3][4]. Such heuristic
approaches to the initial stages of estimating failure rates or other
reliability characteristics of equipment can be most useful. There
are more conventional methods of deriving failure rates by using data
bank information which are well described in the literature[5].

Synthesis methods of prediction. The equipment may be broken
down into parts and by the usual parts count method an overall failure
rate obtained.

A guide figure method has been found useful which from a knowledge
of the design of the equipment a proportion of the overall failure
rate is allocated to a particular mode of failure. A typical set of

Table 1

Summary of ratios of estimated to observed failure-rates

Equipment	Ratio	Ratio range Lowest	Ratio range Highest	Ratio Distribution 0.33	0.33–1.0	1.0–3.0	3.0
INERTIAL NAVIGATION SYSTEM							
1. Gimbal Assembly w/Servo Cards Stable Element	1.1	0.00044	44	16	11	10	17
2. Quantizer	1.6	0.0018	54	13	10	10	20
3. Computer Cards (4 Cards) Computer Interface Cards (3 Cards) Program Memory (2 Cards) Random Access Memory Gyro Bias Memory	1.9	0.020	49	12	4	16	21
4. ARINC Rx/Tx (2 Cards)	4.8	0.0078	63	7	3	15	28
5. D/A Card A/D Converter/A/D Mult. (2 Cards)	1.6	0.0017	34	11	11	12	20
6. CDU (Less Cards)	0.89	0.0015	16	17	11	12	13
CDU Cards (4 Cards)	1.8	0.0044	25	11	9	18	23
7. Overall System							
DIGIPROX SYSTEM							
8. Overall System	1.2	0.0045	18	12	15	14	16

guide figures for electronic equipment is shown in Table 2. This table indicates that for a direct coupled trip amplifier used for shutting down a plant at some predetermined level, then about 50% of the possible component faults could lead to a dangerous failure of equipment. However, if a low trip function is introduced additionally this gives a greater fail safe arrangement so that the fail dangerous rate may be reduced to about 10% of the overall failure rate.

Although such methods may be useful, where the fail dangerous rate is very low it is found that more detailed methods are required to give greater accuracy of prediction.

For detailed safety assessment work, the failure modes and effects analysis with failure rates associated with different failure modes has been found to be most effective.

Recently an investigation has been carried out into the results of a prediction of the reliability of the liquid shut-down system, for the Steam Generating Heavy Water Reactor (SGHWR) which is located at Winfrith. This utilised the detailed failure modes and effects analysis. A comparison has been made between the prediction and actual results over a period of a number of years of operation.

The liquid shut-down system consists of 12 identical loops. The predicted probability of failure on demand for a single loop was 1.55×10^{-2} based on certain assumptions involving testing periodicity. Over the number of years concerned it was found that the liquid shut-down system had operated about 250 times in all 12 loops resulting in

Table 2

Guide Figures

(Fail-dangerous rate as a percentage of total failure-rate)

Type of equipment	Trip functions	Fail dangerous rate, %
Direct coupled amplifier	High or low	50
Direct coupled amplifier	High and low	10
Chopper dc amplifier	High and low	5
Pulse counting channel	High	95
Pulse counting channel	Low	10
Pulse counting channel	High and low	10

3,000 loop firings, which had arisen for various reasons such as experiments and various demonstrations. From the predicted values in the assessment under the conditions appertaining during the number of years in question, it was found that the expected number of failures would be about 5. The practical results showed that under the assumptions made, that 4 failures had occurred in the individual loops. However, because the 12 loops installed gave redundancy there was no question of hazard to the reactor arising from these failures. Nevertheless, this information shows a close agreement between the predicted and actual results and this particular equipment is further discussed later in the paper.

SYSTEM ANALYSIS. In safety assessment the systems often being analysed become available or operational to perform a specific duty when some abnormal event or demand arises. The testing of such a system may occur periodically to prove that the system has a capability of performing and if a failure is found then the system is restored to full working order by a process of repair. On the other hand if the system is observed to have failed at any other time then a process of breakdown maintenance will take place. This failure, testing and repair process is shown in a simplified form in Figure 2. Failure of the system in operation followed by a repair procedure may be a typical change of state. On the other hand, a plant maintenance procedure may be applied to individual equipment or the overall system

at time intervals τ . Changes due to random faults are broadly of
two types, those which are 'revealed' and whose presence are known
immediately and those which are 'unrevealed' and whose presence will
not be known until some routine test procedure takes place.

The model will be based on reliability parameters of interest
such as the mean rate at which the system may fail[6]. In practice,
this requires a statement involving a function of time or environ-
ment such as: $f(t) = \Theta(t) \exp[-\int \Theta(t) \, dt]$ where $f(t)$ is the probability density function for the events in the time domain and $\Theta(t)$ the
function describing the rate of occurrence of events with respect to
time. The cumulative probability of failure $P_f(t)$ up to any given
time t is of interest and is given by integrating the density function
over the time range which gives: $P_f(t) = \int_o^t f(t) dt$.

If the rate of occurrence of events cannot be considered constant
then it may be pertinent to consider the mean time to the first event
or the first failure, which is given by: $\mu = \int_o^\infty [1 - P_f(t)] dt$.

Various combinations of failure may take place in the system such
as a failure occurring and remaining in this state for some period of
time during which a second failure occurs. This requires characteristics of interest such as repair time, replacement or restoration of
the failed device to be considered. Similar methods to those already
described may be applied by replacing $\Theta(t)$ with a repair rate
function.

These techniques were applied to the alternative arrangements for providing electrical supplies to nuclear reactor essential loads

following a reactor trip. The control arrangements were such that

each section of switchboard could be supplied either from the 132 kV

supply, if available, or from any one of three gas turbines.

In this analysis[7] it was found useful to prepare a logic flow

diagram as shown in Figure 3, for a specific reactor requirement.

Each of the blocks in this diagram represent a sub-system and the

diagram combines these in the correct logic pattern to meet the

overall requirement. The probability of failure of each of these

blocks was separately calculated, either manually or by computer

analysis, depending on the complexity of the sub-system and the over-

all system reliability determined. With the calculated value of each

block inserted on the diagram the contribution of each to the overall

power plant reliability could be seen. Table 3 gives the failure

rates used in the assessment. In the initial calculations certain

blocks were found to be dominant. Design modifications and revised

test frequencies were proposed and the calculated improvements in

reliability was shown to meet the required value.

As an example from this assessment box p̄4 Figure 3 represents

the gas circulator logic which gives the final initiating signal to

start the relevant gas circulators, providing all pre-requisites, i.e.

boiler feed, power supplies, etc. are satisfied. The circuit

originally consisted of three control channels of electrical timers

and relays, operating on a two-from-three principle and it

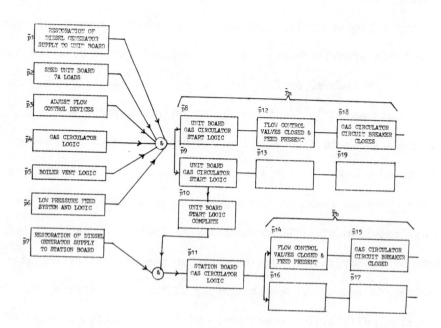

Figure 3 Logic Diagram of Sub-Systems Required for Emergency Core Cooling

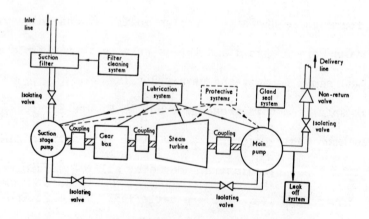

Figure 4   Simplified Scheme of a Pumping System

Table 3

Failure Rates Employed in Assessment

1. Boiler Feed Pumps

REVEALED fault rate 3, 6 and 12 faults/year considered
UNREVEALED fault rate 1% of 12 f/y = 0.12 f/year

2. Circuit Breakers

(a) Failure to close on demand due to unrevealed fault   = 0.01   f/year
(b) Failure to trip on demand due to unrevealed fault   = 0.005 f/year

3. Pressure Switches

Switches fail to open or close on demand due to unrevealed faults   = 0.025 f/year

4. Relays and Timer Units

Sequence Relay Failure Rate   = 0.0055 f/year
Trip Relay Failure Rate   = 0.023   f/year
Contact Failure Rate (per set)   = 0.0015 f/year
                                         (short circuit)
                                       = 0.0015 f/year
                                         (open circuit)
Time Delay Relay Failure Rate   = 0.05   f/year

5. Motorised Isolators and Pressure Control Valves   = 0.05   f/year

6. Non-return Valves   = 0.01   f/year

7. Limit Switches   = 0.02   f/year

8. Diesel Generator Start: Probability of failure to start per demand of 0.02.

represented a dominant area or unreliability. To an extent it
nullified the virtues of redundancy employed in other parts of the
system and set a limiting value on the overall system reliability.
By changing the design to a two-out-of-four configuration and
increasing the frequency of testing to once a month the revised
probability of failure, due to random faults, was greatly reduced.
Thus this common area of dependency no longer represented the sensitive area of unreliability.

This method of approach has been used on various systems and a
typical comparison was made between predicted and actual reliability
values for a composite diesel generator and control system arrangement for a nuclear power station. The reliability tests carried out
demonstrated a .99 probability of successful starting of a diesel-
generator unit at 95% confidence. This compared with a predicted
value of about 0.7×10^{-2} probability of failing to start which shows
a reasonable agreement.

Usually this type of analysis is limited by certain common
assumptions which for their validities depend upon the extent to which
the engineered design has eliminated various dependencies. Such
dependencies are common cause failure, human influence factors in
testing or maintenance and in the design, such as the common specification. These dependencies become important when the safety assessor
is attempting to demonstrate low probabilities of failure.

AVAILABILITY ASSESSMENT. The study of availability arises in nuclear power plant and is obviously particularly geared up to the analysis of maintaining the output of the plant. Consequently most of the plant is involved in some way which normally leads to greater complexity of analysis in undertaking an availability assessment. Whereas in the safety case only particular parts of the overall plant have a significant role to play. Under emergency conditions boiler feed water is often very important in post-trip cooling conditions. However, from a normal operational point of view feed water is also essential for maintaining the generated output of the plant and feed pump availability is usually very important.

Figure 4 shows an example of a pumping system used in a boiler feedwater application. Basically, a turbine drives the main pump and a suction pump is installed to boost the inlet pressure of the main pump. Ancillary services to this equipment include lubrication, gland sealing and protective systems. Other items included in the pipework are a suction filter and, non-return and isolating valves.

In this particular example which was of a new pump design the main objective of the availability assessment was to evaluate the relative significance of the main pump itself together with the repair process. The failure modes and effects analysis was applied in the traditional way. However, important factors which required special study were the influence of human error in certain activities associated with repair, the effects of wear in prolonged running if break-down maintenance were adopted, and the effect of the interaction

of one item with reference to another in the system (e.g. the out of
balance of one rotating item failing another coupled to it).

In the analysis "random" failure was assessed using a simple
exponential model of failure. For human errors the model incorporated the probability of an error occurring during an activity, the
probability of the error being detected and the frequency of the
activity. "Wear-out" failure which was expected to have a substantial time dependency involved more complicated modelling using
mean wear-out life-time estimates, and their distributions,
inspection frequencies and other factors.

From the assessment the overall failure rate of the pump itself
was estimated to be 0.6 faults/year which from subsequent field
experience of the new pump, but in other pump layouts, was shown to
be reasonably comparable. However, the failure rates associated with
the replacement of the pump cartridge were of particular interest
from the point of view of availability. The estimates of failure
rates given in Table 4 show the approportionment between the pump and
other parts of the system and based on given mean repair, replacement
and restoration times the equivalent losses of availability per year
are also given.

It is of interest to note that the assessment shown in Table 4
for the main pump, the overall failures leading to cartridge replacement (0.6 faults/year) constitute the major proportion of outage
failures which confirms the designer's original intention. Further-

Table 4

Failure Rates & Loss of Availability for Pump System Items

Item	Failure Rate faults/year	Loss of Unit Availability hours/year
Main Pump: Cartridge Replacement due to pump components	0.3	10.0
Cartridge Replacement due to interaction faults	0.3	
Other types of repair	0.01	1.5
Ancillary Systems	0.3	5.0
Other Plant Items	5.6	183.0
Pump Protective Systems	0.05	0.3
Total	6.56	199.8

Table 5

Failure Rate by Type for Main Pump Components

Type	Failure Rate faults/year
Random	0.17
Human Error	0.02
"Wearout"	0.11

more the analysis indicates that in a pump of this type that half its
outages result from faults due to interaction with other equipment
e.g. the lubrication system. In addition the pump components themselves contribute only about 5% to the overall incidence of outage.
The analysis of these pump component failures according to type is
shown in Table 5 and indicates that "wear-out" in the break-down
regime proposed for the pump contributes about 35% of the overall
failures. It was shown in the analysis that this could be reduced to
about half this level by introducing an element of planned maintenance
so that the pump is not allowed to operate longer than two years
without an overhaul. In this application the contribution to failure
by human error was estimated to be small.

Whilst the model used for this availability assessment has some
simplifications and assumptions it was found to be useful in the
design decision making process particularly related to standby
arrangements in the pumping system and other associated factors such
as maintenance.

In carrying out availability analysis it may be found that the
field of interest is not limited to the plant and its immediate
surroundings but can extend to activities hundreds of miles away. An
example of this was seen by a study carried out to investigate the
reliability of a data link from a nuclear power plant to a computer
located 500 miles away. It was of particular concern to estimate the
probability that a single job could be successfully transmitted,

received, computed and the computed result returned to the nuclear power plant via the link to the line printer located at the plant. If certain calculations received for such operations as the start-up of the plant were not available as required then a delay may have resulted giving a lack of availability for the plant.

The availability due to hardware faults was estimated to be overall about 90% which at a later date, from field data, was found to be made up of the contributions from major units. This showed good agreement with prediction and gave an increase of confidence in the decisions made at the earlier stages by using a prediction. Although it was appreciated that due to software faults and time scheduling the overall availability would be further reduced, the study concentrated on hardware and showed the problems in the various contributors to non-availability.

In dealing with a complete plant system it is logical to have the availability requirement clearly defined before undertaking the actual availability assessment. Often such a requirement is not explicitly defined with reference to forced outage. Peak winter availability requirement may be defined but may still leave the overall annual requirements not fully defined.

General experience and existing plant history data provide some guide for applying methods to break-down on overall forced outage rate (FOR) target for individual system target allocation. One approach to this problem was the collection of data on a Steam

Generating Heavy Water Reactor (SGHWR) over a period of some eight years, thereby deriving the availability. A prediction of availability was also independently carried out using other sources of data for purposes of comparison.

In this study the evaluation of a mean availability was undertaken on a simple two-state basis, these two states being a nominal working stage and a completely shut-down state. However, it was appreciated that the output of the plant would in practice have a performance value which could vary over a complete range or continuum. With this two-state assumption a simplified definition of mean availability was initially used. i.e. mean availability =

$$\frac{T_1}{T_1 + T_2} = \frac{1}{1 + \theta_f \tau_r}$$

where T_1 = summated time of plant operating in the normal working state

T_2 = summated time of outages due to unscheduled breakdowns from the nominal working state.

θ_f = the mean failure-rate and equivalent to $1/T_1$, τ_r is the mean repair-rate equivalent to T_2,

or mean unavailability = $\dfrac{\theta_f \tau_f}{1 + \theta_f \tau_r} = \theta_f \tau_r$ if $\theta_f \tau_r \ll 1$.

Following the usual methods of theoretical prediction, reliability characteristics for the various sub-systems component parts were expressed in terms of component failure rates and repair times together with some predictions or assumptions of the statistical distribution of these characteristics. The simple series model gave:- mean system availability = A_1 x A_2 - - - - - A_n

where A_1 to A_n are sub-system mean availabilities.

A further stage of analysis was undertaken to investigate the distribution of availability of the system as a whole and of the sub-systems. For this purpose a computer code called PADS (Plant Availability Distribution Synthesis) was used. The code has the strategy of generating a cumulative repair distribution from the component failure and repair characteristics and this uses a Monte Carlo technique to obtain the cumulative distribution of outage time for the system of components.

The results of a computer run based on a 5 year period, 30 year life with a simulation of 5,000 reactor years operation, are given in Figure 5. Curves A, B and C correspond to different mean availabilities calculated from 500 MW or 100-250 MW turbogenerator experience, or from exclusion of turbogenerator data.

It is useful to give the pictorial representation of the combined probabilities of a number of events and the outage time distribution as shown in Figure 6. The height of the hill is a scaled representation of the probability of the predicted number of events and associated outage time , the other axes being the integer number of faults and outage time as a continuous variable. The flag represents the number of faults and the outage time which actually occurred in the liquid shut-down system (previously mentioned in section on safety), over a five year period. It will be seen that good agreement is obtained.

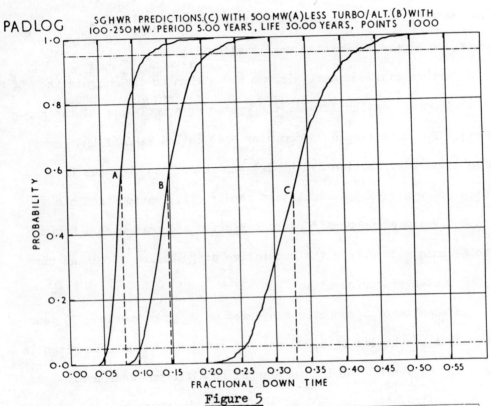

PADLOG   SGHWR PREDICTIONS.(C) WITH 500MW(A)LESS TURBO/ALT.(B)WITH
          100-250MW. PERIOD 5.OO YEARS, LIFE 3O.OO YEARS, POINTS 1OOO

Figure 5

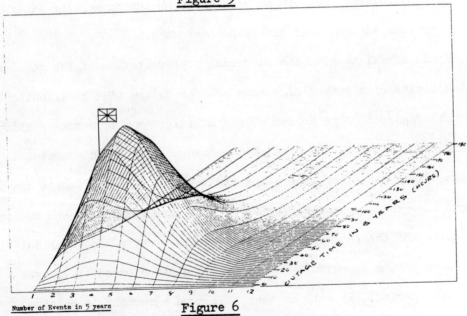

Number of Events in 5 years   **Figure 6**

This method of representation has been used for other sub-systems and the position of the flag relative to the summit of the hill indicates pessimism or optimism in prediction. For a number of sub-systems studied, the general trend is that the number of faults predicted was greater than actual and for predicted repair time, was generally less than that observed.

Although most plants of course do generally provide some form of operating statistics , there are a relatively small number of nuclear plants operating which have provided precise data specifically collected for availability analyses. This means that the appropriate data are not usually easily accessible in areas which have the greatest effect on availability.

Simple methods of estimating overall failure rates are therefore applicable which also involve a high degree of engineering judgement. Flow or activity charts with critical path techniques can be used to estimate overall restoration times. These restoration times may not be adequately dealt with without breaking down the overall restoration process and synthesising the overall restoration time particularly where heavy plant is involved. "Conventional" engineering often presents more difficulties than for specific nuclear components because in many cases the conventional components have not had the intensive design and development process undertaken in many branches of the nuclear industry.

CONCLUSIONS.   The safety and availability studies both involve the analysis of modes of failure, routes to failure and consequences. Loss of availability targets may be between 10% and 25% for large plant but this may represent several orders of magnitude greater than the rare events considered in safety. However, rare events can lead to long and difficult repair processes which are important in availability evaluation. Further work is required in the investigation of modelling both for safety and availability studies and the collection and analysis of the appropriate data. In general, this needs a multidisciplinary approach to the problem involved in order to give the solutions.

Over the past two decades there has been a great impetus to safety investigation and today this has extended into availability analysis and assessment.

The starting point of any assessment whether for safety or availability is to have a properly specified criterion. It is seen that criteria have started to emerge on a risk-consequence basis for safety. In the equivalent case for the Forced Outage Rate it is equally important to have this properly defined. This can then lead to a logical apportionment of targets for whatever purpose the assessment is required.

It emerges that in undertaking safety or availability assessment a generic reliability technology is being applied. The development of this reliability technology can benefit greatly by merging the results of nuclear and non-nuclear applications[8]. Greater discipline together with an improved means of showing whether objectives have been met accrue from the use of quantified reliability techniques of assessment.

REFERENCES

(1)   F. R. FARMER, Siting Criteria - A New Approach, Proceedings of Symposium on the Containment and Siting of Nuclear Power Plants, IAEA, Vienna, April 1967, pp. 303-329.

(2)   A. E. GREEN, Reliability Prediction, Proceedings of the Institution of Mechanical Engineers 1969-70, 184, Part 3B.

(3)   L. E. BOOTH, et al., The Delphi Procedure as Applied to IEEE Project 500 (Reliability Data Manual for Nuclear Power Plants), Proc. 3rd Annual Reliability Engineering Conference for the Electric Power Industry. IEEE & ASQC, Montreal, Sept., 1976.

(4)   M. L. SHOOMAN and S. SINKAR, Generation of Reliability and Safety Data by Analysis of Expert Opinion, Proc. Annual Reliability and Maintainability Symposium, Philadelphia, 1977.

(5)   C. D. H. FOTHERGILL, The Collection, Storage and Use of Equipment Performance Data for the Safety and Reliability Assessment of Nuclear Power Plants, "Reliability of Nuclear Power Plants", IAEA, Vienna, 1975.

(6)   A. E. GREEN, and A. J. BOURNE, Reliability Technology, John Wiley and Son, London, 1972.

(7)   E. R. SNAITH, Optimizing the Design and Operation of Reactor Emergency Systems Using Reliability Analysis Techniques, "Reliability of Nuclear Power Plants", IAEA, Vienna, 1975.

(8)   A. E. GREEN, The Systems Reliability Service and Its Generic Techniques, IEEE Transactions on Reliability, Vol. R-23, No. 3, August 1974, pp. 140-147.

THE ROLE OF RELIABILITY AND RISK ASSESSMENT IN LMFBR DESIGN: IMPLEMENTATION OF RELIABILITY IN LMFBR DESIGN

JOHN GRAHAM AND P. P. ZEMANICK*

Abstract.

This paper concerns future developments in LMFBR licensing technology.

Federal Regulations (10 CFR 50.34) require that the preliminary
safety analysis provide analysis and evaluation "with the objective of
assessing the risk to public health and safety" to determine margins
of safety and the adequacy of the plant. Hitherto, the assessment of
risk has been qualitative but it has become increasingly apparent that
quantitative assessments would provide a better basis for judgement.
Potential future roles of reliability and risk assessment are discussed
in the context of providing additional confirmation of the safety of
LMFBR designs. Potential acceptability criteria for risk evaluations
are outlined.

The reliability implications of designing components to the ASME
Code Section III requirements are discussed. General judgements are
provided as well as the preliminary results of probabilistic studies
of selected specific limits. There is a different reliability significance for the mandatory rules for normal, upset, and emergency conditions versus the non-mandatory rules for normal, upset, and emergency
conditions versus the non-mandatory guidance for faulted conditions.

1. Introduction. In the application for a nuclear power plant

construction license the applicable Federal Regulations[1] require that

the preliminary safety analysis report provide analysis and evaluation

"with the objective of assessing the risk to public health and safety

resulting from operation of the facility" and to determine the margins

of safety during all stages of plant operation as well as the adequacy

of safety related structures, systems and components. This assessment

---
*Westinghouse Electric Company, Madison, PA 15663

of risk has traditionally been made on the basis of deterministic evaluations of conservative plant conditions ranging from anticipated operating modes all the way to accident conditions of exceedingly low probability. The final judgement that no undue risk would result from plant operation is made by the Nuclear Regulatory Commission (NRC) on the basis of this spectrum of evaluations. That this judgement for light water reactors (LWRs) has been reasonably well applied in the protection of the public has been verified by the absence of serious accidents and the absence of hazard in any member of the public even though 70 LWR nuclear power plants are now in operation.

In LMFBRs, licensing review to date has included a further class of accidents, core disruptive accidents, for which a very large range of potential consequences have been postulated. These consequences have ranged from those physically possible to even non-physical envelope conditions. LMFBR design of safety related systems has always sought to minimize the possibility and/or consequences of such accidents and it is proper that the licensing review takes cognizance of this fact. It is clearly possible to do this in the same manner as LWR licensing has been performed but it is becoming increasingly apparent that quantitative assessments of risk would provide an additional valuable basis for judgement where precedent is not clearly applicable. Indeed the publication of the Reactor Safety Study for LWRs[2] has already indicated the desirability of such an additional aid as confirmation even

in this case of a well-established review process. Even those who
seem to reject the value of numerical assessments do so only to
emphasize the value of traditional methods, which should be supplemented rather than supplanted.[3, 3a]

   2. Risk Assessment Criteria. One of the difficulties of using
a quantitative risk assessment guide is that it is necessary to specify
an acceptance criterion, whereas qualitative judgements are usually
made with undefined precision. The Reactor Safety Study did not
specify a criterion but placed the risks in context against equivalent
risks resulting from natural environmental hazards, such as hurricanes
and meteorites, and man-caused hazards resulting from explosions, dam
failures and transportation accidents. The study showed that existing
LWRs, even for a population of 100 plants, were much less hazardous
and therefore, by implication, acceptable. NRC has implied[3] a similar
acceptance criterion for the Clinch River Breeder Reactor Plant by
noting that it "should achieve a level of safety comparable to the
current generation of light water reactor (LWR) plants." The Commission
further notes that in addition to current criteria for evaluation, a
further safety objective is used "that there be no greater than one
chance in one million per year for potential consequences greater than
10 CFR 100 dose guidelines for an individual plant."[3, 4] For a
particular accident, that of an anticipated transient with failure to
scram, a failure rate of one tenth of the overall safety objective
was suggested.[4]

Probability alone cannot constitute a full risk criterion since neither industry nor society are accustomed to viewing a spectrum of potential events without some estimate of the consequences of failure. Nevertheless, in many endeavors probability cut-offs may be employed: either the probability is so high that accommodation must be provided, or the probability is so low that the condition is deemed "incredible". Thus, probability ranges are instructive and confine attention to those areas where different levels of potential consequences may be acceptable.

Table 1 is a tentative listing of reactor plant probability ranges and an indication of an acceptable consequence in each area. Categories 1 through 4 constitute the design bases for the plant and the plant is, therefore, designed with these four categories of events in mind. The maximum consequences are those set by the Regulatory requirements of 10 CFR 20, 50 and 100. The plant operators or owners may, of course, set more restrictive consequence limits to protect their investment and the table shows how these fuel and plant availability limits might also be set in terms of the decreasing probability of the particular event for which the response is calculated.

Normally, these categories are used within the plant design bases with no more than a judgemental separation in terms of probability, so the probability ranges designated are representative only of likely frequencies in each category. The 2.5×10^{-2} value is simply one in forty years and 10^{-7} is chosen by reference to WASH-1270.[4] Since the

Table 1.

CATEGORY (EXAMPLES GIVEN)	FREQUENCY (PER REACTOR YEAR)	ACCEPTABLE CONSEQUENCES	
		PLANT OPERATORS[d]	PUBLIC

1. NORMAL[a]
 	
EVENTS WHICH WILL OCCUR EITHER BECAUSE THEY HAVE BEEN DESIGNED TO OCCUR OR EXPERIENCE HAS SHOWN A HIGH FREQUENCY,
e.g.,   • FULL POWER OPERATION
        • STARTUPS & SHUTDOWNS
        • STOCHASTIC FUEL FAILURES
	≥ 1	• NO INCREASE ABOVE DESIGNATED RADIOACTIVITY ACTIVITY LEVELS IN RESTRICTED CONTAINMENT LOCATIONS
		• ACCUMULATED FUEL PIN CLADDING STRAIN < 0.1%
		• NO INCREASE IN PLANT UNAVAILABILITY
			NONE ($\leq$ 10CFR50, APP. I)

2. ANTICIPATED TRANSIENTS[a]
 	
EVENTS WHICH BASED ON EXPERIENCE ARE EXPECTED TO OCCUR AT LEAST ONCE IN A PLANT LIFETIME,
e.g.,   • LOSS OF OFF-SITE POWER
        • SCRAMS
        • LOSS OF POWER TO ONE PUMP
        • OPERATOR ERROR
	$1 - 2.5.10^{-2}$	• NO INCREASE ABOVE DESIGNATED RADIOACTIVITY LEVEL LIMITS IN RESTRICTED CONTAINMENT AREAS (< 10CFR20)
		• ACCUMULATED FUEL PIN CLADDING STRAIN < 0.3%
		• PLANT UNAVAILABILITY WITHIN PREDICTIONS
			NONE ($\leq$ 10CFR50, APP. I)

3. UNLIKELY EVENTS[a] EVENTS NOT EXPECTED TO OCCUR INDIVIDUALLY BUT WHICH, BASED ON THE TOTAL NUMBER OF SUCH EVENTS, MIGHT OCCUR ONCE DURING PLANT LIFETIME (~ 40 YEARS), e.g., • PUMP SEIZURE • SMALL REACTIVITY ADDITION	$2.5.10^{-2} - \sim 10^{-4}$ [b]	• >10CFR20 • NO LOSS OF FUEL PIN CLADDING INTEGRITY (SAY <0.7% STRAIN) • DOWNTIME FOR PLANT REPAIR	NONE ($\leq$ 10CFR50, APP. I)
4. EXTREMELY UNLIKELY EVENTS[a] EVENTS NEVER EXPECTED TO OCCUR BUT WHICH ARE INCLUDED IN THE DESIGN BASIS OF THE PLANT BY REGULATORY FIAT, e.g., • EXTREME ENVIRONMENTAL CONDITIONS • LARGE SODIUM FIRES	$\sim 10^{-4} - 10^{-7}$ [c]	• NO LOSS OF COOLABLE GEOMETRY • REPAIR OF PLANT FOR CONTINUED OPERATION	$\leq$ 10CFR100
5. HYPOTHETICAL EVENTS POSTULATED EVENTS WHICH ARE NEVER EXPECTED TO OCCUR AND WHICH HAVE NEVER BEEN EXPERIENCED BUT WHICH ARE MECHANISTICALLY WITHIN THE LAWS OF PHYSICS, e.g., • CORE DISRUPTIVE ACCIDENTS	$10^{-7} - 10^{-N}$	• LOSS OF PLANT INVESTMENT	ACCEPTABLE RISK MEASURED AGAINST PUBLICLY ACCEPTABLE CRITERION
6. ARBITRARY CONDITIONS POSTULATED CONDITIONS FOR WHICH NO MECHANISTIC INITIATION EXISTS OR WHICH VIOLATE SOME KNOWN PHYSICAL THEORY, e.g., • CERTAIN CORE COLLAPSE CALCULATIONS • USE OF THERMAL & CHEMICAL CHARACTERISTICS OF SODIUM SPRAY IN AIR WITHOUT CONSIDERING AGGLOMERATION	ZERO	ANY, SINCE THESE EVENTS ARE MERELY AN EVALUATION OF DESIGN CAPABILITY.	ANY, SINCE THESE CONDITIONS DO NOT REPRESENT REALITY.

NOTES

(a)   CATEGORIES 1-4 CONSTITUTE THE PLANT DESIGN BASIS.

(b)   USED AS AN ILLUSTRATIVE NUMBER ONLY.

(c)   CHOSEN BY REFERENCE TO WASH-1270.

(d)   EXAMPLES OF DESIGN ACCEPTANCE CRITERIA ARE GIVEN DEALING ONLY WITH: • RADIOACTIVE RELEASE
        • FUEL BEHAVIOR
        • REACTOR PLANT AVAILABILITY

OTHER CRITERIA MEETING ASME AND IEEE CODE REQUIREMENTS ALSO APPLY IN EACH PROBABILITY CLASS.

Frequency   Ranges

plant is designed to accommodate these conditions, the consequences in terms of the public can be no worse than 10 CFR 100 even for the Extremely Unlikely Events of Category 4. The 10 CFR 100 doses would, however, contribute negligible biological consequences and, therefore, all four categories of events constitute insignificant risk to the public. All events of probability greater than the lower probability limit of Extremely Unlikely Events (about 10^{-7} per reactor year) can thus be omitted from any risk assessment since they contribute insignificant hazard.

Categories 5 and 6 are of lesser probability. Category 6 represents those arbitrary conditions which are sometimes postulated to obtain an upper limit of possible consequences. For example, a distribution of the entire radioactive inventory of the plant across the entire population of the states may be postulated, but since the condition is a non-physical one, the calculated consequences do not represent unacceptable consequences; they simply do not apply. A risk assessment, therefore, should omit all conditions of this type and be restricted to events which are mechanistically possible within the laws of physics.

Category 5 consists of hypothetical events which the plant is not designed to accommodate, but which do constitute a set of events which are physically possible even though never expected to occur. Their probability is less than about 10^{-7} per reactor year. There should

also be a lower probability range cut-off, 10^{-N}, below which the
incredibility of the event is such that it should be omitted without
further ado from the risk assessment. Even though the plant is not
designed to accommodate Category 5 events, nevertheless, it has considerable inherent capability for attenuating the consequences of
extreme events and the resultant risks will be very low.

Figure 1 shows the separation of probability ranges for individual
events and indicates that events which fall within the regulation of
design capability (10 CFR 100) do not contribute in terms of consequences.
The shaded area for Category 5 events constitutes the region of risk
interest. It should be noted that the limit of 10^{-7} for Category 4
events does not define the shaded area since 10 CFR 100 requirements
are below biological significance.[4b]   Therefore, to define the range
of risk assessment three additional parameters are desirable:

> The limit of biological significance;
>
> The probability range cut-off (10^{-9} is suggested
> on the figure), or, alternatively an extrapolated
> acceptance criteria;
>
> The risk criterion envelope for this limited range.

The risk acceptance criterion envelope should be related to a public
perception of acceptable risk, as exemplified by presently accepted
risks, in relationship to alternative risks associated with not constructing nuclear power plants. Various authoritative bodies, including
the ANS 54-10 standards sub-committee, are now working towards proposing
such numerical acceptance criteria.

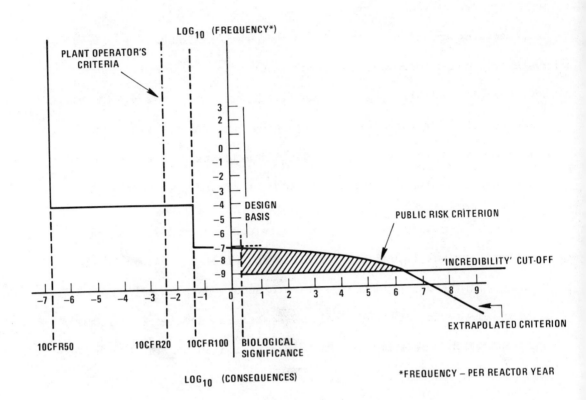

Figure 1.   Frequency Separation Of Risk Criteria (Individual Events)

3. **Implementation of Reliability in Design.** Maintenance of low
plant risk, to be later confirmed in an overall assessment, is provided
by a balanced application of preventative and mitigating measures to
reduce the probability of accidents in the first place and then to
reduce potential consequences.[5]

3.1. **Safety System Reliability.** To ensure a low probability of
significant hazards, it is first necessary to identify those systems
critical to the prevention of accidents such as the shutdown and shutdown heat removal systems, power supply systems and critical structural

components of the plant. Then it is necessary to identify required system redundancy and diversity preferably in a balanced way by reference to reliability goals, and finally it is necessary to maintain the reliability of these systems throughout the design, fabrication and operation of the plant by an identification of critical components within these systems and the institution of proper system controls.

General Design Criteria[6] have provided some guidance as to a necessary level of redundancy for fluid and electric systems by requiring that the plant design be able to accommodate an active or a passive failure in the safety related system in addition to a primary fault condition. This has led to the provision of two largely equivalent shutdown systems, and to at least three paths of decay heat removal in current LMFBR designs. Furthermore, general guidance on a defense in depth philosophy has lead to the addition of backup features such as natural circulation cooling capability, guard tanks, inerted atmospheres and the like. However, detailed comparison of potential concepts have been performed principally on a subjective basis and guidance as to what might constitute necessary or acceptable diversity has been very general.

A recent comparative assessment[7] made on various decay heat removal concepts, all of which complied with the general design criteria, showed that concepts which included a diverse heat removal direct from intermediate loops to air cooled condensers to preclude the necessity of using the steam generators, were nevertheless impacted by possible intermediate loop drainage following some unidentified common cause failure of the steam generator rupture discs. By using a numerical reliability assessment, design changes which allowed failed rupture discs to relieve to intermediate surge tanks were shown to be suffi-cient to raise the reliability of the overall system to a level compar-able to more complex decay heat removal systems connected to the primary loops. Although the calculations were intended to show compar-ative unreliabilities as a trade-off study, in all cases the designs were estimated to achieve acceptable target unreliability figures in the range of 10^{-6} to 10^{-7} per reactor year. Unreliabilities of this order ensure a satisfactorily low contribution to the overall risk from this class of contributors (i.e., failure of the decay heat removal system).[7]

Similarly such a target reliability provides a basis for the assessment of required redundancy in feedwater systems, power supplies, valving, control systems, and pumps. The absolute level of reliability can be verified in many cases by comparison with the proven reliability existing in comparable LWR systems, until sufficient LMFBR experience becomes available.

Although LMFBR reliability studies have now established that most emphasis should be placed on heat removal systems, this is a recent development. The dual shut-down systems were the safety systems of focus and the design trend has been to include diversity between the primary and secondary systems wherever possible. The Clinch River Breeder Reactor Plant (CRBRP) has diverse electronic systems and considerable diversity in mechanical systems (Table 2)[8] which, it is judged, protect the overall scram system from significant common cause failure potential. Further development is aimed at further diversity,[9] although whether or not such complexity of design is warranted cannot be determined without a numerical assessment of the probability of common cause failures due to head movement or core restraint failures. An overall assessment is likely to show that any potential reduction of risk in this area would be small compared to other areas.

Table 2.

Shutdown System Diversity Of Design

CONTROL ASSEMBLY (CA)	PRIMARY	SECONDARY
CONTROL ROD	37 PIN BUNDLE	19 PIN BUNDLE
GUIDE GEOMETRY	HEXAGONAL	CYLINDRICAL
NO. OF CONTROL RODS	15	4
CONTROL ROD DRIVELINE (CRD)		
COUPLING TO CA	RIGID COUPLING	FLEXIBLE COLLET LATCH
CONNECTION TO CRDM	CRD LEADSCREW TO CRDM ROLLER NUTS	CRD ATTACHED TO CRDM CARRIAGE WITH PNEUMATIC ACTIVATION OF CRD LATCH THROUGH SLENDER ROD
DISCONNECT FROM CA FOR REFUELING	MANUAL	AUTOMATIC
CONTROL ROD DRIVE MECHANISM (CRDM)		
TYPE OF MECHANISMS	COLLAPSIBLE ROTOR ROLLER NUT	TWIN BALL SCREW WITH TRANSLATING CARRIAGE
OVERALL MECHANISMS STROKE	37 INCHES	69 INCHES
SCRAM FUNCTION		
SCRAM RELEASE	MAGNETIC, RELEASE CRDM ROLLER NUTS	PNEUMATIC: RELEASE CRD LATCH IN CA
SCRAM ASSIST	SPRING IN CRDM	HYDRAULIC IN CA
SCRAM SPEED VERSUS FLOW RATE	INCREASES WITH DECREASING FLOW RATE	DECREASING WITH DECREASING FLOW RATE
SCRAM ASSIST LENGTH	14 INCHES	FULL STROKE
SCRAM DECELERATION	HYDRAULIC DASHPOT	HYDRAULIC SPRING
SCRAM MOTION THROUGH UPPER INTERNALS	FULL STROKE	0.25 INCH
DESIGNER	WESTINGHOUSE ELECTRIC CORPORATION	GENERAL ELECTRIC CORPORATION

Reliability of power supplies and structural components of the design are both relevant to heat removal. In the former case, added redundancy and diversity can reach a point of small incremental gains in risk reduction since this is eventually governed by the fundamental heat removal path redundancy, but it is desirable to balance the comparative reliabilities of power and fluid subsystems within the overall system reliability. Structural components designed, constructed, inspected and tested to ASME code rules have already an inherent reliability as noted in Section 3.2, and again it is necessary only to balance reliability against significant failure modes to ensure on one hand that weak links do not exist and on the other that over-reliable (and costly) systems are not provided without benefit to the plant.

3.2. System Controls. It is, of course, absolutely vital to ensure that whatever reliability exists in the conceptual design is maintained in the plant systems throughout design, fabrication, construction and operation. This is accomplished by strict procedural controls[10] which ensure that the design engineering groups monitored by an independent reliability group identify failure modes and critical items, perform reliability assessments and provide for tests. Procedures are set to maintain reliability even though the design may change. Component and systems engineers are confined to judgements relative to their own areas of expertise. Special attention is paid to any critical component failure mode which has the potential to produce simultaneous

failure of more than one element of a system or systems. Procedures
become mandatory requirements which ensure no degradation of the plant
reliability. In addition, reliability goals set and evaluated during
the design process ensure an acceptable initial reliability condition.
These goals can be system oriented or may be allocated to specific
sub-systems such as that for a single mechanical control rod drive.

It is important to ensure that reliability activities are strongly
integrated with the design process and that the designer is then able
to improve the design of those components which would otherwise have
failure modes of too high a probability or consequences. The value of
reliability engineering is in the disciplined approach towards screening
the system design for its critical components and its vulnerability to
common cause failures or human errors. Thus a reliable design should
emerge even before quantitative assessments are made.

4. Designing to ASME Code Requirements. Certain structural
components, vessels, piping and support structures, are singular and
non-redundant in their primary functions (even though backup guard
vessels are provided for public protection). Nevertheless, the authors'
judgement is that a significant measure of reliability is incorporated
into their design by adhering to ASME Code design requirements.

The industry experience with pressure vessels and piping has been
marked by only rare instances of major failures during the decades since

introduction of the ASME Boiler and Pressure Vessel Code. Under AEC

sponsorship available service data on non-nuclear vessels were analyzed

to deduce an ASME-Coded pressure vessel unreliability.[11, 12] The

most meaningful block of data was on U. S. vessels designed and con-

structed to the rules of ASME Sections I[13] and VIII[14]. An unrelia-

bility of about 10^{-5} per non-nuclear vessel service year was calculated

for "disruptive failures", major failures with tangible safety conse-

quences. Taking into account an expected improvement in the reliability

of nuclear vessels [ASME Section III[15]], with their mandated enhance-

ment of design, material, construction, and operational quality control,

it was estimated that the likelihood of disruptive failure of nuclear

vessels would be at least one order of magnitude lower, that is, 10^{-6}

per year.

Confidence in the high reliability of Coded components can be

deduced from other considerations than this correlation of non-nuclear

service data. Some of these other considerations are discussed in

Reference (16). For one thing, the past 25 years have been character-

ized by dramatic developments in technical expertise related to pressure

vessel behavior prediction. There have been analytic developments which

include finite element calculational methods which went from concept

to worldwide application over that period and fracture mechanics analysis

models which now have been extended into non-linear regimes. Along with

these analytic advances, there have been important experimental programs

which have provided fresh insight into the mechanics of vessel failure.
Beyond the fact that the latest technical developments are brought to
bear in the vessel design process, the framework within which Codes
and Standards are written provides an added measure of confidence.
The preparation of pressure vessel design rules is supported by the
best available expertise in the country. Superimposed on the specialists' rule-making activity is a system of checks and balances to protect
public safety against any conflicts of interest. It is acknowledged
that the sources of confidence discussed in this paragraph pertain most
directly to LWR-type vessels. Most experience to date relates to this
type of vessel which can be designed to limit primary stresses in all
but local areas to a low enough value to assure shakedown to elastic
action. LMFBR vessels must sustain temperatures within the creep range
of the structural materials and significant non-linear structural
response occurs. Major national effort is being devoted to a better
understanding of the mechanical behavior pertinent to LMFBR materials
and conditions to bring LMFBR Code-qualification to a comparable level
of assurance against failure.

With a failure probability of nuclear vessels at the general magnitude of 10^{-6} per year, the possibility of vessel failure was indicated
to make no important contribution to the likelihood of overall LWR
plant failure with significant public health consequences.[2] The
question may be raised whether the same conclusion can be justified

for vessels typical of LMFBR usage. The LMFBR service is more severe

insofar as higher operating temperatures, with attendant non-linear

material behavior, prevail than is the case for LWRs. Furthermore,

the usual LMFBR vessel material, austenitic stainless steel, is subject

to sensitization, with risk of degraded mechanical performance, due to

welding process temperatures and environmental chemistry during fabri-

cation and operation. On the other hand, austenitic stainless steel

has the advantage of being by nature a more ductile material than

carbon steels. In addition, LMFBR operating pressures are low, creating

conditions of lower mechanical loads and lower stored energy. Signifi-

cant progress has been made in learning to structurally characterize,

analyze, handle, and fabricate LMFBR pressure-retaining components.

However, added development work is required, coupled with the benefits

of broadened operating experience to confirm the judgement that LMFBR

vessel safety and dependability are adequate.

In summary, then, the combination of 1) available non-nuclear Code-

vessel data; 2) the more rigorous standards prescribed for nuclear

vessels, and 3) progress with LMFBR structural design technology, built

as applicable on strong advances in LWR vessel technology, indicates

satisfactory LMFBR pressure vessel integrity, supportive of nuclear

safety objectives.

There is a need for a more refined assessment of LMFBR pressure

vessel reliability [both for designer and regulatory assistance[17]].

If the trend toward further quantification is followed, a further

refinement of Section III pressure vessel reliability prediction would

require "microscopic" evaluation, based on detailed evaluation of the

probabilistic implications of the most important individual Section III

loading limits. Further refinement on a "macroscopic" level, that is,

from observed service experience, is not feasible in the near term be-

cause of the relatively slow accumulation of operational data.

The basic tool of detailed evaluation would be by stress-strength

interference methods, by which a statistical description (expected

value, as well as, uncertainties) of the loads to which a component

is subjected is compared to a statistical description of the capability

of the component to sustain the loads.

Reference (18) describes this methodology which yields a probabil-

ity of failure of the component for the defined loads. The refined

analysis is visualized as following this approach:

> Selection of key Code limits;
>
> Selection of LMFBR characteristic materials, geometries
> (perhaps one for pressure vessels and one for piping),
> environment, and loadings;
>
> Probabilistic analysis of the likelihood of failure for a
> component which just meets the specific Code allowable and
> is constructed and tested in the conformance with Code rules.

The package of results for the key limits could form the basis for a

generic reliability estimate for vessels designed to Section III.

Reference (19) generally follows the proposed approach for the primary membrane stress limit and the fatigue allowables. Although the results are not necessarily representative of Coded component reliability because of simplifications in the analysis, the analysis exemplifies the stress-strength interference process and does provide some insight into these limits. A more detailed analysis along the same lines is presented in Reference (20). The failure mode of interest here is cylinder buckling under external pressure. Again, the probabilistic analysis is exemplary of the techniques applicable to this kind of problem. Preliminary studies were initiated within the LMFBR program to investigate the Code limits against primary membrane stress, fatigue, and buckling under external pressure. Although the studies are not complete, indications are that the primary stress limits against vessel pressure bursting provide a reliability well within the earlier reported estimate of 10^{-6} per year. More preliminary work on the fatigue and buckling limits indicates a smaller margin. These tentative results imply that the Code allowables do not imply consistent reliabilities and that, considering the uncertainty (scatter) in available fatigue and buckling failure measurements, the Code primary stress limit is more conservative than the fatigue and buckling limits.

Three final comments about Code component reliability prediction are made. First, proper evaluation of pressure boundary failure in a risk assessment under upset, emergency, and faulted conditions requires conditional probabilities of failure. That is, the likelihood of the

actual occurrence of these conditions is to be incorporated into the

calculations. Second, failure calculations may need to extend beyond

those mechanisms covered or suggested by the Code. An example is a

fracture mechanics analysis. Even the non-mandatory Appendix G of the

ASME Code which addresses non-ductile failure presents only a rudi-

mentary treatment, probably unsuitable for many situations. This is

not to say that the Code coverage of failure modes requires expansion.

Historical experience implies that the Code coverage generally meets

Code objectives with gradual improvement [e.g., extension into the high

temperature regime[21]] continually occurring. However, for purposes

of risk assessment, potentially dealing with very low probability

caustic factors, broader coverage may be needed. Third, human error

is not entirely eliminated by Code-imposed procedural controls. Quanti-

fication of human error contributions to vessel failure remains unfor-

tunately an elusive element in reliability methodology. Concurrent

with the more manageable evaluation tasks, the human error element,

which is acknowledged to be a real effect, should be introduced on the

most rational practical basis.

In summary, design and construction of LMFBR components in con-

formance to ASME Code provisions are judged to incorporate in the

component a high degree of reliability against failures with serious

safety implications. This conclusion cannot be stated with the same conviction as for LWR vessels because of limited experience with LMFBR

materials/configurations in their different environments and the more

complex structural behavior characteristics of LMFBR temperature levels.

However, the progress to date and the rigorous ongoing program to

advance with LMFBR structural design technology are encouraging.

Additional confirmatory work and broader operating experience will

support a more confident quantification of LMFBR Code-component inherent

reliability and fit naturally into overall progress in the refinement

of risk assessment methods.

5. Special Problems. A major problem inherent in risk assessments

is the demonstration of low probabilities estimated for critical failure

modes which have already, by design, been made very improbable. Parti-

cular examples of these failure modes are those which may arise from a

common cause, also known as common mode failures, and those which may

arise from interfacing human functions, known as human errors.

Since rare events have probabilities of the order of 10^{-6} and

below, it is practically impossible in most cases to obtain classically

significant statistical data, either by test or experience. However,

this is not to say that it is, therefore, impossible to support estimates

of their unreliability, since techniques do exist for their evaluations

which rely on experience with similar systems. Bayesian techniques

establish failure distributions based on all information available as

well as physical understanding of the failure modes within the temporal and physical environment of the component. These failure distributions

may be confirmed or modified by subsequent limited data. Extreme value

techniques or those which depend on decision theory may also be applied

in known conditions. Tails of failure distributions may be estimated

from more abundant data associated with commoner failure modes. Actual

methods used naturally depend upon the specific cases under consideration.

Certain common cause failures, especially those associated with

singular components which cannot be made redundant may be amenable to

calculation. Both in this area and that of human errors data exist

which remain to be screened, and analyzed for particular application.

The Organization for Economic Cooperation and Development (OECD)

has set up a task force on these subjects and from their initial survey

it is clear that a good deal of effort exists, principally in Eurcpe,

directed towards resolution of these particular problems.

6. Summary. It is clear that risk assessment can and will provide

a powerful evaluative tool in assisting the determination of adequacy

of safety of the LMFBR plants. Certainly all trends point to the use

of these techniques in the future. There are problems to be sure:

ensuring that all failure modes have been identified, the possibility

of common cause failures and human errors, the adequacy of data to

support estimated failure rates of rare events, and adequately defined

acceptance criteria. However, there is also an abundance of work proceeding towards a resolution of these deficiencies and there appears to be no insurmountable obstacle to the use of risk assessment techniques.

Moreover nuclear designs are already very reliable by virtue of the use of present day deterministic standards and it is a matter of quantifying this reliability. The ASME Code Section I and VIII application to non-nuclear vessels for example has very meaningful improved vessel reliability even without the application of Section III nuclear Code rules. Further work on the failure criteria beyond Code rules will confirm that LMFBR vessel failure modes which could cause risk to the public are very remote possibilities indeed.

Integration of reliability/risk objectives with the design in a disciplined approach can only improve the design by supplementing the traditional methods of safety assurance. However, this will not occur as part of a regular commercial process unless credit is obtained for its implementation by allowing accredited use of risk assessment as part of the application both for the Construction Permit and Operating License. This infers that present standards work in the area of risk assessment must also be contributing to eventual rulemaking by NRC. NRC is positively contributing to the standards activities so the signs are hopeful that risk/reliability activities will in the future beneficially contribute to the licensing process.

REFERENCES

(1)   Code of Federal Regulations, Title 10, Part 50.34, Paragraph a.4,
      January 1, 1976.

(2)   WASH-1400 (NUREG 75/014), Reactor Safety Study: An Assessment of
      Accident Risks in U.S. Commercial Nuclear Power Plants, Nuclear
      Regulatory Commission, October 1975.

(3)   Letter R. P. DENISE to L. W. CAFFEY, Docket No. 50-537, May 6, 1976.

(3a) O. H. CRITCHLEY, Risk prediction, safety analysis and quantitative
      probability methods - a caveat, J.Br.Nucl. Energy Soc., 15,
      No. 1.18-20, January 1976.

(4)   WASH-1270, Technical Report on Anticipated Transients Without Scram
      for Water-Cooled Power Reactors, Regulatory Staff, USAEC,
      September 1973.

(4a) Code of Federal Regulations, Title 10, Parts 20, 50 and 100,
      January 1, 1976.

(4b) A. AUXIER and W. SNYDER, ORNL, Private Communication, January 1977.

(5)   J. D. GRIFFITH, J. GRAHAM, P. GREEBLER and R. LANCET, Safety-
      related design considerations for large breeder reactor plants,
      ANS International Meeting on Fast Reactor Safety and Related
      Physics, Chicago, October 5-8, 1976.

(6)   Draft ANSI-N214, General Design Criteria for LMFBRs, 1974.

(7)   Letter P. P. ZEMANICK to A. M. SMITH (GE), Large LMFBR shutdown
      heat removal system reliability trade-off study, SNR-77-003,
      January 5, 1977.

(8)   F. J. BALOH, N. W. BROWN, John GRAHAM, A. M. SMITH, and P. P.
      ZEMANICK, Clinch River Breeder Reactor Plant System Reliability,
      IEEE Transactions of Reliability, Vol. R-25, No. 3, August 1976.

(9)   189a, CW077, Inherently Safe Core Design: Self-actuated Shutdown
      Systems, October 1976.

(10) CRBRP Preliminary Safety Analysis Report, Appendix C, Section 1,
      December 1976.

(11) WASH-1285, Report on the Integrity of Reactor Vessels for Light
      Water Power Reactors, Advisory Committee on Reactor Safeguards,
      January 1974.

(12) WASH-1318, Techanical Report on Analysis of Pressure Vessel
      Statistics from Fossil-Fueled Power Plant Service and Assessment
      of Reactor Vessel Reliability in Nuclear Power Plant Service,
      Regulatory Staff USAEC, May 1974.

(13) ASME Boiler and Pressure Vessel Code, Section I, Power Boilers,
      ASME, 1974.

(14) ASME Boiler and Pressure Vessel Code, Section VIII, Pressure
      Vessels, ASME, 1974.

(15) ASME Boiler and Pressure Vessel Code, Section III, Nuclear Power
      Plant Components, ASME, 1974.

(16) W. E. COOPER and B. F. LANGER, Nuclear vessels are safe, Mechanical
      Engineering, p. 18, April 1975.

(17) R. C. DEYOUNG and D. C. EISENHUT, Current plans of the Regulatory
      staff for the use of probabilistic assessment, presented at ANS
      Winter Meeting, San Francisco, November 1975.

(18) E. B. HAUGEN, Probabilistic Approaches to Design, John Wiley,
      New York, 1968.

(19) H. G. ARNOLD, Pressure vessel reliability as a function of allow-
      able stress, ASME paper No. 73-WA/NE-15, August 13, 1973.

(20) J. M. DUKE, M. MAZUMDAR, W. J. MORALES, Reliability of slightly
      oval cylindrical shells against elastic-plastic collapse,
      Reliability Engineering in Pressure Vessels and Piping, ASME, New
      York, 1975.

(21) ASME Boiler and Pressure Vessel Code Case 1592-7, Class 1
      Components in Elevated Temperature Service, Section III, approved
      by Council December 22, 1975.

THE ROLE OF RELIABILITY IN THE LMFBR INDUSTRY

J. R. PENLAND*, A. M. SMITH† AND D. K. GOESER†

Abstract. This mission of a Reliability Program for an LMFBR
should be to enhance the design and operational characteristics relative to safety and to plant availability. Successful accomplishment
of this mission requires proper integration of several reliability engineering tasks--analysis, testing, parts controls and program controls.
Such integration requires, in turn, that the program be structured,
planned and managed. This paper describes the technical integration
necessary and the management activities required to achieve mission
success for LMFBR's.

1. Introduction. The development and application of reliability

engineering techniques in the nuclear power industry to date have been

primarily directed at evaluation of safety-related systems and events.

These evaluations have utilized not only quantitative aspects of reli-

ability analysis but also significant areas of qualitative analysis.

These studies have addressed subjects such as common cause failure and

human operator error. The study objectives have been to enhance safety

design features and related decisions and, thus, to improve licensabil-

ity. The Reactor Safety Study was performed to enhance the ability to

make decisions. Numerous examples of studies aimed at licensability

exist (see references 2, 3 and 4, for example).

---

*TVA, CRBRP Office, Oak Ridge, TN 37830. †General Electric Company.
+Westinghouse Electric Corporation.

The Liquid Metal Fast Breeder Reactor (LMFBR) industry has pro-
ceeded toward commercialization only very recently and has had the
ability to learn from the experience of the Light Water Reactor (LWR)
industry and from other industries utilizing reliability engineering
techniques. These experiences indicate that the role of reliability
engineering should also include design and operation enhancement to
realize the most effective use of the technology.

The goal of design and operation enhancement is applicable to
licensability, safety, and plant availability and is being used in
varying degrees in each of the areas by the LMFBR industry. Addressing this goal effectively requires the use of the full spectrum of
reliability engineering techniques. For reliability engineering to
achieve enhancement of design and operation, organizational elements
other than reliability specialists must be an integral part of the
reliability program.

This paper describes a suggested approach to utilize the full
scope of reliability engineering technology in an integrated manner
with the total design and operational cycle. This approach reflects
the results of the current learning process (including trial and
error) in the LMFBR industry as well as the experience in other
industries.

In the following sections, the objectives of an LMFBR Reliability

Program are described, the basic approach to achieving the objectives
are delineated, key technical elements are defined and conclusions are
reached.

2. Objectives of an LMFBR Reliability Program. The objective is
to achieve reliability enhancement of design and operation for safety,
plant availability and licensability. Although functional and technical
interdependencies exist, these areas are described below.

In the area of reactor safety, the greatest emphasis is applied to
prevention of the Core Disruptive Accident (CDA). Accidents of lower
consequence but of higher probability are not excluded; but at this
stage of LMFBR development, primary emphasis must be placed on the CDA.
Breeding of new fuel and utilization of a fast neutron spectrum produces core configurations which have the potential, albeit probabilistically remote, for the Core Disruptive Accident. It is theoretically
possible that the LMFBR CDA can cause core disassembly, partially
vaporize fuel and produce severe structural loadings.

In applying an approach to reduce the probability of the CDA, the
LMFBR industry is taking additional precautions relative to normal
reactor safety practice. Although LWR safety indirectly treats reliability[5,6], the principal LWR emphasis has been placed upon mitigation
of consequences of hypothetical accidents. This approach is used for
LMFBR's[7] but is augmented by direct programs to decrease the probability

of such occurrences.

To minimize the CDA probability, major emphasis is placed upon two systems--the Reactor Shutdown System (RSS) and the Shutdown Heat Removal System (SHRS). Although other systems are involved in CDA event sequences, all involve failure of one or both of the RSS and SHRS[8]. The most effective treatment of the CDA is through concentration on these systems.

Treatment of RSS and SHRS reliabilities is directed at reducing their failure probabilities in fact, and establishing via numerical analysis that these probabilities are vanishingly small for all known failure modes. Although regulatory authorities endorse this approach[9], the lack of convincing statements hampers the use of reliability technology directly in the licensing process. Quantitative reliability estimates are an important part of the process but other aspects of qualitative analysis, testing program results, reliability controls and engineering judgement must be an integral part of the program.

The LMFBR reliability program approach to licensability is, therefore, to concentrate on those activities and items which will produce enhanced safety. Many of these activities, by their very nature, are not amenable to generating absolute proof of conservatism. As these activities mature and confirm the ability to enhance safety, the licensing process must endorse those activities and give appropriate credit.

An equally important application of LMFBR reliability technology
is plant availability. The current utility atmosphere requires high
plant availability as the most effective manner of achieving high
utilization of capital. The LMFBR is a new type of plant for commercial power production and, as such, must demonstrate the capability to
produce reliably. While safety and availability have many areas of
common interest and interface, availability encompasses additional considerations and, therefore, must be explicitly addressed in a comprehensive Reliability Program.

Reliability engineering is a vital ingredient in assuring economic competiveness. Without a sizeable base of operating plants from
which to learn, reliability technology must be more inclusive in
scope for LMFBR's than would be appropriate for LWR or fossil plants.
Analysis, testing and reliability related design controls must perform
functions which use of operating plant data would perform for LWR's
and fossil plants.

3.0. Basic Approach.

3.1. Program Strategy. The need for product reliability is
universally accepted. In viewing an overall product cycle from
inception through operation, reliability is often thought of as every-
body's job. It is true that a large percentage of the personnel

involved in designing, producing and using a product contribute directly to its reliability, but to set specific reliability objectives--and
achieve them--requires a deliberate and structured effort. As the
product complexity increases, and as the economic and/or safety consequences of unreliability become more severe, the necessity for a
structured reliability effort becomes a primary consideration.

Structuring a formal reliability effort for a complex product
is no mean task. Given that specific reliability objectives have been
selected, a cost-effective reliability program that permeates the product cycle must be developed. While different approaches can be
envisioned, a strategy that is simple and effective is illustrated in
Figure 3.1. An organized Reliability Engineering operation provides
the focus and technical expertise required to accomplish specific
reliability tasks. These tasks, in turn, involve two categories of
activities:

1. Reliability Engineering provides a <u>support role</u> to
   the line organizations responsible for design, fabrication, test and operation. In this support function, their role is to <u>influence</u> the product to
   achieve the specified reliability objectives. This
   is done in several ways, ranging from evaluations
   for system configuration selection to detailed reliability analyses that may be essential to deci-

sions for inspection techniques, sampling plans and
the formulation of operational procedures.

2. As a direct and prime responsibility, Reliability
    Engineering is charged with the independent and
    objective <u>measurement</u> <u>and</u> <u>evaluation</u> of reliability
    status versus the specified reliability objectives.
    It is essential that the responsible design organi-
    zations are not held accountable for such measure-
    ments; they plan an important support role in this
    case, but cannot be held both responsible for design
    and independent of its assessment.

(Figure 3.1)   **ROLE OF RELIABILITY ENGINEERING**

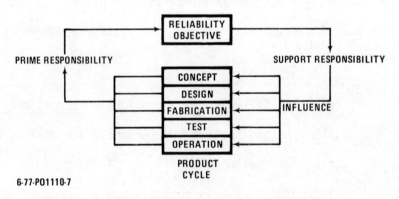

Thus, Figure 3.1 shows a typical closed loop situation. The
reliability objective guides the degree of influence that should be
fostered by Reliability Engineering. The measurement activity provides the status of reliability achievement. Feedback occurs when the

status and goal are not properly matched. This is a very important cycle and is the essence of the cost-effective reliability program. When further attention is required, the cycle so indicates; or conversely, it becomes clear when further effort (and dollars) are unwarranted. Without this definitive cycle, management must, at best, intuitively judge the necessity for further effort.

3.2. Program Elements. Implementation of the strategy requires the development of key program elements which can be used to effectively structure a formal reliability effort. Four such elements are suggested.

1. Reliability Objectives. Specific reliability objectives are essential to the conduct of a meaningful reliability program. These objectives most often are stated in quantitative terms and may be numerical goals defined at the plant level. These numerical values must be allocated down to specific systems and components so that each responsible design engineer will know that part of the reliability objective which is his to meet and satisfy. These objectives may have qualitative aspects to provide proof or confirmation of reliability factors which have been appropriately treated. Without these quantitative and qualitative objectives,

reliability studies tend to be meaningless.

2. Reliability Program Plan. This is the <u>road</u> <u>map</u> of
    how the objectives will be achieved. It is a spe-
    cific definition of the tasks to be performed, the
    level of effort required to do them, the schedular
    milestones for their accomplishment, and the assign-
    ment of responsibilities to do them. It includes
    organizational definitions and interface relation-
    ships (including those with co-contractors, custom-
    ers and vendors, as appropriate). Any good manage-
    ment approach requires a meaningful planning effort
    and a reliability program is no exception.

3. Reliability Control. Since reliability tends to be
    everybody's job, the question of control becomes
    very difficult. While the measurement and evalua-
    tion role of Reliability Engineering discussed in
    Section 3.1 is an important facet of overall con-
    trol, other, more discrete, control points through-
    out a product cycle should also be exercised. Four
    such control points where Reliability Engineering
    becomes an integral part of the mainstream process
    are noted.

a. Reliability approval of Final Design Review
before drawings are released for prime hardware
manufacture. If the Reliability Engineering support role has been properly accomplished, there
should be no problem with such approval. If a reliability problem does exist, that problem must not
proceed downstream into the prime hardware where
its correction becomes more expensive and time
consuming.

b. Reliability control and approval of a master
parts, materials and process (PMP) list. The PMP
list controls should require that all product specifications and drawings adhere to the disciplined
usage of only those items that have been scrutin-
ized and technically adjudged as suitable for the
product. The haphazard usage of parts, materials
and processes has proven to be the Achilles heel of
more than one product.

c. The placement of reliability objectives (or
their allocation) on all first-tier vendors. It is
virtually impossible for a prime manufacturer to
achieve his reliability objectives if he cannot
obtain the same degree of reliability achievement

from his first-tier vendors. To affect this element

of control, Reliability Engineering should have ap-

proval or concurrence authority on procurements to

assure the continuity of reliability objectives to

the first-tier vendors. Reliability requirements

placed on vendors must reflect the technical feasi-

bility of measuring compliance with the requirements.

d. Reliability control of failure analysis and

reporting activity on all reliability-critical com-

ponents. Data on failures of critical components is

perhaps the most important information that will

develop over the course of the product cycle. It

indicates the true reliability status and must be

addressed and resolved with meticulous attention to

detail. Reliability Engineering is not necessarily

the group embodied with the expertise to perform this

technical evaluation, but they are in the proper posi-

tion to control and administer this function indepen-

dently.

4. Reliability Visibility. Management, engineer and

technician alike must know what the objectives are,

where the product stands against those objectives,

and what their role is. Reliability Engineering

must institute written and oral reporting procedures
to assure that the reliability message is properly
conveyed. As a part of the Reliability Program
Plan, a structured reporting cycle should be established. This should be augmented throughout the
cycle with special presentations and reports to
keep management well advised of the reliability
status and activities.

A fifth element should be noted to avoid any
misconceptions. All the strategy and planning in
the work is for naught unless there is solid top
management backing and support for an effective
Reliability Program.

4. Key Technical Elements. Although safety, availability and
licensability differ in reliability missions, the applicable tasks to
the three areas have much in common.

To identify the required tasks, consider the ideal--having systems
which never fail to perform their required functions. For the system
never to fail either (a) the system must be defect free, or (b) the
system must be able to perform its function with failed components.

If the system is to have no defects, then it must embody 1.)
perfect design, and 2.) perfect parts. For the system to function in

spite of faults it must 3.) be fault tolerant, and 4.) there must

exist the ability to monitor the system and repair or replace failed

components.

None of the items 1 through 4 are absolutely achievable. To pro-

duce a system approaching the ideal, items 1 through 4 must each be

pursued.

In the following sections, each of these areas are examined. The

two areas related to design--assuring the design can perform its func-

tion and assuring that the design can tolerate failures--are combined.

4.1 Design Considerations. Reliability techniques appropri-

ately combined with design activities are an essential element of the

conceptual design phase. Their role is to analyze the plant concept to

identify (a) critical systems for safety, (b) critical systems for

availability, and (c) interrelations of systems for each of the safety

and the availability missions. This analysis is qualitative and semi-

quantitative in nature, first in the form of system level failure modes

and effects analysis (FMEA) and, second, the the form of quantitative

evaluations to select between various alternate configurations. Inputs

are deterministic systems analyses, initial systems requirement and

engineering specifications, LWR history and data and engineering judg-

ment. An example of the impact such an evaluation may have is that

the need for two separate, redundant and diverse shutdown systems may

be identified. Evaluations leading to this conclusion rely on opera-

tional histories from other reactors and estimates of the effect of
such failures. In addition to forming an early identification of
critical systems, results of the analyses can be fed back into the
design process as design requirements. At this early stage in which
no detailed design information exists, the impact, and therefore the
importance, of reliability evaluations is significant.

A top-level analysis of identified critical systems should be
performed as a cooperative effort of design and reliability engineers.
This analysis should identify system configuration problems and define
a Reliability-Critical Items List (RCIL). The RCIL is a tool to maintain visibility and cognizance of the status, actions required, results
of tests, etc. of components, subsystems, or systems which are either
critical to success of the overall system or which are subject to
significant uncertainties.

System analysis should be performed by reliability engineers and
design engineers acting as a team. This coordination is necessary
since: 1) The reliability engineer is trained to evaluate in a systematic manner from a potential weakness viewpoint as opposed to the
designer's more direct "here is now it works" perspective. 2) Without the detailed knowledge of the design engineer, the results of the
reliability evaluation may be incomplete or, in cases, incorrect.
3) The combined effort assures reliability feedback into design at
the earliest possible time.

The top level system evaluation should identify areas for more detailed evaluation. From this definition, the reliability engineer should develop guidelines or criteria for design of these areas. Typically, these will take the form of qualitative requirements such as modified single failure criteria[10]. Formal mechanisms for imposition of these guidelines into design should exist.

After completion of the initial system evaluation, more detailed analysis and initiation of testing programs should begin. This second level of analysis is generally quantitative in nature. To impact design the most realistic portrait of the system is necessary in order to accurately pin-point system weakness. If overly conservative data or assumptions are utilized, the introduced conservatism may mask real problems.

Based upon qualitative and quantitative reliability studies, testing programs may be advisable. It is often not economically feasible to test to demonstrate failure probabilities at a given confidence level. Effective testing should generally be directed at identification of failure inducing parameters, definition of failure modes, characterization of failure effects and experimental searching for unsuspected failure factors. Appropriate statistical treatment of the data should be performed to achieve full utilization of experimental information.

Due to the large costs, testing must be concentrated on unproven systems. Furthermore, advantage must be taken of existing tests. For nuclear plant components these could include design verification tests, prototype tests, acceptance tests and pre-operational tests. In certain instances specific reliability testing beyond these tests may be a good return on investment. This should be considered in the case of new designs or where potential failure modes warrant the accumulation of more detailed data. In addition, Reliability Engineering should have strong input to all phases of testing.

4.2. Parts Controls. Reliability analysis and the initial phases of testing will identify components critical to system reliability. These parts should then be entered into the Reliability-Critical Items List (RCIL). The RCIL by itself accomplishes nothing, but the controls applied to RCIL items are the key to an effective means of improving reliability.

Controls for RCIL components should address initial specifications through system operation. Major points at which parts controls come into play are specifications, procurement, development testing, acceptance testing, installation, pre-operational testing and operations.

For electrical parts, a detailed structure for grading of parts exists from military and aerospace experience. Grading levels reflect not only manufacturing procedures but also differing levels of inspection and burn-in. Data[11] indicate that reductions in component

unreliability of factors of five to ten are achievable by more rig-

orous specifications. Although these data primarily relfect random

failure rates, reduction in susceptability to random failures also

reduces sensitivity to common cause failures. More rigorous parts

specification impacts safety and also aids plant availability by

reducing the probability of spurious activation of back-up systems.

While existing specifications cover several aspects of electron-

ics reliability, mechanical component controls are more complicated.

For mechanical components, reliability engineering should have a

direct, formal input into detailed engineering specifications, manu-

facturing procedures, installation procedures and QA requirements.

It may become necessary to perform qualitative analyses of manufac-

turing and installation procedures to identify and eliminate elements

which could degrade the inherent design reliability. Further effort

to evaluate historical failures in like equipment may be necessary as

a reliability grading system comparable to that for electrical parts

does not exist.

Procurement of RCIL components must be monitored closely by

reliability engineering. Although specifications may be established

to perfection, the real world of procurement produces exceptions to

the specifications. These exceptions must be analyzed to determine

their effects on component reliability.

A failure reporting, analysis and corrective action program must be implemented. This system consists of the following elements.

1. A procedural system must detect and report failures which occur during development, acceptance, pre-operational testing and operations.

2. Sufficiently detailed information must be collected on reported failures to guide failure analysts.

3. In many cases, special failure analysis must be performed to identify the failure causes.

4. Parts traceability must exist in order to implement corrective actions or like or similar components.

5. A positive corrective action system must exist.

The failure reporting, analysis and corrective action system is one of the most critical aspects of a reliability engineering program. This system is the dynamic means for insuring feedback into future equipment.

4.3. Maintainability and Repairability. For even a system of high order redundancy to achieve high reliability, that system must be capable of being effectively maintained and efficiently repaired. From a quantitative analysis point of view, the repair time is as important a parameter as failure rate.

Maintainability and repairability considerations should permeate design and operational procedures. The following summarizes key tasks

and elements which should be considered prior to operation.

1.   Design. The design should be analyzed to evaluate abilility
     to perform maintenance or repair when the plant is "on-line".

2.   Diagnostics. Analysis should be performed to assure that key
     failures are annuciated to the operations staff. Considera-
     tion should be given to special diagnostic systems for incipi-
     ent failure detection.

3.   Accessability. Analyses and tests, if feasible, should be
     performed to assure that repair or replacement can be perform-
     ed with a minimum of strain on personnel or equipment. Plant
     scale models are an invaluable tool for this effort.

4.   Spare Parts. Analysis and historical experience should be
     used to determine what spare parts should be on-site to expe-
     dite repair.

5.   Tools. Special tools required for maintenance and repair
     activities should be identified.

6.   Manpower. Human resource requirements should be identified.
     Not only the number of men, but also their necessary skills
     should be identified and training programs established.

7.   Time Line Studies. For critical operations such as refueling,
     detailed time line studies should be performed to identify
     possible problems and to identify means of accelerating the
     operation. Contingencies for generic types of problems should

be identified. As a part of the time-lining efforts, logistics problems should be analyzed.

After operation of the facility, repairability and maintainability programs should stay in effect. Key elements for this phase are maintenance and analysis of repair operations and feedback of this information into on-going efforts.

5.   Conclusions. This paper has presented descriptions and recommendations based upon current LMFBR and other related industrial experience. Based upon this information, the following conclusions are reached.

1. The need for reliability programs directed at safety and at availability exists in the real world and should become a more direct part of the licensing process.

2. Haphazard approach to a reliability program will not produce desired results. In order to achieve design and operation enhancement, many complex tasks must be integrated to achieve the synergistic benefits. An overall "systems" approach to building reliability into nuclear plants must be adopted.

3. For a large project a formalized and structured approach is necessary and should be appropriately considered in the contractual aspects of the project. This is required to assure uniformity, viability and management support and also to assure that the results of reliability technology are fed

back into design.

4. A balanced program relative to design, parts and maintenance
   is required. Omission of one down grades the effectiveness
   of the others.

REFERENCES

(1) Reactor Safety Study: An Assessment of Accident Risks in U.S. Commercial Nuclear Power Reactors, WASH-1400 (NUREG-75/-14) U.S. Nuclear Regulatory Commission, October 1975.

(2) J. SCOTT, V. GALAN, L. CARTIN, W. HELDENBRAND, J. PENLAND, Analysis of Anticipated Transients Without Trip, BAW-10016, Babcock and Wilcox, September 1972.

(3) W. GANGLOFF and W. LOFTUS, An Evaluation of Solid State Logic Reactor Protection in Anticipated Transients, WCAP-7706, Westinghouse Electric Corporation, July 1971.

(4) B. TASHJAN, Sensitivity Analysis of a Two-Out-of-Four Coincidence Logic Reactor Protection System, Nuclear Science Transactions, vol. NS-18 (Feb. 1971), pp. 455-464.

(5) IEEE Standard 279-1971 (ANSI N42.7-1972), Criteria for Protection Systems for Nuclear Power Generating Stations.

(6) Technical report on anticipated transients without scram for water-cooled reactors, WASH-1970, US. Atomic Energy Commission, 1973.

(7) Third Level Thermal Margins in the Clinch River Breeder Reactor Plant: Updated Analysis for the First 24 Hours, November 1976. An enclosure to ERDA Letter S:L:1820, A. R. Buhl to R. S. Boyd, CRBRP - 24-Hour Non-Venting Criteria, November 5, 1976.

(8) Clinch River Breeder Reactor Plant Preliminary Safety Analysis Report; Appendix C: Safety-Related Reliability Program, Docket No. 50-537.

(9) Letter, R. DENISE to L. CAFFEY, Docket No. 50-537, May 6, 1976.

(10) IEEE Standard 379-1972 (ANSI N41.2-1972), Guide for the Application of the Single Failure Criterion to Nuclear Power Generating Station Protection Systems.

(11) Reliability Prediction of Electronic Equipment, MIL-HDBK-217B, September 20, 1974.

A METHOD FOR QUANTIFYING LOGIC MODELS
FOR SAFETY ANALYSIS

R. C. ERDMANN, J. E. KELLY, H. R. KIRCH, F. L. LEVERENZ AND E. T. RUMBLE[*]

Abstract. The accomplishment of any detailed reliability or risk analysis task involves both engineering judgement and accurate analytical procedures. In this paper procedures are described which have been programmed so that a variety of information concerning reliability, availability, risk assessment, and cost impact can be evaluated quickly and accurately. Utilizing a common input deck the "WAM" codes efficiently and accurately provide information about systems modeled by any Boolean function. This information includes:

1. Point estimates of the system (top event) reliability (or unreliability) together with the reliability of any event within the system (WAM-BAM code).

2. A reevaluation of the system as described in (1) with changes made to the probability of occurrence of basic events (WAM-TAP code).

3. Qualitative assessment of the system in terms of failures (cut-sets) which cause the system to fail and which cause any event within the system to occur (WAM-CUT code).

4. Qualitative assessment of the system and events within the system together with the first, and if desired, second moment of the probability of the events being analyzed. This allows modeling the basic system components as random variables with a mean and standard deviation included in the model (WAM-CUT code).

5. Qualitative assessment of the system which is displayed in terms of cut sets and the probability polynominal (WAM-CUT code). This can be stored for use by a Monte Carlo code which allows determination of the entire distribution of the system reliability as a function of component distributions (SPASM code).

6. A drawing of the fault tree as input to the evaluation codes (WAM-DRAW).

This paper will describe the development of these codes and present example problems which illustrate the codes' capabilities.

---

[*]Science Applications, Inc., Palo Alto, California 94304. This work was supported by the Electric Power Research Institute.

1. Introduction. The WAM series of computer codes has been
designed to provide flexibility as well as accuracy in the analysis of
system reliability models utilizing any or all Boolean functions. The
use of these codes requires a minimum amount of user interaction after
the problem has been logically modeled (fault tree, event tree, Boolean
expression, etc.).

These codes were developed to take advantage of techniques used in
existing codes and to make available capabilities not found in existing
codes. Codes examined included PREP-KITT [1], GO [2], MOCUS [3],
ALLCUTS [4], and SAMPLE [5].

Subsequent sections of this paper discuss the interrelationships
of the WAM codes, their capabilities, and how they work. Finally,
problems solved by the WAM codes are compared to PREP-KITT and MOCUS
solutions.

2. WAM Code Interrelationships and Capabilities. Figure 1
depicts the functional relationship between code modules. The preprocessor named WAM allows the system analyst to easily communicate
with the BAM, CUT and DRAW codes. WAM accepts the logic tree in which
components and gates are input with alphanumeric names, and allows up
to eight inputs per AND* and OR* gate. A multitude of checks are
performed to advise the user of mistakes in his model. In addition,
the input to the desired evaluation module is optimized to reduce
running time and maximize accuracy.

---

*The AND, OR, NOT gates represent the intersection, union and "not"
  logical operations between sets. Also see reference [6].

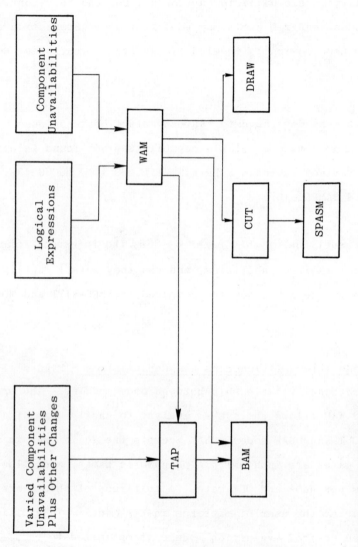

BAM = Boolean Arithmetic Model

SPASM = System Probabilistic Analyses by Sampling Methods

TAP, WAM, CUT, DRAW = Not anacronyms

Figure 1. Code Module Interrelationships.

The inputs to WAM allow models which includes all sixteen Boolean operations [6] between two variables. Basically, this extension is accomplished by the incorporation of NOT* gates. This makes possible the explicit modeling of dependent events. WAM also allows the input of combinational failures; i.e., then a system fails because at least N of M branches fail. The input to WAM to describe the combinational failures is the name of each M branch and how many (N) must fail to cause system failure. Up to eight branches are allowed as inputs to the "combination" gate.

The BAM model uses Boolean algebra minimization techniques to find intermediate and final logic expressions for the input tree and then calculates the associated point unavailabilities. BAM forms a truth table that describes each gate as a function of all possible event combinations. Details of BAM will be presented later.

At the user's option, the input to BAM can be saved from a WAM run and subsequently called by WAM-TAP. This saves re-running WAM and allows probabilities to be changes for specific components identified by common alphanumeric characters in the component name. For example, if the event code used letters "HE" as the first two letters of every Human Error component, WAM-TAP can search for these and change the failure rate for each component starting with "HE" by a multiplication factor or set them to a specific value. Following this, the tree is reevaluated via the BAM code. This capability allows sensitivity studies or common mode studies to be easily accomplished.

---

* The AND, OR, NOT gates represent the intersection, union and "not" logical operations between sets. Also see reference [6].

Three options are available through the utilization of CUT. Inputs processed through WAM are logically interrogated to determine the minimum cut (path) sets in all three options. Option one prints these cut sets for the requested events. In the second option, the first and second moments about the origin, the mean and variance, and the 95% Chebyshev bound of any event including the top event are calculated and printed with the cut sets. The third option determines the arithmetic probabilistic statement (probability polynominal) from the cut sets.

The probability polynominal obtained from CUT is used in the SPASM module. This module determines the top event distribution through Monte Carlo sampling of the probability polynominal.

The DRAW module accepts logical input data processed through WAM and draws a fault tree pictorial representation for consistent graphical display.

Thus, utilizing the WAM codes, logic models can be quantified effectively, in a variety of ways, by specifying the right module combinations, consistent with the size and complexity of the problem.

3.   Principles of Fault Tree Evaluation by WAM-BAM. The BAM code uses Boolean algebra minimization techniques to find the resultant logic expressions from an input tree and then calculates the associated point unavailability. BAM first forms all possible combinations of basic events and them forms a truth table that describes each event and gate as a function of these combinations.

This basic methodology is computationally optimized, based on techniques used in the GO computer code [2].

The combinations of events are called "P Terms" as they are the product terms of the canonical P form [6] of the equivalent Boolean expression for the fault tree being analyzed.

In this application, however, the P terms not only correspond to the events for the logical construction of the truth table, but, also, are assigned their numerical values in the probability space. This can be done since the logical variables represent statistically independent events and the P terms are all mutually exclusive events. Thus, the probability value for any P term is the product of the specific variable probabilities.

In addition, the probability of the union of the P terms is the sum of the P term probabilities. Thus, the BAM module resolves the logical expression as represented by the fault tree by: forming the P term events; calculating the P term probabilities; representing the logical expression by the canonical P term expression using a truth table; and, summation of the applicable P term probabilities to give the desired event probabilities.

As an example of the operation of WAM-BAM, consider the following example fault tree, Figure 2 .

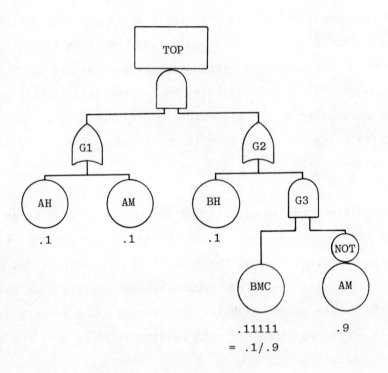

Figure 2.

This tree would be described by the following input statements to WAM-BAM:

TOP	AND	G1	G2
G1	OR	AH	AM
G2	OR	G3	BH
G3	ANOT	BMC	AM

WAM then resolves the inputs into gates with two inputs (in this case the input is already in this form) and turns the tree over.

G3	ANOT	BMC	AM
G2	OR	G3	BH
G1	OR	AH	AM
TOP	AND	G1	G2

BAM then forms the truth table and P terms starting at the first of these statements. (The "bottom" of the tree.) BAM deletes gates or components not needed in further operations and also deletes product terms less than a user specified minimum (PMIN).

The general flow of BAM is shown in Figure 3. As an example of the processing of one gate, consider the evaluation of:

G1 = AH   OR   AM

At the evaluation of G1, G3 and G2 have already been evaluated and the resultant truth table is as shown in Table I. Note that the table contains only events needed in future evaluation, (i.e., evaluation of G1 and TOP).

PRODUCTS	AM	G2
.01	1	1
.09	1	0
.18	0	1
.72	0	0

Table I.

Processing of the next gate (G1) continues as shown below, illustrating one pass through the flow chart of BAM, Figure 3 .

1) Include AH (P = .1) in the truth table: multiply each existing product by P(AH) and P($\overline{AH}$) leaving out products less than user specified minimum: assume PMIN = .005.

PRODUCTS	AM	G2	AH
.009	1	1	0
.009	1	0	1
.081	1	0	0
.018	0	1	1
.162	0	1	0
.072	0	0	1
.648	0	0	0

Table II.

2) Calculate next the gate as a function of existing gates and components: G1 = AH. OR. AM

PRODUCTS	AM	G2	AH	G1
.009	1	1	0	1
.009	1	0	1	1
.081	1	0	0	1
.018	0	1	1	1
.162	0	1	0	0
.072	0	0	1	1
.648	0	0	0	0

Table III.

3) Delete components of gates not appearing in the tree in future steps: AM and AH in this case.

PRODUCTS	G2	G1
.009	1	1
.009	0	1
.081	0	1
.018	1	1
.162	1	0
.072	0	1
.648	0	0

Table IV.

4) Simplify the truth table by adding products where truth table rows are identical.

PRODUCTS	G2	G1
.027	1	1
.162	1	0
.162	0	1
.648	0	0

Table V.

5) Process next gate.

3.1. Capabilities and Limitation of WAM-BAM.   WAM-BAM will accept
trees with up to 1500 basic events and the equivalent of 1500 two-input
gates.   This limitation allows the code to operate any CDC machine
having 65K small core memory.   The size of the tree could be larger if
more than 65K of memory were available.   Conversely, it could also be
modified to accept smaller trees and fit in less core.   A general rule
is that seven locations of memory are required for each additional gate
or basic event above the current limitations.   For example, if a
capability of 2000 basic events and 2000 gates were required then 7K
of additional memory must be made available.

During the evaluation of the fault tree, probabilities of any basic
event or gate can be saved for the final output.   This allows numerical
evaluation of portions of the tree "below" the top event.   It is also
possible via WAM-TAP to repeat the evaluation of the tree with changes
in the probabilities assigned to the basic events.   Thus, sensitivity
studies or the evaluation of the effect of a parameter such as time
could be accomplished.

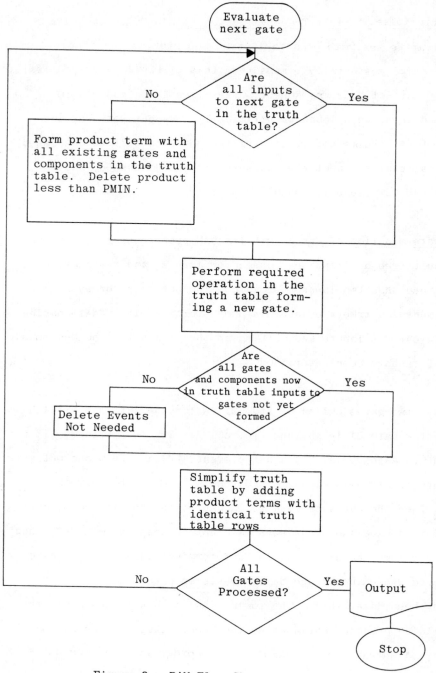

Figure 3.   BAM Flow Chart

4. Principles of Fault Tree Evaluation by WAM-CUT. In order to
allow qualitative analysis of a fault tree and provide for formulation
of the fault tree most easily transformed to a probability polynominal,
an efficient cut set code was sought. First, existing codes were
examined, and then, when found insufficient, a new cut set code was
developed. First the method of finding the cut sets will be discussed,
then numerical evaluation of the cut sets will be discussed, and
finally the WAM-CUT capabilities discussed.

4.1. Determination of Cut Sets. The WAM preprocessor is used to
read the fault tree descriptions, check for errors, and restructure the
tree to no more than two inputs per gate. The cut sets for each gate
of the restructured tree are determined by starting with a gate having
only basic events as inputs and working up the tree until the cut sets
of the top gate have been determined.

Consider two gates called G1 and G2. After the cut sets of G1 have
been found, the gate G1 is an input to, call it G2, is examined. If the
second input to gate G2 is: (1) a basic event; (2) a gate whose cut sets
have been previously determined; or (3) the NOT of a gate previously
determined, then the cut sets of gate G2 can be found. If the other
input to G2 is a gate whose cut sets have not yet been determined, then
the code moves down the branch of the undetermined input until a gate
is encountered whose inputs are basic events or gates whose cut sets
have already been determined. The logic flow is then reversed and the
cut set building process continues up the tree. This process can be
more clearly illustrated by an example. The order in which the cut sets
are determined for the fault tree in Figure 4 is indicated by the

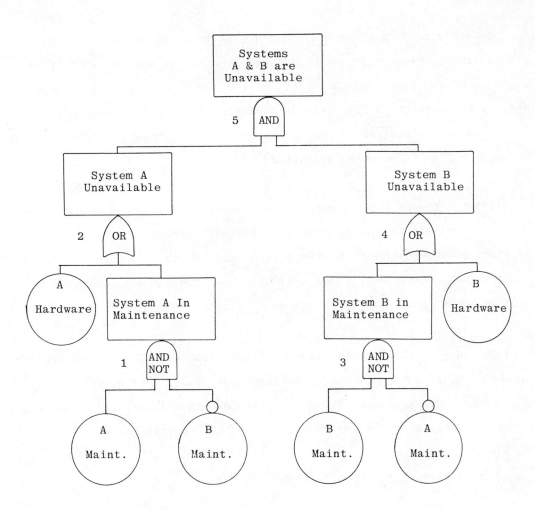

Figure 4.   Example Fault Tree Showing Order of WAM-Cut
            Operation.

numbers to the left of the gates.

In order to form the cut sets of a gate from the cut sets of its inputs, three operations must be defined; the ANDing and ORing of the cut sets of two gates and the NOTing of the cut sets of a gate. These three operations are a functionally complete set of logic operations, i.e., any of the sixteen Boolean functions of two variables can be determined from these three operations [6].

A cut set of a gate is defined as a set of basic events which, when simultaneously failed, cause that gate to fail. Hence, a cut set can be written in Boolean algebra as the intersection of the basic events in that cut set. Since a gate may have more than one cut set, the Boolean expression for that gate is the union of its cut set.

Consider a gate A having two sets, the first containing two basic events a AND b, and the second containing one basic event c; and a gate B with three cut sets containing the basic events d AND e, f AND c, and a AND b. These gates can be represented by the Boolean equations:

$$A = (a \cdot b) + c, \quad B = (d \cdot e) + (f \cdot c) + (a \cdot b).$$

The operations of ANDing and ORing the cut sets of two gates are simply the applications of the Boolean operators to these two equations:

$$A + B = a \cdot b + c + d \cdot e + f \cdot c + a \cdot b,$$

$$A \cdot B = (a \cdot b + c) \cdot (d \cdot e + f \cdot c + a \cdot b) = (a \cdot b \cdot d \cdot e) + (a \cdot b \cdot f \cdot c) + (a \cdot b \cdot a \cdot b) + (c \cdot d \cdot e) + (c \cdot f \cdot c) + (c \cdot a \cdot b).$$

At this point some minimization can be performed to delete duplicate events in a cut set and to delete cut sets that are supersets of other cut sets; i.e., the cut sets are reduced to minimal cut sets. This minimization yields:

$$A + B = (a \cdot b) + c + (d \cdot e) \ ,$$
$$A \cdot B = (a \cdot b) + (c \cdot d \cdot e) + (c \cdot f) \ .$$

The NOTing operation is simply the application of DeMorgan's Theorem [7] to the cut set equation; e.g.,

$$\overline{A} = \overline{(a \cdot b + c)} = \overline{(a \cdot b)} \cdot \overline{c} \ \ (\overline{a} + \overline{b}) \cdot \overline{c} = \overline{a}\overline{c} + \overline{b}\overline{c} \ \ .$$

These Boolean operators along with the minimization technique described above have been implemented into the WAM-CUT computer code.

4.2. Numerical Evaluation of the Cut Sets. The evaluation option of WAM-CUT allows estimation of the effect of treating the fault tree components as random variables. In essence, the cut sets of the desired events are evaluated by calculating the first and second moments about the origin based on component first and second moments about the origin. The process of determining the first and second moment of an event requires first determination of the first and second polynominals, and then evaluation of these polynominals using input component data.

For example, assume the following cut sets result from a fault tree:

A

BC

D

This is equivalent to the following probability expression for the top
event, T:

P(T) = P(A U BC U D)

= P(A) + P(B)P(C) - P(A)P(B)P(C) + P($\overline{A}$)P(D)

- P($\overline{A}$)P(B)P(C)P(D).

This expression assumes all events are independent.

For this discussion, each event is a random variable and the
expected value operation on the polynominal will yield polynominals for
the moments of the top event which are functions of the moments of the
components.

Thus, the nth moment polynominal becomes:

$$E\left[T^n\right] = E\left[(a + bc - abc + \overline{a}d = \overline{a}bcd)^n\right] .$$

Then, dropping the expected value operation for convenience, the top
event first moment is obtained from the following expression:

T = a + ab - abc + $\overline{a}$d - $\overline{a}$bcd.

In the above expression, a, b, c, and d are first moments of the components. Note:

$$E[\bar{a}] = E[1-a] = 1-E[a]$$

or the first moment of "NOT a" is one minus the first moment of a.

Similarly, the second moment polynominal is obtained (again the expected value operator is dropped):

$$T^2 = a^2 - abc - a^2bc + a\bar{a}d - a\bar{a}bcd$$
$$+ bca + b^2c^2 - ab^2c^2 + bc\bar{a}d - \bar{a}b^2c^2d$$
$$- a^2bc - ab^2c^2 + a^2b^2c^2 - a\bar{a}bcd + aab^2c^2d$$
$$- a\bar{a}bcd - \bar{a}b^2c^2d + a\bar{a}b^2c^2d - \bar{a}^2bcd - \bar{a}^2b^2c^2d^2.$$

In the above expression, a, b, c, and d are the first moments of component unavailability and a^2, b^2, c^2, and d^2 are the second moments of component unavailability. Other terms are calculated:

$$\bar{a} = 1 - a ,$$
$$a\bar{a} = a - a^2 ,$$
$$\bar{a}^2 = 1 - 2a + a^2 .$$

Where a is the first moment, and a^2 is the second moment.

After the polynominals are determined, component data are used to calculate the mean and standard deviation of the desired event. Should a complete definition of the top event distribution be desired (assuming complete component distributions are known), an option in WAM-CUT writes

a subroutine for SPASM. The subroutine is essentially the first moment
polynominal. SPASM provides values for the terms in the polynominal
from the component distributions for up to 40,000 trials. Thus, the
top event distribution is determined via Monte Carlo methods.

4.3. Capabilities and Limitations of WAM-CUT. WAM-CUT is designed
to utilize specific features of the CDC 7600. Conversion to any other
system may be costly, and much of WAM-CUT's versatility could be
sacrificed. The system dependent features used are directly addressible
large core memory, available with CDC's FTN 4 compiler, and the random
access disk used for temporary storage. At execution WAM-CUT requires
65K of small core memory and 130K of large core memory.

The size of the fault tree input to WAM-CUT is limited to 1500
two-input gates and 1500 basic events. WAM-CUT can process up to 2150
cut sets per gate with no limitation on the number of basic events per
cut set. An estimation of the probability of each cut set is calculated
and those cut sets with probability less than a specified minimum are
discarded. This minimum probability is gate dependent so that valuable
information for gates other than the top gate is not lost and,
conversely, more information than needed is not generated to clutter up
the cut set determination process. WAM-CUT can also operate in the more
familiar mode of finding all the cut sets up to a specified number of
basic events; i.e., all singles, doubles, triples, etc.

As the cut sets of each gate are generated, they can be printed,
ordered by size and by probability, the first and second moments
calculated and the first moment polynominal generated for later use in a

Monte Carlo calculation. These options are all controlled by input
switches.

When the cut sets of a gate are determined, they are stored on a
random access disc if that gate appears more than once in the tree.
Thus, each gate is resolved into its cut sets only once, eliminating
the duplication which appears in many of the other cut set codes. This
elimination of duplicate analysis and the implementation of the minimum
probability cut-off have greatly decreased the execution time.

5. Automatic Fault Tree Drawing. WAM-DRAW provides a drawing of
the input tree on a Calcomp plotter. The same deck used for evaluation
is the input deck for WAM-DRAW. This drawing can be used as a formal
drawing for documentation purposes or as a visual representation of
the tree being analyzed for purposes of verifying that the original tree
was correctly transmitted for evaluation. WAM-DRAW presently will only
run on the Berkeley CDC computers as its Calcomp routines are unique
to that facility.

6. Code Operation on Problems. Table VI shows a comparison of
execution for evaluation of various fault trees using WAM codes. The
execution times for WAM-BAM and WAM-CUT (cut sets only) are dependent
not only on the size of the fault tree but also on the logic structure
of the tree, the probabilities of the basic events and the minimum
probability. The execution time for the moments calculation is
dependent not on the size of the tree but on the number of cut sets
and their probability.

Number of Two Input Gates	Number of Basic Events	Number of Cut Sets	Minimum Probability	WAM-BAM	WAM-CUT (Cut Sets Only)	WAM-CUT (First Moment)
18	11	25	10^{-10}	0.74	0.09	0.32
224	171	11	10^{-5}	9.82	1.47	1.62
		143	10^{-7}	10.08	8.73	9.62
298	124	20	10^{-5}	10.96	2.43	3.76
		297	10^{-7}	17.78	8.55	22.09
559	411	19	10^{-5}	24.86	6.76	8.09
		159	10^{-7}	25.10	10.39	72.66

Table VI. CDC 7600 Execution Times (Sec.) for Various Trees Solved via WAM Codes.

Finally, Table VII compares the execution times of WAM-CUT, PREP-KITT and MOCUS on a sample problem. The problem was varied by asking for first "singles", then "doubles", etc. All times are CDC 7600 execution times for these codes.

	SINGLES	DOUBLES	TRIPLES	QUADRUPLES
MOCUS	1.96	41.1	–	–
PREP-KITT	.78	1.23	14.8	–
WAM-CUT	.38	.68	13.4	22.0

Tree Size: 118 Basic Events
123 Two Input Gates

Table VII.   Comparative CDC 7600 Execution Times for MOCUS, PREP-KITT and WAM-CUT.

REFERENCES

[1] W.E. VESLEY and R.E. NARUM, PREP and KITT Computer Codes for the
     Automatic Evaluation of a Fault Tree, Idaho Nuclear
     Corporation, Idaho Falls, IN-1349, 1970.

[2] W. GATELY, D. STODDARD, and R. WILLIAMS, GO A Computer Program for
     the Reliability Analysis of Complex Systems, Kaman Sciences
     Corp., KN-67-704 (R), April, 1968.

[3] J. FUSSELL, E. HENRY, N. MARSHALL, MOCUS - A Computer Program to
     Obtain Minimal Cut Sets From Fault Trees, Aerojet Nuclear Co.,
     Report ANCR-1156, August 1974.

[4] W. VAN SLYKE, D. GRUFFING, ALLCUTS, A Fast Comprehensive Fault Tree
     Code, Atlantic Richfield Hanford Co., Report ARH-ST-112
     (Draft), July 1975.

[5] Reactor Safety Study, An Assessment of Accident Risk in U.S.
     Commercial Nuclear Power Plants, U.S. Nuclear Regulatory
     Commission, WASH-1400 (NUREG-75/014), October 1975.

[6] Y. CHEE, Digital Computer Design Fundamentals, McGraw-Hill, New
     York, 1962, pp 101-108.

[7] H. GSCHWIND, E. McCLUCKEY, Design of Digital Computers, Springer-
     Verlag, New York, 1975, p.33.

Topic 7
AUTOMATED LOGIC MODEL CONSTRUCTION

Topic 7
Automated Logic Model Construction

The papers in this final section all address automated system logic model construction. The subject is relatively new; the first published work in this area appeared in 1973. The papers present progress in a subject area that is not yet in production status.

J. R. Taylor and E. Hollo present a review of the work that has been carried out at RISØ in the first paper. The methods presented involve both the automatic construction of fault trees and cause-consequence diagrams. The resulting algorithms have been tested on academic sample problems. Presently, comparisons with conventional analysis of systems encountered in practice is under way.

In the second paper, S. A. Lapp and G. J. Powers present their algorithm for constructing fault trees. The algorithm uses a directed graph (digraph) as input. Some controversy exists concerning this algorithm. Professor E. J. Henley of the University of Houston writes "Following up on our conversation at Gatlinburg, we tried very hard to use Powers' nitric acid cooler plant (IEEE Transaction on Reliability, April 1977) as a test problem for comparing fault tree synthesis methods, and we failed". Henley goes on to state, "There are undoubtedly a set of assumptions that will lead to the tree. They are, however, hard to pick out". The editors feel that, in spite of this

controversy, Lapp and Powers have made significant advances in the area
of automated logic model construction.

In the next paper, by J. S. Wu, S. L. Salem and G. E. Apostolakis,
a method is presented for the construction of fault trees based on
tables that describe each possible output state of system components.
The method is stated to be applicable to mechanical, electrical and
hydraulic components. An example fault tree that is generated by the
methods is presented for a nuclear reactor containment spray recirculation system.

R. L. Williams and W. Y. Gateley present a technique for determining system minimal cut sets without manual construction of a logic
model. However, like the method of Lapp and Powers an intermediate
diagram is required. This diagram, intermediate to the system schematic
and the minimal cut sets, is called a "GO Chart". Considerable experience using GO charts as an aid to reliability analyses exists.

EXPERIENCE WITH ALGORITHMS
FOR AUTOMATIC FAILURE ANALYSIS

J. R. TAYLOR* AND E. HOLLO†

Abstract. The following pages present a brief review of the work carried out at Risø from 1973 to 1977, with the objective of constructing fault trees and cause-consequence diagrams automatically.

1. **Introduction.** The motives for automatic process plant failure analysis are

- to reduce the cost; at present, safety analysis costs may be several percent of the plant capital cost.

- to make as effective use as possible of the plant engineers' time; the engineers in posession of detailed information about plant construction often bear a very heavy workload.

- to improve the completeness of safety analyses.

- to develop the methods as a basis for alarm analysis and failure diagnosis.

Others have worked on problems of automatic fault tree construction (Fussell 1973, Powers & Tompkins 1973, Lapp & Powers 1976, Salem & Apostolakis 1976). Our work has especially concentrated on problems of event sequence - that is, on situations in which the order in which failures

---

* Research Establishment Risø, 4000 Roskilde, Denmark.
† Permanent address: Institute for Electrical Power Research, H-1051 Budapest, Zrinyi Street 1, Hungary.

occur is significant. This makes the methods suitable for treating
safety systems which operate in several phases, such as nuclear power
plant shut down systems; for batch production systems; for treating
problems of human error; and for alarm analysis, where the sequence of
alarm arrival provides additional information.

Central to the methods are: the concepts of an event - an event is
a significant change in some process plant variable or group of variables; the concept of a plant component transfer function - when an
event occurs at the input to a component, further events may occur at
the component outputs; and the concept of a condition - a condition is
a logical requirement which may be fulfilled by plant variables. Which
output events occur at the output of a component in response to a change
in an input variable may depend on which conditions hold within the
component, or on other component input variables.

2. Plant component descriptions. There is a wide range of possibilities for describing plant components. The simplest is to use a finite
state model, in which each plant variable may take only a restricted
number of values. Finite state transfer functions then describe the
change in component output variables (output event), or in component
state variables, in response to a change in input variables (input event)
(Fig. 1, 2). In such a model, failures can be represented by a "working
state" variable for each component, with values such as "OK", "failed
ON", "failed OFF". Working state variables change in response to "spon-

taneous events", which may occur even in the absence of input events,
or in response to an input event with large variable values, corresponding to an "overload".

To use such models, a plant block diagram is required, describing
the way in which components are interconnected. Then, starting with an
initial event at the input to one component, the event can be "matched"
to different terms in the component transfer function, and corresponding
output events "deduced". The process can be repeated treating output
events from one component as an input event to adjacent components in
the plant block diagram. In this way complete sequences of events can
be built up.

Finite state models are adequate for many types of plant components,
but for detailed modelling of quantitative phenomena, can be over restrictive. An extension of the finite state transition function is to allow
numeric expressions to represent events - an event establishes a new
equality or inequality between a plant variable and a numeric expression.
Transfer functions can then allow conditions expressed in terms of numerical equalities or inequalities, and output variables can be expressed
as a numeric function of input variables.

If equalities and inequalities using simple numbers are allowed,
then plant models are still somewhat restricted. A further relaxation
is to allow events to be described in terms of algebraic functions,
involving other plant component variables, and involving functions of
time. For example, a comparator might be described as IF INPUT1 > INPUT2

CONDITIONS	WATER IN	PRESENT				ABSENT			
	WORKING STATE	WORKING		FAILED OPEN		WORKING		FAILED OPEN	
	CONTROL	OPEN	CLOSED	OPEN	CLOSED	OPEN	CLOSED	OPEN	CLOSED
EVENTS	WATER IN → PRESENT					WATER OUT → PRESENT		WATER OUT → PRESENT	WATER OUT → PRESENT
	WATER IN → ABSENT	WATER OUT → ABSENT		WATER OUT → ABSENT	WATER OUT → ABSENT				
	WORKING STATE → FAIL OPEN		WATER OUT → PRESENT						
	WORKING STATE → FAIL CLOSED	WATER OUT → ABSENT							
	CONTROL → OPEN		WATER OUT → PRESENT						
	CONTROL → CLOSED	WATER OUT → ABSENT							

Fig. 1. Finite state transfer function table.

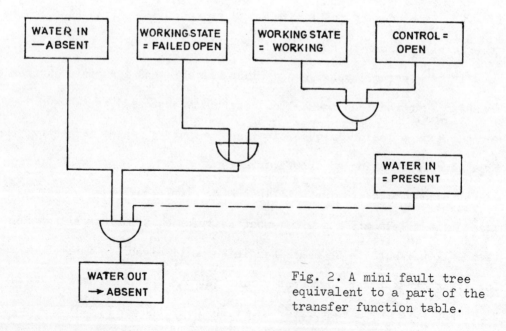

Fig. 2. A mini fault tree equivalent to a part of the transfer function table.

CONDITIONS	STATE	ON				OFF			
	R	ON		OFF		ON		OFF	
	S	ON	OFF	ON	OFF	ON	OFF	ON	OFF
EVENTS	S→ON						STATE → UNDEFINED		STATE → ON Q → ON Q̄ → OFF
	S→OFF								
	R→ON			STATE → UNDEFINED	STATE→OFF Q → OFF Q → ON				
	R→OFF								

Fig. 3. Transition functions for component with an internal state variable and time delay - an RS flip flop.

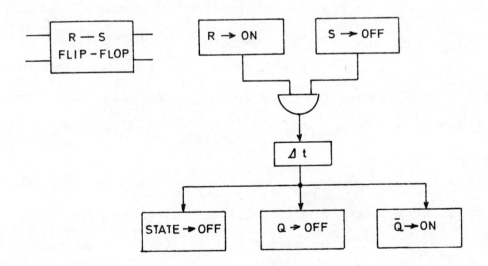

THEN OUTPUT = ON.

When such freedom of expression is allowed it introduces some problems. Matching an input event to a particular condition in a transfer
function becomes difficult in general. For this reason, our efforts
have so far been limited, allowing only piecewise linear models of plant
components. In this case, a straightforward matching algorithm exists
(Meserve 1955).

3. Causality. To analyse a process plant in terms of events and
conditions is to accept the logic of causality. While this provides a
very convenient method of describing changes in a plant (and particularly
of describing reliability models), causal models are less general than
models described in terms of equations. Some kinds of components can
allow causality to work in either direction. For example a motor connected to a variable voltage power supply may have a transfer function
from voltage to output torque. Connected differently, for example to a
water turbine, the same motor might serve as a generator, with a transfer function from torque to voltage. The direction of causality in this
case depends on how the motor is connected into a net-work of components.

It is desirable to be able to build up a library of component models
which is independent of a particular plant construction. If this can be
done, then all that is required as input to a fault tree construction
program, for example, is a list of component names, modification of a
few components parameters, perhaps, and a list of physical intercon-

nections between components.

The direction of causality in a process plant (which variables are dependent on which) can be determined by providing equations to describe each component, and by applying the following rules:

- Each variable may be determined by at most one equation.

- Each equation may determine only one variable.

- When time delay is involved the direction of cause and effect is from the variable which changes earliest.

- When variables are related by integration/differentiation, causality is assigned in the direction of integration.

- When the value of one variable corresponds to several values of another variable in a deterministic way (a many - one mapping) then the direction of causality is in the direction of the many - one mapping.

- For systems in which the rules given above are not sufficient, it is convenient if the gain around any loop in a network is less than one. That is, if a change occurs at one point in a loop, and propagates around the loop, then further changes at the same point are smaller in magnitude than the original change.

These rules are justified in (Taylor 1977).

A simple way of applying these rules is to construct a graph in which each equation is represented by one type of node (square), each variable by another type (circle) (Fig. 4). Arcs are constructed between variable and equation nodes, and indicate the equations in which variables appear. Directions are then marked on arcs according to the time

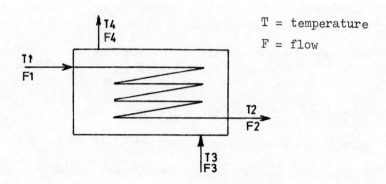

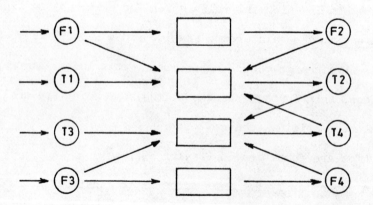

$$F_3 = F_4$$

$$F_1 = F_2$$

$$C_1 \frac{dT_2}{dt} = T_1F_1 - T_2F_2 - K(T_2 - T_4)$$

$$C_2 \frac{dT_4}{dt} = T_3F_3 - T_4F_4 - K(T_2 - T_4)$$

Fig. 4. Causal diagram for a (very) simple model of a heat exchanger.

delay, many one mapping, and integration rules. After this, a backtracking algorithm is applied to complete the marking of directions, choice
of dependent variables being made on a trial and error basis, until a
consistent allocation of causal directions is achieved.

Causal directions must be calculated dynamically during event
tracing, in general, because of the problem of reversal of causality.
For example, for an amplifier, the direction of causality is generally
from amplifier input to output. However, if a short circuit occurs at
the input to the amplifier, events may propagate "backwards", affecting
components "earlier" in the circuit. In practice it is more effective
to precompute the possible sets of causal directions for each component
and the conditions under which these causal directions may apply. The
causal directions are expressed most simply by storing a list of
alternative sets of output variables for each component.

4. Consequence analysis. Algorithms for consequence analysis
have been presented earlier (Taylor 1974a, 1974b). The algorithm shown
in Fig. 5 is the basic algorithm for developing event sequences in time
order. A list of events is maintained, and events are selected from this,
the earliest first. The selected event is applied as an input event to
its relevant component, and the output events for the component are
deduced. The new events become input events for the following components,
and are added to the events list.

Just which output events are deduced for a component will depend

1) Start with initial event description, place it in OE. Find component affected by initial event, and place it on current components list. Go to step 4.

2) Select component with earliest event from active components list (select at random, if there are several components with same event time). If there are no such components, exit. The selected component becomes SC.

3) Find the list of following components of SC. Find the earliest event description on SC's future events list, and call it OE. Find which of the following components are affected by OE, and place them on the current components list.

4) Select the next component from the current components list and call it CC. If there is no such component, go to step 2.

5) Use OE to update the input variable values of CC. Use CC's input variable values, state variable values, and transfer function, to deduce a new output event, call it NE. If NE is null, go to step 4.

6) Split NE into an immediate event IE, a delayed event DE. Apply the changes of variables described in IE, to update the input and state variables of CC. Add both DE and IE to CC's future events list. Place CC on the active components list, marked with the time for DE, if DE exists, and with the time for IE, if IE exists. Go to step 4.

    OE - output event variable

    SC - selected component variable

    NE - new event variable

    CC - current component variable

    IE - immediate event variable

    DE - delayed event variable

Fig. 5. Basic consequence analysis algorithm.

on the component state, and on the conditions obtaining at the inputs
as well as on which input event occurs. The first additional complication
to the basic algorithm is to allow each component to have several alternative possible states. Then, in treating an input event, one of the states
is selected arbitrarily. The event chains which are a consequence of this
choice are evaluated.

When the event chains terminate, the evaluation steps are retraced,
in a classic backtrack programming fashion, and an alternative component
state is chosen. In this way, event chains incorporating alternative
possible event sequences can be plotted.

A further level of complication is to include the possibility of
variable, or random delays between input and output events for a component. When such a delay is reached, a first alternative time is chosen
- a zero delay is assumed. The event is also stored on a stack. When
the event chains following a zero delay have been evaluated, the evaluation steps are retraced, and a new delay period is chosen, equal to the
smallest difference between event times in the previous event chain construction. The process is repeated, using successively the second
shortest event time difference, the third shortest, and so on.

A third level of complication is to allow backtracking in cases
where incompatibilities are found in the causality structure of the
plant models, as described in the previous section.

These various versions of the basic algorithm all require a backtrack search procedure. A system for storage of component state infor-

mation has been programmed, which allows the state of the entire plant
to be recorded using just a single program instruction. Then, when steps
are to be retraced, the old plant state can be restored, again using
just a single instruction. A further finesse is that it is not necessary to store all of the plant state information in each "generation" of
the "plant data base". Only information concerning differences between
generations is stored, with a consequently large saving in storage space.

5. Fault tree construction. A fault tree construction algorithm,
similar in principle to those described by Fussell (1973) and by Powers
and Tompkin (1973) is given in Fig. 6. It works by tracing the course
of events backwards through a component network. At each stage it takes
a given event established in the fault tree and looks among component
transfer functions for a combination of an input event and component
variable conditions, which can lead to the already established event.

Because, at each stage, the transfer functions involve an input
event in combination with an earlier established condition, the AND
gates incorporated into the tree are priority AND gates. The relative
ordering of events is thus preserved in the fault tree.

It will be seen from Fig. 6 that when an attempt to complete a
branch of the fault tree fails, the dummy event FALSE is added to the
tree (step 8). This means that a further stage is then required, in
which any AND gate with a FALSE event is removed, and any OR gate with
no input events is removed.

1)   Get an initial output event (TOP event) and make it the root of
     the fault tree. Form a set of output event pointers, with a pointer
     to the TOP event as its only member, and place this on the "current
     combination stack".

2)   Take one event pointer OE from the set at the top of the "current
     combination stack". If there are no more pointers in the set, go
     to step 5. If the pointer points to a spontaneous event, or a nor-
     mal component condition, go to step 2.

3)   Search in the component transfer functions for the input event and
     conditions which can produce the selected output event, OE. If
     there are no such conditions, go to step 7. Convert the input events
     and conditions to output events of the preceeding components.

4)   Add an OR gate to the fault tree, and an AND gate for each event/
     condition combination found in step 3. Place a pointer to each AND
     gate on the "alternative combinations stack".

5)   Take the top combination indicated on the alternative combinations
     stack. If there is no such combination, go to step 6. Construct a
     set of pointers, one to each event or condition in the combination,
     and place it at the top of the current combinations stack. Go to
     step 2.

6)   Remove the current entry from the "current combinations stack". If
     the stack is empty, stop. Otherwise go to step 2.

7)   Replace the selected output event on the fault tree by FALSE. Go
     to step 5.

Fig. 6. Fault tree construction algorithm.

Lapp and Powers (1976) have pointed out the importance of feedforward and feedback loops in a plant, in introducing mutually exclusive
events into "anded" branches of a fault tree. Such events must be checked
to ensure that inconsistencies are not introduced into the fault tree.

The algorithm of Fig. 6 does not do this checking, and so, by itself could not be used for construction of correct fault trees, if feedback or feedforward loops were present. To overcome this limitation, we
have adopted the expedient of evaluating the "sequential failure sets"
of the fault tree ("cut sets", but with individual failure events placed
in time order), and checking these by using them as initial failure
event input to the consequence analysis algorithm. This enables inconsistent sequential failure sets to be eliminated.

The algorithm of Fig. 6 is efficient in finding possible causes of
failure, but by failing to eliminate exclusive events, can produce an
excessive computational load. An alternative algorithm, which develops
events in time sequence, and which takes into account the state of the
plant at each event time is given in Fig. 7. Whenever a new event is
added to the tree, the question is asked "if there has been an earlier
event in this component, is this new event compatible with the state
arising from the earlier event?" (step 3). If the answer to this question
is negative the entire set of conjunctive terms involving the new event
and the earlier event may be removed from the fault tree. An alternative
question could be asked "have there been any intervening events which
could have changed the state of the component following the earlier

1) Enter the program by treating the TOP event as the input event, IE, of a national "final" component. Go to step 5.

2) Take the latest input output event pair from "event pair stack". If the stack is empty, go to step 7. Make IE the selected input event, OE the selected output event, and CC the component affected by these events.

3) Find whether the conditions necessary for OE to follow from IE are fulfilled by one of the possible states of the current component CC. If not, go to step 8.

4) Record the current state of the component by creating a new generation of the component state data base. Update the list of possible states of CC and also the state of any other component affected by IE, using the consequence analysis algorithm, and recheck that OE is feasible, if not, go to step 8.

5) Find the previous component PC, for which IE is an output event. Find the input event/condition combinations for PC which can lead to IE. Put these combinations on the alternative combination stack. Add corresponding OR gate, and a set of AND gates to the fault tree with the combinations of events and conditions as input.

6) Take one combination from the alternative combinations stack, and extract its input event PE, if the stack is empty, STOP. Calculate the event time for PE from the transfer functions of PC. Store the event time, PE, IE, the previous component name PC, and the condition under which IE may occur, on the event pair stack. Go to step 2.

7) Return the component state data base to the previous generation, and go to step 6.

8) Change the event on the fault tree corresponding to OE to FALSE, and go to step 7.

Fig. 7. Algorithm for developing sequential fault trees in time order.

IE - input event
PE - previous event
CC - current component
OE - output event

Fig. 7. Continued.

event, making the current event feasible?". An algorithm incorporating this kind of test involves a heavy additional computation load, without in practice discovering very many new failure modes.

6. Experience with the algorithms. These algorithms have been coded in a dialect of LISP and run on both an 8k PDP8 computer and a Burroughs B6700 computer. The algorithms use from 1000 to 4000 LISP function calls in their coding. The LISP programs are compiled to an intermediate code, and then interpreted by a small program, written in FORTRAN. This makes the programs readily portable. The execution time for the programs is of the order 0.2 - 2 sec. per event, on the PDP8 computer. This is sufficiently fast to allow human interaction during failure analysis.

Human interaction is an important feature in the fault tree and cause consequence diagram construction - the analyst becomes involved in construction, can direct attention to areas of special interest, increase the level of detail of failure models etc. One important aspect of the analysts task is to stop the development of event sequences, when several identical sequences are developed. This is especially the case when stochastic delays are incorporated in the plant model.

Programs which take account of the sequence of events, are inevitably slow when compared with programs treating event combinations only. What are the advantages of the "sequential" algorithms? Some answers can be given:

- the algorithms can readily treat excessive delay as a cause of failure;

- failure modes which are event order sensitive can be described (for example sequence errors in plant operator procedures);

- for problems which are essentially sequential such as operator procedures, plant shut down, or batch processing, the results can be presented in a much clearer fashion.

The algorithms have been tested on toy examples drawn from the literature. Current tests involve comparison with practical safety analyses constructed by hand, and in the construction of an alarm analysis system for a pressurised water reactor.

REFERENCES

(1) J.B. FUSSELL, Synthetic Tree Model, a Formal Methodology for Fault Tree Construction, Report ANCR-1098, March 1973.

(2) E. HOLLO and J.R. TAYLOR, Algorithms and Programs for Consequence Diagram and Fault Tree Construction, Report Risø-M-19o7, 1976.

(3) S.A. LAPP and G.J. POWERS, Computer-Aided Fault Tree Synthesis, Dep. of Chemical Eng., Carnegie-Mellon University, Febr. 1976. (to be published).

(4) B.E. MESERVE, Decision Methods for Elementary Algebra, American Mathematical monthly, Vol. 62, pp. 1-8, 1955.

(5) G.J. POWERS and F.C. TOMPKINS, Fault Tree Synthesis for Chemical Processes, AICHE Journal Vol. 2o No. 2, March 1974, pp. 376-387.

(6) J.R. TAYLOR, Sequential Effects in Failure Mode Analysis, Conference on Reliability and Fault Tree Analysis, Berkeley 1974. Ed. J. Barlow and J.B. Fussell, NO Singapurwalla, SIAM 1975.

(7) J.R. TAYLOR, A Semiautomatic Method for Qualitative Failure Mode Analysis, CSNI Specialist Meeting on the Development and Application of Reliability Techniques to Nuclear Plant, Liverpool 1974.

(8) J.R. TAYLOR, Automatic Failure Analysis, (to be published).

(9) SALEM, APOSTOLAKIS and WU, The Use of Decision Tables in the Systematic Construction of Fault Trees, this volume.

THE SYNTHESIS OF FAULT TREES

STEVEN A. LAPP AND GARY J. POWERS*

Abstract. In order for fault tree analysis (FTA) to be useful in
the assessment and control of risk, the synthesis of the trees should
be: 1) Routine -- so that project engineers, etc. will use it. 2)
Rapid -- so that it won't slow down the project. 3) Accurate -- so
that the results will have some meaningful relationship to the actual
risk. 4) Flexible -- so that the synthesis procedure could be used on
a wide range of systems. We present an algorithm which uses directed
graph (digraph) models of the cause and effect relationships between
variables and events. Given the models, the algorithm will deduce the
combinatorially correct fault tree for the system. The models and the
algorithm are described and tested on two examples.

1. Introduction. Fault tree analysis commonly takes a large

amount of high quality manpower. If fault trees are going to be use-

ful in finding the important modes of failure (overlooked modes, com-

mon-cause, etc.) in a system, the scope of the analysis has to be both

wide and deep. The width is required to reveal interactions between

systems of different function (utilities, control, maintenance, etc.)

and of different modes of interaction (propagation of disturbances

both inside and outside the pipes, wires, etc.). The depth of analysis

is required to reveal the commonalities. An analysis which is both

wide and deep is destined to be large, hence the obvious need for an

algorithm to organize the bookkeeping and searching. We have been

working on this problem for the last five years.

*Department of Chemical Engineering, Carnegie-Mellon University,
Pittsburgh, Pennsylvania 15213

Dr. Frederick Tompkins started the use of input/output models for describing the local cause and effect relationships between variables and failure events for a single component of a system [1]. Steve Lapp developed the multiple edged digraph method of modeling of cause and effect. He also invented an algorithm which converts the digraph into a fault tree [2]. A number of other workers have contributed to the study of fault tree synthesis. J.B. Fussell developed the Synthetic Tree method for fault tree construction [3]. R. Taylor has developed an Algol-like language for describing the cause and effect relationships for a system component [4]. Steve Salem has used decision tables to aid in the generation of fault trees [5].

1.1. Requirements for Fault Tree Synthesis Models.

1. It must be modular in the modeling of behavior (component models).

2. All possible modes of behavior must be describable. The models need to represent failure modes which cause deviations in output variables and failure modes which change relationships between variables. Classes of devices include capacitors, inductors, resistors, switches, latches, amplifiers, etc., and macros made from these elements. Sequential relationships must also be easily represented.

3. Loops should be modeled naturally. Feedback and feed-forward loops should be described by the component

models which make up the loops.

4. The range of the variables must be greater than 0, 1.
    To describe the continuous interactions between vari-
    ables and events will require at least a multilevel
    logic.

5. The fault tree (Boolean equation) must be deduced from
    the models. (The models should not imply all AND/OR
    relationships.)

1.2. Digraph Models. We have developed multiply edged directed
graphs models as a means to meet the needs of fault tree synthesis.
The features of digraphs are illustrated in Figure 1.

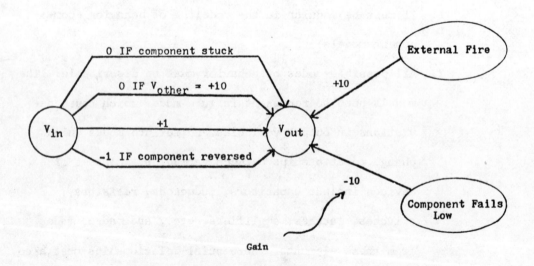

Figure 1. A General Digraph

1. The nodes may be variables or events.

2. A variable may have input edges.

   a. A variable may take on values -10, -1, 0, +1, +10. These values are deviations from the normal conditions.

   b. Some variables may take on only selected values (deviations) (e.g. only positive deviations for a valve that is normally closed).

3. An event has no input edges.

   a. Events may take on values -10, -1, 0, +1, +10.

   b. Some events may take on only selected values (e.g. only positive deviations for the event fire).

4. An edge connects two nodes.

   a. Each edge has a gain, a first-order time constant and a dead time.

   b. $\text{Gain} = \dfrac{\Delta \text{ Output/Output Value (Normal)}}{\Delta \text{ Input/Input Value (Normal)}}$.

   c. Gains may have values -10, -1, -.1, 0, +.1, +1, +10. These gains may be computed from expressions derived directly from the mass, energy, momentum and electrical balances or from data. The actual gains are discretized to these values.

   d. Gains may be conditional on the value of variables or events.

   e. The time constants and dead times are derived from the system equations or data.

5. A system digraph is constructed from component digraphs. The system digraph is constructed by working strictly backwards from the output variable (defined by the top event) through the component digraphs.

6. Loops are found by tracing node-edge-node paths in the system digraph.

a. Negative feedback loops are loops from a node back to itself with a net negative gain.

b. Negative feedforward loops involve two or more branches which fan out from one node and converge at another. One or more paths must have a gain which is the negative of the other paths.

c. Loops may exist through the conditional variables on an edge.

d. Multiple loop situations exist when more than one loop passes through a single variable.

7. Potential common cause events are detected by fan out of edges from a single event.

Figure 2 illustrates a digraph for a control valve.

1.3 The Lapp-Powers FAULT TREE GENERATION Algorithm.

Characteristics:

1. Directly deduces the fault tree from the digraph.

2. Is based on local conversion of the digraph variables, events, and edges into a partial development of the fault tree.

3. There are three basic conversions of digraphs into fault trees.

a. They are based on whether negative feedback loops or negative feedforward loops pass through the digraph variable.

b. They depend on the value of the deviation of the output variable. The range of operation of the loops is used to select the operator.

c. The fault tree "operators" are shown in Figure 3.

4. The operators are applied recursively until all nodes in the digraph have been developed.

5. The consistency of loop variables are checked against conditions developed in the tree. (e.g. Previous conditions

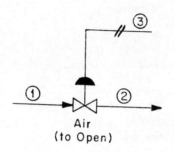

Air
(to Open)

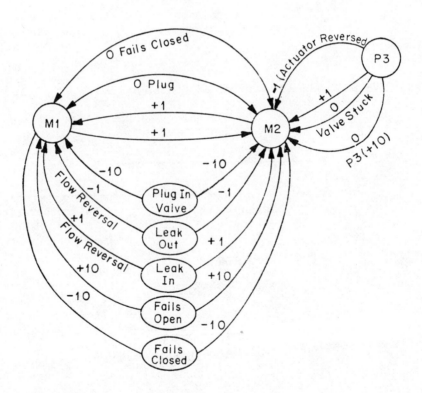

Figure 2.   Digraph for a Control Valve

FOR NEGATIVE FEEDBACK LOOP VARIABLES

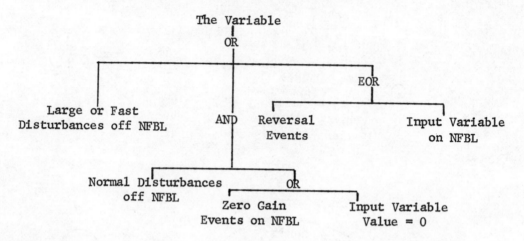

FOR NEGATIVE FEEDFORWARD LOOP VARIABLE JUST BEFORE START OF LOOP

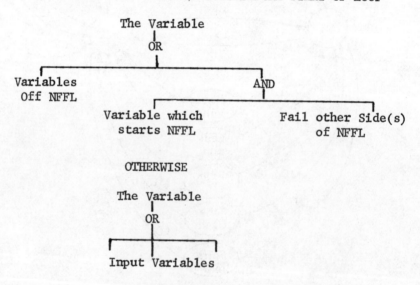

Figure 3. Boolean Expressions for Digraph Variables

and the domain of AND gates.)

Example 1. Consider the flow control loop shown in Figure 4.
The system senses the flow and adjusts the valve position to maintain
flow at the set point value. The controller may be placed on manual.

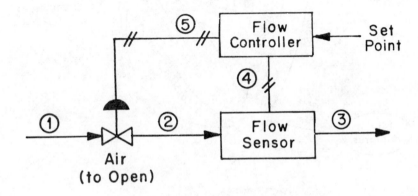

Figure 4. Flow Control System

When on manual the controller maintains the output of the controller
at its last value. Changes in the input to the controller do not
cause changes in the output when in manual operation. The controller
may also be made reverse acting by either explicitly changing the action by activating a switch on the controller or by increasing the loop
gain or dead time (or reset rate or derivative rate if integral or derivative action is present in the controller). The digraph for the
flow control system is shown in Figure 5. The digraph was constructed
by starting at the top event (in this case flow out of the system M3)

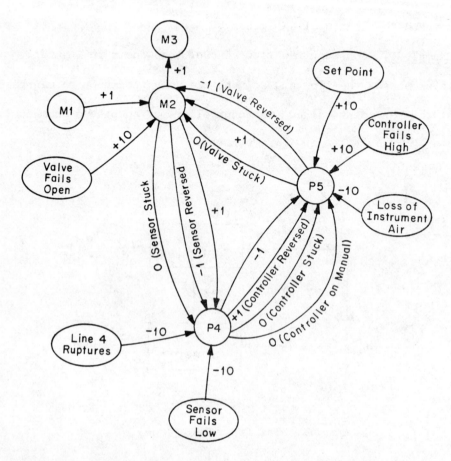

Figure 5.   Digraph for a Flow Control System

and working backwards through the digraph models for each component in
the system.   The rules for constructing the system digraph from the
component digraphs are:

1.   Start at the top event (output) variable.

2.   Get the component model from which the variable is the output.

3.   Work backwards through the component models to its inputs as-
     sembling the system digraph.   (Do not trace any of the output
     variables to other variables at this time.)

4. For each input variable on the resulting digraph, repeat step
    3 until variables are encountered which have no inputs (sys-
    tem boundary or failure modes).

5. The variables in the component digraph models are labeled by
    their location in the system flow diagram. External vari-
    ables in the same location are labeled as the same variable.
    Components manufactured by the same company are given the
    same variable for certain failure modes. This leads to dis-
    covery of common-cause situations. Extension of the analysis
    to common systems such as power supply, instrument air, or a
    single operator results in numerous common variable situa-
    tions. These are deduced directly from the interconnections
    shown in the flow diagram.

6. If loops exist in the process, it is possible to pass through
    the same component digraph twice. The same rules given in
    steps 3 and 4 should be followed. Do not trace variables
    that have already been developed.

7. Variables that are conditions on edges are developed in the
    same manner as input variables.

The resulting system digraph can be used for the output variable having
any value (+10, +1, -1, or -10). The only part of a system digraph
which explicitly denotes variable values are variables on conditional
edges.

1.4. The Lapp-Powers Fault Tree Synthesis Algorithm. This algorithm is explained in detail in reference 2. The algorithm is based
on the logical (AND, OR, EOR) combinations of digraph variables (and
their values) which could cause a particular deviation in an output
variable. The Boolean expressions for a digraph variable given in
Figure 3 were based on a detailed study of how negative feedback loops
and negative feedforward loops might fail. The negative feedback loop
Boolean expression has three major terms. The leftmost branch shown
in Figure 3 indicates that a large or fast deviation in an input variable to a negative feedback loop will pass through the loop. That is,
an OR gate is indicated. The second term (center branch) denotes the
fact that if a normal deviation in an input variable enters a negative
feedback loop that it is also necessary (AND) to fail the feedback loop
by inactivation. The third term (right branch) of the Boolean expression for a negative feedback loop indicates that the loop can directly
cause the deviation in output variable by becoming a positive feedback
loop. As noise is always present in these loops we assume that reversal of the loop is all that is necessary to cause the loop to go unstable. However, if two reversals occur they cancel each other and
the loop remains negative. Hence the exclusive OR gate indicates one
reversal event or another but not both.

   The negative feedforward loop Boolean equation has two major terms
(branches). The leftmost branch indicates that if the disturbance
variable entering the loop does not send signals down all sides of the

negative feedforward loop, the loop will not cancel out the disturbance (hence an OR gate). The right hand term indicates that if a deviation in an input variable activates all sides of the negative feedforward loop, it is necessary to have the disturbance AND fail the
other sides of the loop.

If the digraph variable is not on any negative feedback or feedforward loops, use an OR gate.

The fault tree derived from the flow control digraph is shown in
Figure 6. Gates 3, 4, and 5 are due to the feedback loop operator.

Example 2. Consider a heat exchanger with feedback from the hot
stream outlet temperature to the cold stream flowrate and feedforward
from pump shutdown to the hot stream flowrate. (See Figure 7.)

This example is drawn from part of a process to nitrate benzene
to mononitrobenzene. If the temperature T4 shown in Figure 7 gets too
high (i.e. T4(+1)) the reaction may accelerate at an uncontrollable
rate. The digraph shown in Figure 8 was constructed from models of a
temperature sensor, heat exchanger, control valve, temperature controller and pump. The fault tree shown in Figure 9 was derived using
the fault tree synthesis algorithm. Gates 2, 5, and 6 were developed
using the negative feedback loop operator. Gates 13 and 14 were developed using the negative feedforward operator.

2. Discussion. We have developed fault trees containing over
500 gates using this algorithm. Trees containing over 1000 gates are

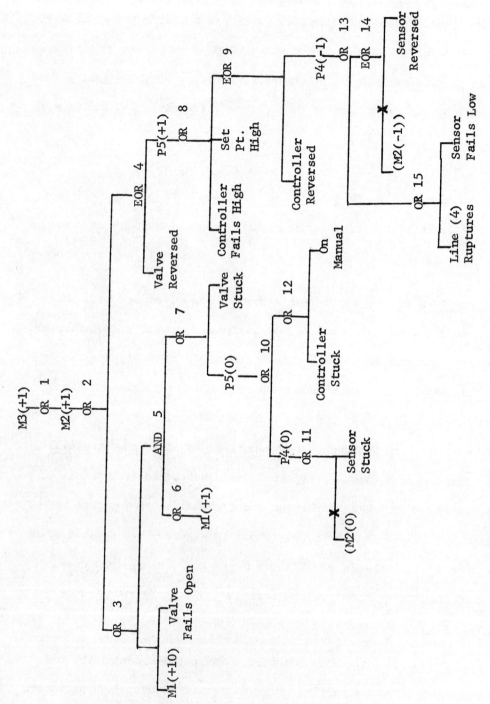

Figure 6. Fault Tree for a Flow Control System

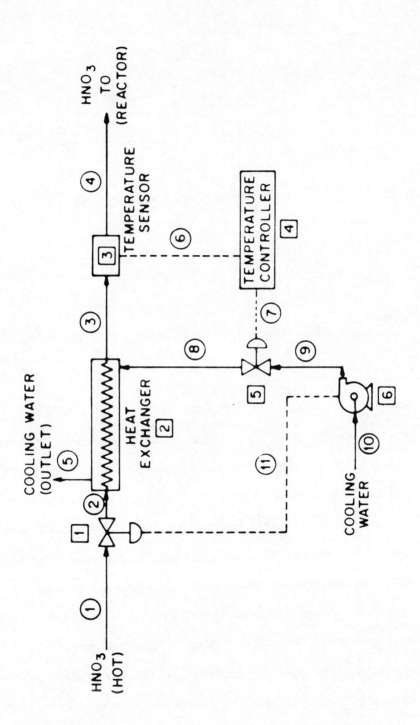

Figure 7. Heat Exchanger System

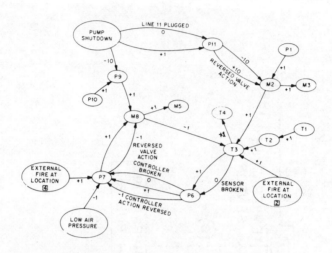

Figure 8. Heat Exchanger Digraph

well within the reach of a computer program which would use this
algorithm.

Question: Can this approach work for human operator interactions?

Answer: Yes, if we can describe how the operator interacts with the
          process. For example, consider the flowsheet shown in Fig-
          ure 10. When the high pressure alarm is on, the operator
          is supposed to move the valve switch into the closed posi-
          tion. This action will close the valve and hopefully re-
          duce the pressure in the system. A digraph for this pro-
          cedure is shown in Figure 11. A fault tree could be di-
          rectly deduced from this digraph. The challenge is to
          understand the possible operator actions to describe them
          with a system digraph. The operator becomes a control loop

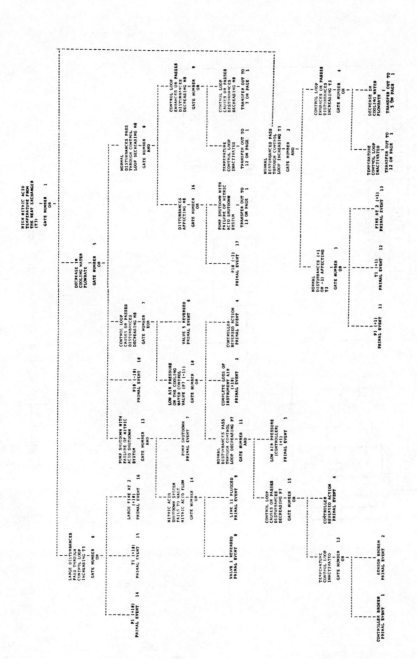

Figure 9. Heat Exchanger Fault Tree

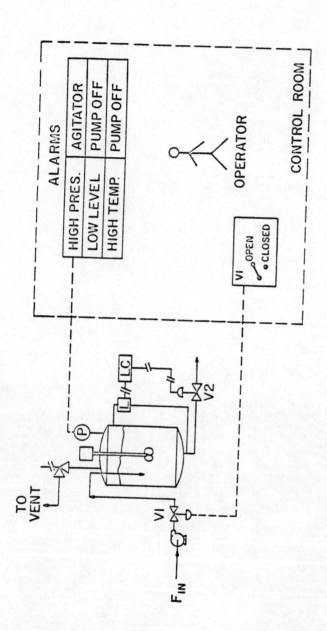

Figure 10. Operator-Process Flowsheet

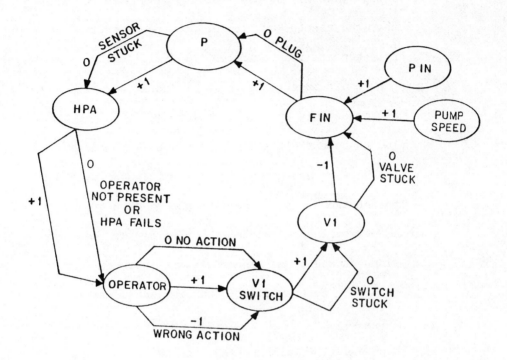

Figure 11.   Operator-Process Digraph

in the process. He senses variables (alarms, indicator,

etc.), "computes" some action, and then takes actions by

opening and closing switches or valves, or changing set

points. Of course, he may also be a random cause of failure.

Question: Will this system find all failures?

Answer: Obviously not. There are several ways to go wrong.

1. An incorrect flow diagram might be analyzed. (This

raises the challenging problem of how to control pro-

cess modifications.)

2. A top event may be overlooked. If you do not know
    that the reactor may rupture due to a reaction between
    hydrogen and iron oxide on the inside of the vessel,
    then fault tree analysis is of little use. We have
    addressed this problem in reference 6.

3. The component digraph models may be incomplete. If
    the physics and chemistry of a particular system com-
    ponent are unknown, then the propagation of cause and
    effect will be incomplete. For example, steam explo-
    sions caused by molten metals are just now being
    studied. Certain power excursion accidents in nuclear
    cores are extremely difficult to simulate and have not
    been experimentally tested. In addition, a model may
    simply overlook a known class of cause and effect.
    Most commonly these include interactions "outside" of
    the pipes and the wires. These external interactions
    include fire, flooding, vibration, shorting of wires to
    each other, impingment of leaks from one pipe onto
    another pipe or structure, impact from one component
    onto another (pipe whip), maintenance, installation,
    sabotage, etc. If they are not in the component di-
    graph, they will not be in the fault tree. Our hope
    here is to develop and continuously update component
    models to capture experience with each class of device.

At the functional level of modeling the effort can approach completeness. That is, a valve can only fail open or closed, leak in or out, stick in position, be reversed, etc. However, these generic functions may not be detailed enough to reveal common cause variables or to remind the system designer or operator to take specific action. For example, valve stuck may be caused by a large number of different variables including corrosion, leak of a material which will freeze on the valve stem, packing overtightened, etc. Each of these causes would require different design and operation responses. In addition, the leak may have been caused by high pressure which could be traced to other causes.

If the above three items are not in error, we believe that the algorithm will generate the "correct" fault tree. If the problem is mainly one of uncertainty in the flow diagram, definition of top events, or component models, then this algorithm will be of use only, perhaps, in getting the uncertainty out in the open. If the problem is due to complexity, this approach should help considerably.

Question: Is the fault tree unique?

Answer: Yes, if a given system, top event, and set of models are used the fault tree derived by the algorithm is unique. The fault tree algorithm is completely deterministic.

Question: What about dynamics?

Answer: We are now adding dynamics to the digraphs. Each edge is
            now going to be given two dynamic parameters in addition to
            its steady state gain. The parameters are the dead time
            and net first order lag between the input and output vari-
            able. With these values, it is possible to categorize the
            loops in the process by their time domain response. If the
            transit time for a disturbance is much faster than the re-
            sponse of a negative feedback loop, then the loop is not
            required to fail to allow the propagation of that variable.
            These classes of variables are shown entering on the left-
            most gate of the negative feedback loop operator, that is,
            larger or faster than the loop can respond to.

Question: What about sequential systems?

Answer: They are a lot harder. A paper (reference 6) addressing
            this problem will appear shortly. Briefly, the edges be-
            tween variables are made dependent on variables that depend
            on time. In this way the consistency checking is handled
            by the algorithm. The example presented in reference 6 had
            two four-way valves and one three-way valve that were op-
            erated at four different times during a process cycle.
            Using this approach, time appears as a primal event. That
            is, Time = Phase 1 is a primal whose probability would be
            0.33 if phase 1 is 1/3 of the normal mission or cycle.

Failure modes must be added to the time or control mechanism to fail to switch, switch too soon, etc.

Question: Has this method been tested?

Answer: We have tried about twenty different problems to date. They have ranged from simple single component problems to one which contained 75 components, 12 feedback loops (6 through the human operator) and 3 negative feedforward loops. We are still learning rapidly but it looks like it might be able to handle very large (i.e. 500 components cases). Of course, the interfacing of this algorithm with computer programs, actual users, and project schedules is still a major problem.

REFERENCES

[1] G.J. POWERS and F.C. TOMPKINS, Fault Tree Synthesis for Chemical Processes, AIChE J., 20(1974), p. 376.

[2] S.A. LAPP and G.J. POWERS, Computer-Aided Synthesis of Fault Trees, IEEE Transactions on Reliability (to appear).

[3] J.B. FUSSELL, G.J. POWERS and R.G. BENNETTS, Fault Trees -- A State of the Art Discussion, IEEE Transactions on Reliability, R-23(1974), p. 51.

[4] J.R. TAYLOR, A Formalisation of Failure Mode Analysis of Control Systems, Danish Atomic Energy Commission, RISO-M-1654, September (1974).

[5] S.L. SALEM, A Computer-Oriented Approach to Fault Tree Construction, UCLA-ENG-7635 UCLA School of Engineering and Applied Science, April (1976).

[6] J. SHAEIWITZ, S.A. LAPP AND G.J. POWERS, Fault Tree Analysis for Sequential Systems, Industrial and Engineering Chemistry -- Process Design and Development (to appear).

THE USE OF DECISION TABLES IN THE SYSTEMATIC CONSTRUCTION OF FAULT TREES

J. S. WU,* S. L. SALEM** AND G. E. APOSTOLAKIS*

Abstract. A systematic methodology for the construction of fault trees is presented based on the use of decision tables. These tables are used to describe each possible output state of a component as a complete set of combinations of states of inputs and internal operational or failed states. The methodology has been implemented by the development of a computer code, which, following several stages of editing, produces a fault tree in conventional format. The analysis of an actual reactor system using this methodology is presented which demonstrates the applicability and flexibility of the approach.

1. Introduction. Fault tree analysis is currently one of the most important and useful methods of safety and reliability analysis in the nuclear industry. As a logical model of system operation and failure, it offers a well developed methodology for evaluating the failure modes and failure probabilities associated with complicated systems (Apostolakis and Lee [1], Lambert [3], WASH-1400 [9]). At the present time, however, the process of constructing the fault trees themselves is still a time-consuming task for the analyst. Although a number of fault tree analysis programs have been developed, automated

---

*Chemical, Nuclear and Thermal Engineering Department, School of Engineering and Applied Science, University of California, Los Angeles, Ca. 90024.
**Atomics International, Canoga Park, California 91304

This research was supported by the Electric Power Research Institute under Contract RP297-1.

fault tree construction methodologies have only recently been formulated, such as those of Fussell [2] for electrical systems, and Powers
and Tompkins [5] for chemical process systems.

The current approach of Salem, et al.[6,7] is a general,
computer-oriented method of modeling complex systems of mechanical,
electrical and hydraulic components, allowing for human interactions
and common cause effects as well. This paper presents the basics of
the decision-table methodology used for component modeling, which
allows component operational and failure modes to be conveniently described in tabular form. Furthermore, the application of the methodology as implemented in the CAT code (Computer Automated Tree, [7]) is
developed. Finally, a fault tree for the Containment Spray Recirculation System of a PWR is produced by the CAT code. This application necessitated the modeling of inhibit gates in order to include
in the analysis the possibility of one of the redundant trains of the
system being disabled due to maintenance as well as the different
logic of the system for different periods of time (two-out-of-four
before twenty-four hours following a LOCA and one-out-of-four after
twenty-four hours).

   2. Methodology.

   2.1. Decision Table Modeling. In order to construct fault trees
for general systems, information is required both to describe the system itself, and the operation of the specific components within the
system. It is desired to develop component models which are

independent of any specific system, and which depend only upon a set of inputs from other components within the system. With a group of such component models available, the description of the system configuration simply requires a specification of the interconnections of inputs and outputs of various components, along with a suitable set of initial conditions and a TOP event in order to begin construction of the tree.

Since any system may include a number of similar components, it may be only necessary to develop a single model, or component "type," in order to describe the behavior of any number of system components. Such a component type will be defined as a unique description of all possible output states of a component in terms of all possible input states and internal operational and failure modes. Notice that any two components which can be described by the same model are defined as the same type. For example, simple fuses and resistors, which have the same failure modes and effects, would be of the same type.

The methodology proposed in this paper employs an extension of truth tables, known as "Decision Tables" (Pollack [4]). Although truth-table (binary) logic is especially easy to utilize in computer programming, such binary logic is often insufficient for use in nuclear systems. For example, a three-state variable might be required to describe the water pressure in certain light water reactors which have both low and high pressure systems. It would then be possible to model such effects as the introduction of high pressure coolant into a low pressure system using a variable with the states "no pressure", "low pressure", and "high pressure." Another desirable feature of the present approach is the capability of both current and voltage (or flow and pressure) modeling. This allows a recirculation-type system to be analyzed in terms of flow, and other types of systems in terms of signal or pressure levels, as appropriate.

The present decision table approach is a generalization of truth-table methods in that any number of states may be used for each entry in the table. It is only necessary to associate each state value with a physical meaning. The following numbering scheme was developed for failure modes:

-1	"don't care"	(internal mode irrelevant)
0	good	
1-1000	general faults	
1001-2000	electrical	(shorts, surges, etc.)
2001-3000	mechanical	
3001-4000	fluid	(leak, rupture, plugged)
4001-5000	electronic	(logic errors, etc.)
5001-6000	human	
6001-7000	environmental	(temperature, pressure, stress,etc.)

Within these categories, a representative number of basic failure modes were defined. For example, some of the states in the category of general faults are:

1   failed open (fails to close; fails to transmit signal)

2   failed closed (shorted, fails to open; welded shut)

3   internal failure (general, undefined)

4   fails to start (fails to actuate or change position)

5   fails to run (fails during operation)

6   operates prematurely (starts without signal to start)

In addition, several signal states were defined as follows:

-1   = "don't care" (signal state irrelevant)

0   = none or too low

1   = normal signal

2   = overload (or overload too long)

3   = low

100   = ground (zero) or short to ground

101   = floating (open, undefined)

Decision tables can now be constructed which will describe each possible output state of a component as a complete set of combinations of states of inputs and internal states. As an example, a decision table for a pump is presented below.

Number of inputs   : 2

    Input 1 : Main flow input (flow or pressure)

    Input 2 : Power input

Number of internal modes : 1

Number of outputs   : 1

Table 1

DECISION TABLE FOR PUMP

Row	Input 1 Main Flow	Input 2 Power	Internal Mode	Output
1	0	-1	-1	0
2	-1	0	-1	0
3	-1	-1	4	0
4	-1	-1	5	0
5	1	1	0	1

Thus, Row 2 of the table shows that there will be no output from the pump if there is no power, regardless of the main flow and the internal mode of the pump.

The use of such decision tables in constructing fault trees will now be illustrated, assuming that the event "no output from pump" is desired as an event to be analyzed. This might be the TOP event of a tree, or some intermediate event which would be required to produce a zero input to a succeeding component.

Given the desired output state, a search is made for rows with the correct state, in this case rows 1 through 4 of the table. Since any one of these rows has the correct output, they are connected by an OR gate, each row being a single input. Since in all rows two of the three signals are of the "don't care" type, each row is replaced by a single event. Thus, row 2 is replaced by the event "no power," which must be developed further with the use of another decision table. Row 4 is also replaced by the event "pump fails to start", which is a primary failure and thus becomes a direct primary input event.

If the desired output state were "normal output from pump", then a search of the rows reveals that only row 5 gives the correct output.

Here there are three states defined, all of which must be true for the
output state to be 1. The result, then, is an AND gate with the three
appropriate inputs.

2.2. System Nodes. The method of decision tables has allowed the
modeling of the operation and failure of each individual component in
a system. Now it remains to determine a method of coupling components
and tracing through the system in order to construct the fault tree.
This can be done by defining a set of "nodes" throughout the system.

A node will be defined as any point in the system at which the
output of a component is connected to the inputs of one or more succeeding components. That is, a node, or junction, represents the
point of interconnection between component inputs and outputs. In
order to define a node, one assigns it a unique number, and the same
node number used for several inputs implies a common signal source for
all such inputs. In assigning such node numbers, one restriction
occurs which will be defined by the following rule: Only one output
may be connected to any input or inputs; however, one output may be
connected to any number of inputs. That is, inputs may be connected
in parallel, but outputs may not. A common example of parallel inputs
is the frequent occurrence of a single power supply connected to multiple components in parallel.

2.3. System States. A system state consists of a specific condition of the system, defined either at a node, or as an internal mode of
a component. "No signal at node i" is the definition of a particular

state, at a specific point in the system. "Pump fails to start" defines the internal state of a component. Thus, in order to determine a system state, both the state itself, and the node or component, must be specified. However, once done, states of both nodes and components are treated identically.

The states at any points may either have been defined by previous events, or left undefined. All events being traced at a particular moment must be compatible with system states already defined, but are completely free with respect to undefined states.

2.4. Boundary Conditions. In addition to system states which are defined as the tree is generated, other states may have been defined as "initial" or "boundary" conditions. Both are states of the system which are defined initially and continue to exist throughout the entire fault tree; quite often they will be used to qualify the TOP event description.

In general, a boundary condition may be defined at any node of the system, or as any pre-existing component failure mode or operational state. It will continue to exist as the fault tree is constructed. In addition, however, certain component types will also have "initial positions." These are internal modes which may be preset, but which allow the final state to change, depending on further events. Thus, a switch or valve which was "left open" may close if the appropriate operator action or signal is present. Aside from the fact that boundary conditions cannot be reset, they are handled the same way as any other system states.

2.5. TOP Events. In order to construct a fault tree, the final
information required is the TOP event. The only requirement for the
TOP event is that it be defined in terms of system states which can
serve as starting points for the fault tree. That is, the TOP event
is essentially a set of boundary conditions which have the added function of starting the construction process. In addition, the TOP
events must have some logical relationship to each other in order to
structure the tree beneath.

For example, the TOP event for the failure of an Emergency Core
Cooling System which is comprised of two legs might be defined as:
"flow in neither leg 1 nor 2." Such a TOP event would have the following effects. First, the event "no flow in leg 1" would be developed,
under the boundary condition that no flow occurs in leg 2. Then, the
event "no flow in leg 2" would be evaluated, using "no flow in leg 1"
as a boundary condition. These, then, would be combined under an AND
gate at the top of the final tree.

2.6. Editing Concerns. In this work, the editing is divided into
three phases:

1) Editing while in the process of constructing a gate. This includes
checking events for consistency with pre-defined system states and
deleting them as required.

2) Removing excess, redundant or contradictory gates or events after
each gate or group of gates has been completed.

3) Final editing concerns after the tree has been constructed. This includes checking for transfers, renumbering gates and other operations.

In the editing during construction of a gate, the basic concerns are that no events be developed, or allowed to exist, which contradict or duplicate existing events. The method of editing is based upon the utilization of component decision tables in conjunction with the system states as already defined. As an example, let us assume that the event "no output from pump" is desired in the presence of a system state of one (signal present) at the input to the pump.

Referring to Table 1, an output state of zero is satisfied by rows 1 through 4. However, row 1 does not satisfy the condition of normal signal present at input. Thus row 1 is eliminated by the requirement of satisfying a pre-existing system state and would be edited out of the fault tree.

Several general principles can be developed for checking decision table entries against existing system states. Three situations can arise in searching for a specific system state:

1) The system state has not yet been defined. If no other entries in a specific row are contradicted by preset states, the row is thus found to be compatible with the system states.

2) The system state contradicts the state required in the row. That is, the system is in a state such that the row cannot occur. In this case, the row is not allowed.

3) The system has already been defined to be in the state required by
the input to the row. This implies that the system state is "sure to
occur," and thus need no longer be considered.

Two principles are used in this methodology to edit out "cannot
occur" events.

1) If a contradictory (cannot occur) event appears beneath a multi-
input OR gate, that event is deleted from the gate.

2) If the event occurs beneath an AND or a single input OR, that gate
is removed as well as all AND and single input OR gates above, up to
the lowest multiple input OR gate.

The general procedures for dealing with events which have been
preset to the correct state are:

1) If the event occurs beneath a multi-input AND gate, delete that
event only.

2) If the event occurs beneath an OR or a single input AND, completely remove that gate and all OR and single input AND gates above, up to
the lowest multiple input AND gate. These rules are the complements
of those for removing "cannot occur" events.

The above procedures also provide a means of dealing with many
types of feedback loops in a system; that is, a path through one or
more components by which a signal can be traced back to its starting
point. Alternately, one may consider a component, one of whose outputs eventually returns to one of its inputs.

The way in which a loop is dealt with is the following. First, a specific event at some point in the loop will be required as an event to be developed. The specific signal at that point will be set as a system state, and the backtracking of events will begin. If a loop does exist, this process will eventually return to the starting point, and the state of the system will be checked in connection with the specific decision table in use at that node. If the state originally defined at that point is compatible with the state required by the decision table (either as the same state, or a don't care entry), then the loop is consistent and the checking ceases. No further backtracking is required, since the discovery of a predefined state terminates that branch, either with a "sure to occur" or "cannot occur" event.

On the other hand, the original state may contradict the state needed by the current decision table. That is, the originally defined state does not produce a consistent result when traced through the loop. In this case, the specific row of the decision table being used at that point cannot occur. However, another row of the table may indeed produce the required consistency. That is, under a different set of events, the initial state might be compatible with the new row of the table. If no row allows the needed state, however, then the loop cannot be completed under the initially assumed state, and the original event thus cannot occur. All gates and events back to that point, then, will be deleted, using the previous editing methods.

2.7. Intermediate and Final Editing. In addition to the editing which takes place during the construction of each gate, there are several situations which call for further editing once each gate has been completed. This editing generally consists of two phases. First is the elimination of single input gates which have resulted from the deletion of other inputs during previous editing steps. Then follows a search for events which may have been rendered redundant or contradictory due to the removal of the single input gates. This "post-gate" editing will take place each time all branches beneath a specific gate have been completed and comprises a significant portion of the CAT code.

The first phase of this editing is the removal of any single input gates and the transfer of that single input directly into the gate above. However, once this has occurred, further editing may be required since the elimination of an OR gate may lead to contradictory events within a branch. This can be seen, for example, when a single OR gate between two AND gates is removed, leaving a series of AND gates, which now behave as a single AND gate. Alternately, the elimination of a single input AND may produce a direct chain of OR gates which then behave in the manner of a single OR.

Once the top gate of the fault tree has been completed and intermediate editing finished, one final form of editing involves the searching for transfers with the tree. This involves the removal of identical gates within the tree and their replacement by numbered transfer symbols. That is, when a particular sub-tree appears more than once, only its first appearance will be shown explicitly, and will include a numbered "transfer in" symbol which terminates that branch. After the insertion of these transfers, the tree is in its final form and is ready for further analysis.

3.0. Applications. The preceding methodology has been programmed in the computer code CAT, and has been used to investigate a number of systems in the past two years. These include a pressure tank system and reactor Residual Heat Removal System (Salem, Apostolakis, Okrent [7]), a reactor scram system, and a containment spray recirculation system. This latter system will be presented here.

3.1. Containment Spray Recirculation System. The Containment Spray Recirculation System (CSRS) is described in WASH-1400 [9]. For purposes of illustration, this system has been simplified slightly and its flow diagram is shown in Fig. 1. The intended function of the CSRS is the recirculation of the containment sump water through the heat exchangers of the Containment Heat Removal System to spray headers inside the containment, thus removing energy and fission products from the containment in the event of a LOCA.

The following are important features of the system:

1. The system is comprised of four trains. During the first twenty-four hours following a LOCA, the logic of the system for successful operation is two-out-of-four and it becomes one-out-of-four after that period.

2. Each train consists of a pump, a heat exchanger and a spray header. Two of the pumps are inside the containment.

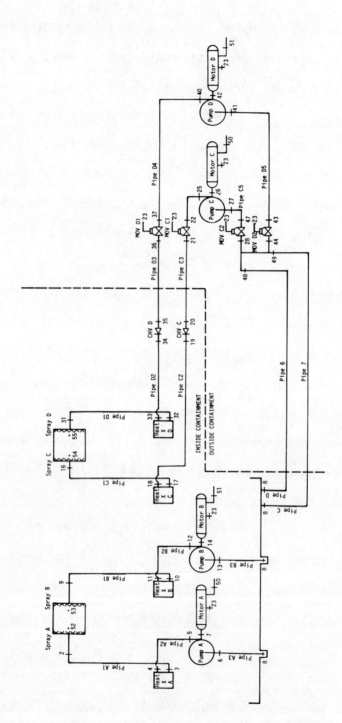

Figure 1. Simplified Flow Diagram for the Containment Spray Recirculation System

3.   All valves in the two trains with the pumps outside the containment are normally open.

4.   The CSRS is actuated by a signal from the Consequence Limiting
Control System which turns on the pumps and also opens the motor-
operated valves if they had been inadvertently left closed after maintenance.

5.   The electrical power supply is common to one inside and one outside train. The other pair of trains have also a common electrical
power supply to the motors of their pumps.

6.   A train can be disabled due to maintenance. However, only one leg
is allowed to be down for maintenance at any time.

   3.2. TOP Event and Preliminary Considerations. In order to construct the fault tree for this system, the decision tables and TOP
event definition had to be developed. Many of the tables had been used
for previous examples [7]. However, new models were developed for the
spray headers, heat exchangers and pump motor. Furthermore, the model
used for the motor-operated valve [7] was modified by deleting a slip
clutch failure mode.

   The model for the pump motor was included in order to separate the
failures of the pump and pump motor into two components. Thus, the
motor itself received power from the power supply, and a signal to turn
on or off. The "power" for the pump (Table 1) was then actually the
mechanical coupling from the motor.

   The TOP event for this system was defined as failure of the

recirculation system to supply adequate spray cooling to the containment following a LOCA. This requires operation of at least two of the four legs for the first twenty-four hours, and one-out-of-four legs thereafter. Thus, the logic of the TOP event changes after twenty-fours, and this situation must be developed explicitly. This was done using inhibit gates modeled by the following decision table (Table 2) as part of the TOP event:

Table 2.

Logic Model for Inhibit Condition

Row	Leg A	Leg B	Leg C	Leg D	Inhibit Condition	System Output
1	0	0	0	-1	1	0
2	0	0	-1	0	1	0
3	0	-1	0	0	1	0
4	-1	0	0	0	1	0
5	0	0	0	0	2	0

In this table, zeroes represent the condition "no flow" from the appropriate leg, or "insufficient flow" as a system output. The inhibit condition is treated as an internal mode, with state 1 representing the condition $0 < t < 24$ and state 2 as $t > 24$. Only those rows leading to the system state "insufficient flow" are shown, that is only those combinations in Table 2 will allow system failure. Thus, it is seen that for $0 < t < 24$, either three or four leg failures will lead to TOP failure, while, for $t > 24$, four failures are required.

An additional complication in this system is the inclusion of maintenance. Any single one of the four legs is permitted to be under repair for a period of up to twenty-four hours. Thus, should the system fail while one leg is under maintenance, during the first twenty-four hours, only two subsequent failures would be needed, or three subsequent failures thereafter. Since only one leg may be under maintenance at a time, this situation represents an "exclusive OR" type of logic, and is modeled analogously to the inhibit conditions in Table 2. The TOP event is defined as "insufficient output with no leg under maintenance OR insufficient output with one leg under maintenance"; "one leg under maintenance" is further divided into "leg A under maintenance", "leg B under maintenance", etc. Since each of these maintenance conditions is treated as if it were an inhibit condition, the "exclusive" nature of the TOP OR gate is retained without explicitly requiring such a gate. Thus, the resulting tree can be analyzed by any of the existing computer codes (such as PREP-KITT [8]). This structure is seen in the upper level structure of the completed fault tree, Figure 2.

Finally, as an example of the input used in the CAT code, Table 3 reproduces the component node-numbering table for this system. Referring to this table and Figure 1, and using heat exchanger A as an example (component 3 in the table), the input defines this component as "type 2" (i.e., using Decision Table number 2), with input node 4 and output node 3. This output node is connected to the input of the pipe (component 2), whose output (node 2) is ultimately connected to spray header A (component 1). Components 47-52 represent the inhibit gate

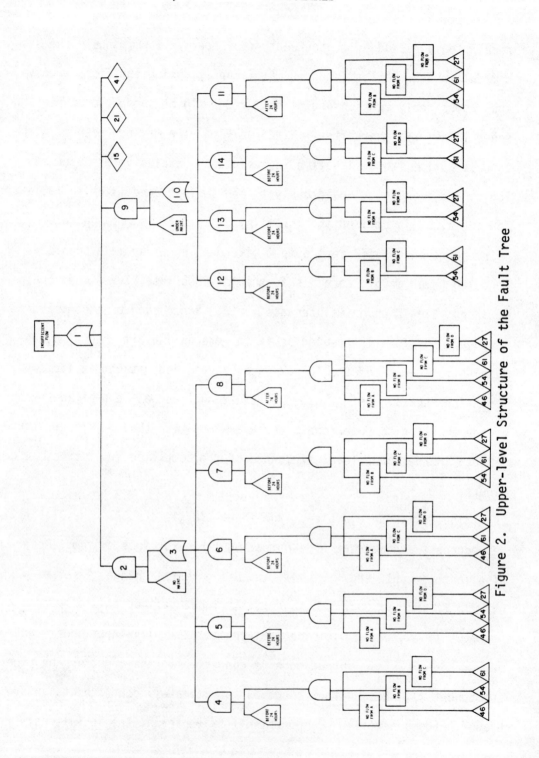

Figure 2. Upper-level Structure of the Fault Tree

Table 3. COMPONENT INDEX INPUT PRINTOUT

CONTAINMENT SPRAY RECIRCULATION SYSTEM

COMPONENT INDEX INPUT PRINTOUT

COMPONENT INDEX	CARD CODE	PRINTOUT NAME	TYPE	INPUT/OUTPUT NODES				
1	COM1	SPRAYA	1	2	52			
2	COM2	PIPEA1	3900	3	2			
3	COM3	HTEXA	2	4	3			
4	COM4	PIPEA2	3900	5	4			
5	COM5	PUMPA	4300	6	7	5		
6	COM6	PIPEA3	3900	8	6			
7	COM7	WATER	3	8				
8	COM8	MOTORA	4	50	23	7		
9	COM9	SPRAYB	1	9	53			
10	COM10	PIPEB1	3900	10	9			
11	COM11	HTEXB	2	11	10			
12	COM12	PIPEB2	3900	12	11			
13	COM13	PUMPB	4300	13	14	12		
14	COM14	PIPEB3	3900	8	13			
15	COM15	MOTORB	4	51	23	14		
16	COM16	SPRAYC	1	16	54			
17	COM17	PIPEC1	3900	17	16			
18	COM18	HTEXC	2	18	17			
19	COM19	PIPEC2	3900	19	18			
20	COM20	CHVC	5	20	19			
21	COM21	PIPEC3	3900	21	20			
22	COM22	MOVC1	6810	22	23	21		
23	COM23	CONT SIG	3	23				
24	COM25	PIPEC4	3900	25	22			
25	COM26	PUMPC	4300	27	26	25		
26	COM27	MOTORC	4	50	23	26		
27	COM28	MOVC2	6810	28	23	47		
28	COM31	SPRAYD	1	31	55			
29	COM32	PIPED1	3900	32	31			
30	COM33	HTEXD	2	33	32			
31	COM34	PIPED2	3900	34	33			
32	COM35	CHVD	5	35	34			
33	COM36	PIPED3	3900	36	35			
34	COM37	MOVD1	6810	37	23	36		
35	COM40	PIPED4	3900	40	37			
36	COM41	PUMPD	4300	41	42	40		
37	COM42	MOTORD	4	51	23	42		
38	COM43	PIPED5	3900	43	41			
39	COM44	MOVD2	6810	44	23	43		
40	COM47	PIPEC5	3900	47	27			
41	COM48	OR1	6	48	49	28		
42	COM49	OR2	6	48	49	44		
43	COM50	PIPE6	3900	8	48			
44	COM51	PIPE7	3900	8	49			
45	COM52	POWER1	3	50				
46	COM53	POWER2	3	51				
47	COM54	TOPC	101	52	53	54	55	15
48	COM55	TOPD1	102	53	54	55	56	
49	COM56	TOPC2	102	52	54	55	57	
50	COM57	TOPC3	102	52	53	55	58	
51	COM58	TOPC4	102	52	53	54	59	
52	COM59	TOTAL	103	15	56	57	58	59	1

END

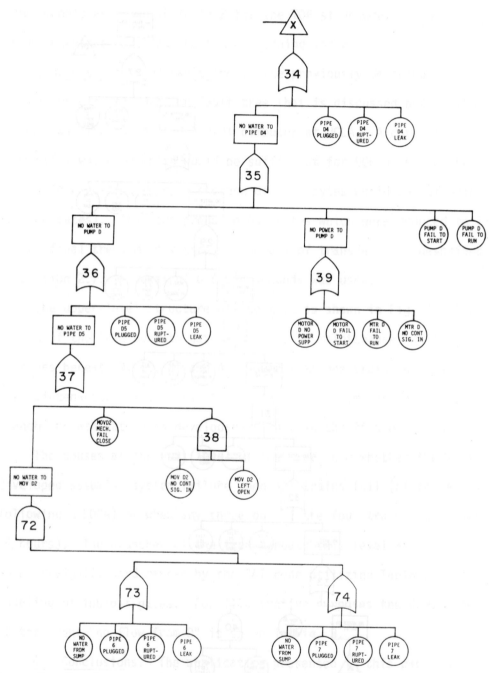

Figure 3.   Development of Lower Events for System Fault Tree (page 2)

fault tree construction. In particular, the handling of inhibit gates as developed here shows a general technique for dealing with mutually exclusive events, such as are encountered in test and maintenance schemes of redundant systems. This approach has also been shown to be useful in situations where the basic configuration of the system changes as a function of time. This is especially important in reactor applications where decay-heat levels, and therefore cooling requirements, are reduced with time.

Finally, it should be emphasized that approaches such as this can only be used to aid the analyst and not replace his understanding of the system.

REFERENCES

[1] G. APOSTOLAKIS, and Y. T. LEE, Methods for the Estimation of Confidence Bounds for the Top-Event Unavailability of Fault Trees, Nuc. Eng. and Des., 41(1977), pp. 411-419.

[2] J. B. FUSSELL, Computer Aided Fault Tree Construction for Electrical Systems, in Reliability and Fault Tree Analysis, R. E. Barlow, J. B. Fussell and N. D. Singpurwalla, eds., SIAM, Philadelphia, 1975.

[3] H. E. LAMBERT, Fault Trees for Decision Making in System Analysis, Lawrence Livermore Laboratory, UCRL-51829, October 1975.

[4] S. L. POLLACK, Decision Tables: Theory and Practice, Wiley-Interscience, New York, 1971.

[5] G. J. POWERS, and F. C. TOMPKINS, Jr., Fault Tree Synthesis for Chemical Processes, AIChE J., 20, pp. 376-387 (1974).

[6] S. L. SALEM, G. E. APOSTOLAKIS and D. OKRENT, A New Methodology for the Computer-Aided Construction of Fault Trees, Annals of Nuc. Energy, to appear.

[7] S. L. SALEM, G. E. APOSTOLAKIS and D. OKRENT, A Computer-Oriented Approach to Fault Tree Construction, EPRI NP-288, Palo Alto, November 1976.

[8] W. E. VESELY, and R. E. NARUM, PREP and KITT Computer Codes for
     the Automatic Evaluation of a Fault Tree, Idaho Nuclear
     Corporation, IN-1349, Idaho Falls, Idaho, 1970.

[9] WASH-1400 (NUREG-75/014), Reactor Safety Study, U.S. Nuclear
     Regulatory Commission, October 1975.

USE OF THE GO METHODOLOGY TO DIRECTLY GENERATE
MINIMAL CUT SETS

R. LARRY WILLIAMS* AND WILSON Y. GATELEY

Abstract. The set of computerized algorithms called "GO" for operating upon the logic models created in system reliability analyses has
been established as an efficient and economical procedure. Recent developments have extended its capability to generate minimal cut sets.
This is accomplished by retaining the original sequential computer
generated model information, selecting an output event of interest,
assigning response modes to each operator and inverting the original GO
process to ascertain operator response mode combinations inducing the
selected event. The original GO model can thus be used to generate
minimal cut sets to the fourth order which lead to system events of
interest.

1. The GO Procedure. The GO computerized reliability analysis

technique was originally conceived and reported by researchers at Kaman

Sciences Corporation, Colorado Springs, Colorado in 1968. [1] It has

been utilized extensively during the last decade to perform probabilistic

studies of the SPRINT [2], SPARTAN [3], POLARIS-POSEIDON and TRIDENT

missile systems, nuclear reactor SCRAM systems [4], coal gasification

plants [5], NASA [6] and FAA equipments, etc.

The GO procedure utilizes a set of standardized operators or com-

ponents which describe the logical operation, interaction and combination

of physical equipments. The logic to properly combine the inputs to

each such operator is defined in a series of algorithms contained in the

computer program. The standardized set of operators identified by their

"type" numbers are shown in Figure 1, GO Building Blocks.

Modeling a system then consists of selecting the operators by

"type" numbers which characterize the elements of the system and inter-

relating their inputs and outputs. The specific probabilities (point

estimates) of component operation applicable to the component are iden-

tified by the "kind" number associated with the operator. These are

---

* Kaman Sciences Corporation, Colorado Springs, Colorado 80933. This
research was sponsored by the Electric Power Research Institute.

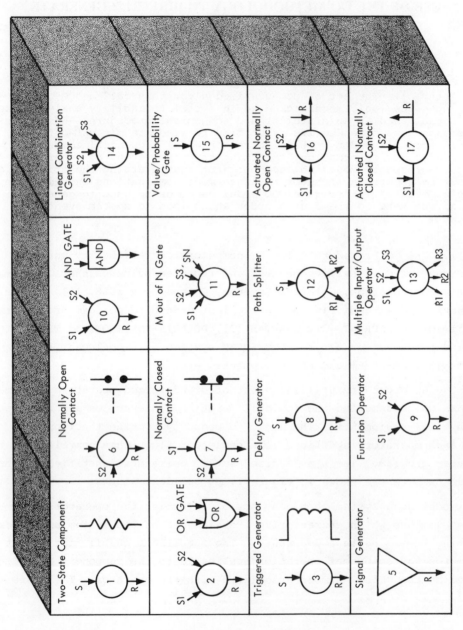

FIGURE 1.   GO BUILDING BLOCKS.

defined separately and follow the system configuration description as
inputs to the computer program.

The GO procedure differs from typical fault tree construction in
that the normal operating sequence is modeled and all possible system
states including premature and dud fault modes, as well as system success,
are generated. With this total approach, events of interest can be
subsequently examined automatically.

Presume that the following subsystem, Figure 2, is an integral part
of a larger system requiring two distinct electrical outputs precisely
sequenced in time. The GO reliability diagram of this subsystem using
the standardized GO building blocks is shown in Figure 3.

In Figure 3 the various operators are identified by the sequential
number assigned by the processing sequence, by the standardized GO type
number and the specified kind number. The signals themselves representing electrical, mechanical, etc., responses are likewise numbered
uniquely.

Data from the GO chart is punched on cards in the format shown in
Table 1. The first entry of the table indicates that operator #1 is
type 5, kind 4 and generates signal 1. The second entry defines operator 2 as type 5, kind 5 generating signal 2. Signal 3 is the output
from the relay to actuate the normally open switch contact and is denoted as a standardized type 3 operator with kind 1 probabilities of
operation having signal 2 as an input, etc.

Table 2 lists the probabilities of component responses specified
for this execution. Kind 1 is a type 3 operator and requires three
probabilities for the success, failure and premature operational modes
of the relays modeled. Kind 2 is a type 1 operator (the resistor of
Figure 2) and requires probabilities for the success and failure of the
resistor.

The kind data for the type 5 operators is somewhat more complex
(kinds 4,5,6 and 7). For this problem, eight signal values (time points)
are specified from 0 to 7 inclusive. A time point of 0 would be premature and a time point of 7 that the signal failed to arrive or never
arrived. The kind data for the type 5 operators must therefore specify
at which time points the signals are expected to arrive and the likelihoods of arrival.

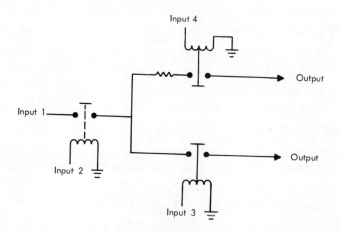

FIGURE 2.   EXAMPLE SUBSYSTEM

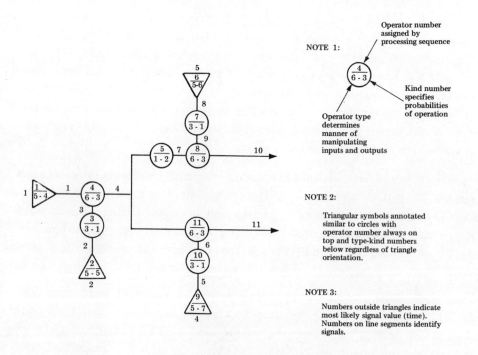

FIGURE 3.   EXAMPLE GO RELIABILITY DIAGRAM.

Operator Data

OP	DATA
1	5 4 1 $
2	5 5 2 $
3	3 1 2 3 $
4	6 3 1 3 4 $
5	1 2 4 7 $
6	5 6 8 $
7	3 1 8 9 $
8	6 3 7 9 10 $
9	5 7 5 $
10	3 1 5 6 $
11	6 3 4 6 11 $
12	0 10 11 $

TABLE 1.   GO INPUT OPERATOR
            DATA.

Record Kind Data

1	1,3,0.8,0.1,0.1$
2	2,1,0.9,0.1$
3	3,6,0.8,0.1,0.1$
4	4,5,2,1,0.9,7,0.1$
5	5,5,2,2,0.9,7,0.1$
6	6,5,2,4,0.9,7,0.1$
7	7,5,2,5,0.9,7,0.1$

TABLE 2.   GO INPUT KIND
            (PROBABILITY) DATA.

For this example each type 5 operator was defined to generate a signal with high likelihood at one time point and fail to generate a signal with low likelihood at time point 7 (infinity or never). Hence the data for kind 4 specifies that there are two time points; that a signal is generated at time point 1 with likelihood 0.9 and at time point 7 with likelihood 0.1. (The sum of the likelihoods must be unity). Similar kind definitions were made for the other type 5 operators.

The computer output for this sample problem is shown in Table 3, indicating that with the arbitrarily defined component probabilities of operation depicted in Table 2 that the event of greatest likelihood of occurrence is "signal 10 occurs at time point 7 and signal 11 occurs at time point 7" ($10_7 11_7$), with likelihood 0.373. All other operational states (joint events) and their likelihoods of occurrence are likewise printed. Time point 7 represents failure to occur, or never occurred. Thus the most likely event, $10_7 11_7$ represents failure of both signals 10 and 11 to arrive, etc.

If one were interested in the marginal distributions (individual signal behavior instead of the joint behavior), the entries in Table 4 appropriately sum the entries of Table 3 to characterize the individual output signal probability distributions. Thus the probability

Final Event Table (Infinity = 7)

Signals and Their Values

Probability	10	11
.0047239200	1	1
.0064035360	1	7
.0093195360	7	1
.0151165440	1	5
.0151165440	4	1
.0151165440	2	2
.0204913152	2	7
.0298225152	7	2
.0483729408	2	5
.0483729408	4	2
.0860635238	4	7
.1252545638	7	5
.2031663514	4	5
.3726592250	7	7

Total Probability = 1.0000000000
Total Error =   .0000000000

TABLE 3.   GO EXAMPLE FINAL EVENT TABLE.

Val	10	11
1	.0262440000	.0291600000
2	.0839808000	.0933120000
4	.3527193600	0.0000000000
5	0.0000000000	.3919104000
7	.5370558400	.4856176000

TABLE 4.   INDIVIDUAL SIGNAL PROBABILITY DISTRIBUTIONS.

of the event "signal 11 arrives at time point 1" (a premature) is 0.02916 without regard to what happens to signal 10, etc.

The signals, of course, are in fact random variables taking on the discrete values representing time of occurrence in this model. The output variables have been properly conditioned by the intermediate operators and random variables which have not been explicitly retained.

From the list of possible system events suppose it were desired to identify the minimal cut sets for the dud event $10_7 11_7$ and the premature event $10_1 11_1$. Using the convention (1) to indicate a dud operator response mode and (2) to indicate a premature operator response the Fault Finder routines were exercised with the results shown in Tables 5 and 6.

From Table 5 the first order cut sets leading to event $10_7 11_7$ are operators 1, 2, 3, and 4 in a dud response mode. The twelve second order cut sets are also shown beginning with both operators 5 and 10 in a dud mode, etc.

To generate event $10_1 11_1$ only third order sets were obtained. The eight sets each require triple premature operator combinations. Thus operators 10,7 and 3 all prematured produce event $10_1 11_1$ as do operators 11,7, and 3, etc.

The purpose of this treatise is to explain how the GO methodology identifies these cut sets. The balance of the paper outlines the specifics of the fault finder process.

NO 0-ELEMENT FAULT SETS EXIST.

1-ELEMENT SETS.

NO.	OP(MODE)
1	3(1)
2	4(1)
3	2(1)
4	1(1)

2-ELEMENT SETS.

NO.	OP(MODE)	OP(MODE)
1	10(1)	5(1)
2	11(1)	5(1)
3	9(1)	5(1)
4	10(1)	7(1)
5	11(1)	7(1)
6	9(1)	7(1)
7	10(1)	8(1)
8	11(1)	8(1)
9	9(1)	8(1)
10	10(1)	6(1)
11	11(1)	6(1)
12	9(1)	6(1)

TREE VANISHED AT OP 11.   NO 3—ELEMENT SETS
NO 3—ELEMENT FAULT SETS EXIST.
TREE VANISHED AT OP 11. NO 4—ELEMENT SETS
NO 4—ELEMENT FAULT SETS EXIST.
HIGHER-ORDER FAULT SETS MAY EXIST

TABLE 5.   MINIMAL CUT SETS FOR EVENT $10_7 11_7$.

1	10(2)	7(2)	3(2)
2	11(2)	7(2)	3(2)
3	10(2)	8(2)	3(2)
4	11(2)	8(2)	3(2)
5	10(2)	7(2)	4(2)
6	11(2)	7(2)	4(2)
7	10(2)	8(2)	4(2)
8	11(2)	8(2)	4(2)

TABLE 6.   MINIMAL CUT SETS FOR EVENT $10_1 11_1$.

2.   Computer System Flowcharts. An overview of the computer
programs which comprise the GO system is provided in Figure 4, <u>GO</u>
<u>Computer</u> <u>System</u> <u>Flow</u> <u>Chart</u>. The GO1 program processes the operator deck
information to produce the operator file which will be used in exercising
the GO2 and GO3 programs. It also generates some printout of operator
information. The GO2 program is then exercised to generate the kind
file and a printed kind summary. Subsequently the GO3 program uses the
operator and kind files to generate the analysis results in accord with
various parameter choices made by the user. If the SAVE parameter is
set equal to 1, all intermediate signal distributions will be written on
the distribution file for subsequent utilization by the fault finder
programs in determining the minimal cut sets for selected events.

The Fault Finder computer system flowchart of Figure 5 depicts the
interaction of the three fault finder programs, the required input data
and files and the generated output. The program SYSFILE first operates
on the three temporary files from the prior GO execution and restructures
the information for use by the Fault Finder 1 (FF1) program.

Once the output event distribution has been determined by the
application of the three GO programs, the user has the option to select
any output event or any combination of the output events and determine
the inducing operator response combinations. The selected event, or
combination of events, is chosen by the analyst from the GO3 printout
and is input into the FF1 program. It produces the selected event tree
file which is a subset of the total tree pertinent to ascertaining the
minimal cut sets generating the specific selected event.

Finally the FF2 input data deck specifying any user-supplied response mode definitions is read and the FF2 program determines the
minimal cut sets for the specified selected event.

3.   Fault Finder Theory. To illustrate the manner by which the
Fault Finder programs operate, a simple example will be thoroughly discussed. The program printouts are included at the conclusion of the
text and may be referenced by the reader as he desires.

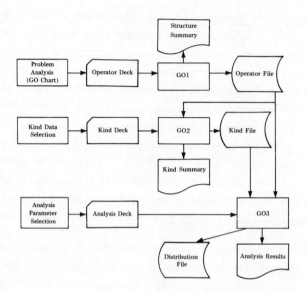

FIGURE 4.   GO COMPUTER SYSTEM FLOWCHART.

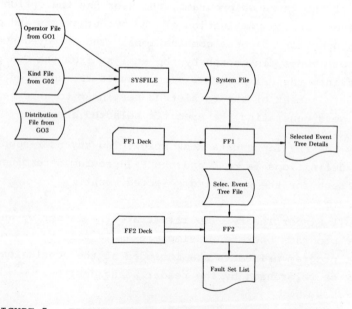

FIGURE 5.   FAULT FINDER COMPUTER SYSTEM FLOWCHART.

The GO chart for our example is:

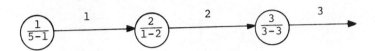

FIGURE 6.   EXAMPLE GO CHART

The numbers in the upper half of the circles are the sequential operator numbers as assigned by GO1 and those in the lower half are the type-kind numbers assigned by the analyst. The type 5 operator (#1) has two output values, 0 and 1. In the value domain the largest possible value (specified as ∞) is set equal to 2. The specific kind probability data and PMIN are such that no pruning takes place in GO3. Signal #3 is the only final output signal.

The complete theoretical tree is shown in Figure 7. Instead of indicating the twig probabilities, we have shown the response states of the various operators. For this the following abbreviations have been used:

    ① , ② : first and second values (0, and 1) of operator #1; ① indicates a good response and ② a failure

    g : good

    f : fail

    p : premature

The numbers at the end of each twig are the number and value of the signal at that point (value is subscripted).

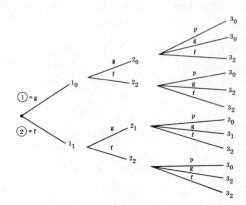

FIGURE 7.   THEORETICAL TREE.

The fault finder process for this example can be conceived by listing the twelve operator response mode combinations where the operator response modes (0=success, 1=dud, 2=premature) are placed in parentheses. These are shown in column 1 of Table 7. The corresponding system events, where the subscripts represent the output signal values (time points), are shown in column 2 and the number of faults for each operator response mode combination are shown in column 3.

TABLE 7.   OPERATOR RESPONSE MODE COMBINATIONS.

Operator Response Mode Combinations	System Events	No. Faults
1(0),2(0),3(0)	3_0	0
1(0),2(0),3(1)	3_2	1
1(0),2(0),3(2)	3_0	1
1(0),2(1),3(0)	3_2	1
1(0),2(1),3(1)	3_2	2
1(0),2(1),3(2)	3_0	2
1(1),2(0),3(0)	3_1	1
1(1),2(0),3(1)	3_2	2
1(1),2(0),3(2)	3_0	2
1(1),2(1),3(0)	3_2	2
1(1),2(1),3(1)	3_2	3
1(1),2(1),3(2)	3_0	3

If it is of interest to know the fault set combinations which generate system event 3_2, from Table 7 the fault sets having zero, one, two or three elements can be identified as follows:

Zero-Element Fault Sets Generating Event 3_2

    None

1-Element Fault Sets Generating Event 3_2

    3(1)

    2(1)

2-Element Fault Sets Generating Event 3_2

    2(1),3(1)

    1(1),3(1)

    1(1),2(1)

3-Element Fault Sets Generating Event 3_2

    1(1),2(1),3(1)

The 1-element fault sets that generate system 3_2 are identified as 2(1) and 3(1), that is, operator 2 in a dud response mode or operator 3 in a dud response mode would result in system event 3_2. The 2- and 3-element fault sets identified are seen to be supersets of the 1-element sets, therefore are not minimal and are not generated in the computer output listings.

The zero-element fault set may seem to be contradiction. It is entirely possible, however, that the event selected for scrutiny could be generated by operator combinations having no operator responses which are termed critical or faulty. For example, if system event 3_0 were to be examined the operator response mode combination 1(0),2(0),3(0) induces it with zero faults.

The mechanics of identifying the minimal fault sets using the GO procedure will now be discussed. The actual tree as created by GO3 as contrasted with the theoretical tree of Figure 7 is shown in Figure 8. It differs from the theoretical tree because at each stage GO3 combines branches which differ only in details which are no longer relevant.

The information created and saved on a mass storage file by GO3 for use by the Fault Finder programs consists of all of the probability distributions - one for each operator. Each distribution consists of a

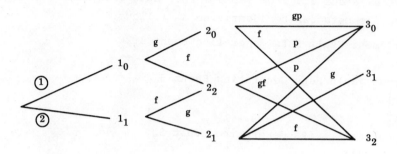

FIGURE 8.   ACTUAL GO3 TREE.

set of terms (where each term contains the number and value of each
currently active signal) and the probability associated with each term.
Thus our example produces three distributions, the terms of which are
shown below (we do not include the probabilities here as they are
irrelevant except for pruning purposes). The notation D_i means the
distribution produced by the ith operator.

TABLE 8.   OPERATOR DISTRIBUTIONS.

D_1	D_2	D_3
1_0	2_0	3_0
1_1	2_2	3_1
	2_1	3_2

Note that the final distribution, D_3 in this case, is the primary printout data from GO3.

If all signals have been retained at each step, the three distributions would have been as follows:

TABLE 9. COMPLETE OPERATOR DISTRIBUTIONS

D_1	D_2	D_3
1_0	$1_0 2_0$	$1_0 2_0 3_0$
1_1	$1_0 2_2$	$1_0 2_0 3_2$
	$1_1 2_2$	$1_0 2_2 3_0$
	$1_1 2_1$	$1_0 2_2 3_2$
		$1_1 2_2 3_0$
		$1_1 2_2 3_2$
		$1_1 2_1 3_0$
		$1_1 2_1 3_1$
		$1_1 2_1 3_2$

Since information concerning the values taken on by random variables 1 and 2 are not of interest they are deleted to generate Table 8 previously noted. It is also noted that there are twelve possible operator response mode combinations but only nine would be generated in a GO exercise. This is so because either a g or p response of operator 3 coupled with event 2_0 induce the single event 3_0. Similarly a g or f response of operator 3 coupled with event 2_2 induce the single event 3_2 thus reducing the set of possible events from 12 to 9.

3.1. Program FF1 Processing. For this example the selected event 3_2 is chosen. The fault set combinations which induce it will be determined. The function of the program FF1 is to create a new set of operators which can be used by FF2 to create the subtree which contains all of the branches terminating in the selected event. Data for each new operator contains a set of input terms and for each such term one or more output terms with a connective response mode for that operator.

FF1 starts by reading the final distribution D_3 and the user-defined selected event (3_2). Each term of D_3 is scanned to see if it is included in the selected event. If it is not, it is ignored; if it is, it is placed in a temporary "target list".

The (old) operator #3 and associated kind data are now read along with D_2 which is the input to operator #3. FF1 replicates the processing done by GO3, operating upon each term of D_2 with the logic of operator #3. As each new term is created by this process, it is compared against the target list. If no match is found, the new term is ignored. Otherwise, it is stored in a new target list (which will be used for the processing of operator #2), and the old term, the new term, and the response mode are placed in the definition of the new operator #3. Finally, after all terms of D_2 have been processed in this manner, the complete definition of the new operator #3 is written on a mass storage file record.

The above process is then carried out for old operator #2 (operating on D_1) and old operator #1 (operating on the single-element "dummy" distribution which always starts a GO tree).

Table 10 shows the definitions of the three new operators. Operationally this table is created from right to left - i.e., operator #3 first, etc.

OP#1			OP#2			OP#3		
Input Term	Response Mode	Output Term	Input Term	Response Mode	Output Term	Input Term	Response Mode	Output Term
	1	1_0	1_0	g	2_0	2_0	f	3_2
	2	1_1		f	2_2	2_2	gf	3_2
			1_1	f	2_2	2_1	f	3_2
				g	2_1			

TABLE 10.   NEW OPERATOR DEFINITIONS.

The table may be constructed by hand by simply tracing the tree in Figure 8 backwards from the final node 3_2.

In FF1 these definitions are recoded in order to simplify their use in FF2. The response modes are given numerical values (as described in the Fault Finder Manual) [8]. In this example the g is replaced by

1, f by 2, and gf by 1. In addition we note that the information contained in a term - i.e., signal numbers and values - is really no longer
needed. It is sufficient to be able to match up the input terms of one
operator with the corresponding output terms of the previous one. It
turns out that a satisfactory and simple code for a term is furnished
by the location of that term in the appropriate target list. Thus there
is only 1 output term of interest for operator 3, event 3_2, the selected
event. For operator 2, however, there are three events of interest, the
first being 2_0, the second 2_2 and the third, 2_1, etc.

This coding procedure provides considerable economy in the use of
central memory because a single uncoded term may require up to five
computer words whereas its location can always be identified by a single
word. These locations are always sequential. With these codings in
mind, Table 10 can be rewritten to produce Table 11.

OP #1			OP #2			OP #3		
Input Term	Response Mode	Output Term	Input Term	Response Mode	Output Term	Input Term	Response Mode	Output Term
1	1	1	1	1	1	1	2	1
	2	2		2	2	2	1	1
			2	2	2	3	2	1
				1	3			

TABLE 11. CODED OPERATOR DEFINITIONS.

3.2. Program FF2 Processing. The first objective of FF2 is to
replace the response mode values in the operator definitions by critical
response mode values, using either the default values or values supplied
by the user. In this example the default values for operators #2 and
#3 are used and for operator #1 the response modes 1 and 2 are replaced
by the critical response modes 0 and 1 respectively; thus 1(0) is non-
critical but 1(1) is critical. With these changes the operator definitions become as shown in Table 12.

OP #1			OP #2			OP #3		
Input Term	Critical Mode	Output Term	Input Term	Critical Mode	Output Term	Input Term	Critical Mode	Output Term
1	0	1	1	0	1	1	1	1
	1	2		1	2	2	0	1
			2	1	2	3	1	1
				0	3			

TABLE 12.   REVISED OPERATOR DEFINITIONS.

The complete tree generated by these operators is shown in Figure 9. FF2 will seldom generate this entire tree but it will be useful to compare it with what FF2 actually creates. The nodes of this tree contain the following information separated by colons: (1) the term identifier, (2) the number of critical operators, and (3), if any, the critical operator numbers and their critical mode values (the fault set). The number on each twig is the critical mode value associated with that twig.

FF2 creates a subtree of this complete tree for each fault set order. The subtree results from pruning the complete tree by two rules. The first rule is that a branch is pruned if and when the number of critical operators at a node exceeds the fault set order being currently processed. The second rule is that if the fault set of a node contains a proper subset which has been previously found to be minimal fault set, that branch is pruned thus removing nonminimal fault sets.

Let us now create the FF2 tree using these two rules. A pruned branch is indicated by a circle containing either a 1 or 2 which refers to a rule #1 or rule #2 pruning respectively. We first start with the zero[th] order tree- that is, we are searching for zero order fault sets. Rule 2 cannot be applied here because no lower order fault sets have yet been found. We find that all branches are pruned. Consequently, no zero order fault sets exist.

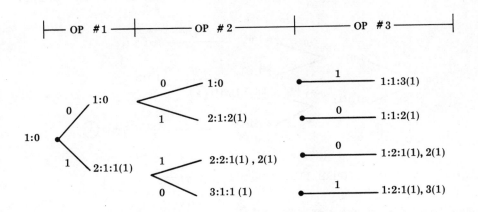

FIGURE 9.   COMPLETE FAULT SUBTREE GENERATING EVENT 3_2.

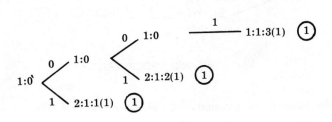

FIGURE 10.   ZERO ORDER TREE.

Now we do the first order tree. Again rule 2 is not operative.

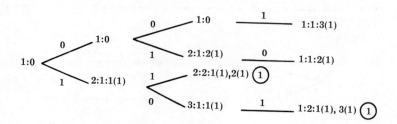

FIGURE 11.   FIRST ORDER TREE.

We have found two first order fault sets: 2(1) and 3(1).

For the second order tree, rule 2 can now be applied because of the existence of the two first order fault sets.

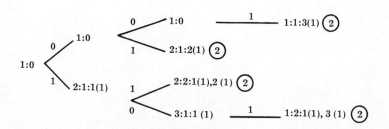

FIGURE 12.   SECOND ORDER TREE.

Again all branches have been pruned. Consequently, no second order fault sets exist. In addition, observe that none of the pruning was by rule 1. This implies that no higher order fault sets exist, and there is no need to look at higher order trees (they would all be identical to the second order tree).

Note that branches in FF2 trees are not combined as they are in GO3. This feature tends to produce relatively larger trees in FF2 but also permit one to split a tree at any point and analyze the parts separately. FF2 carries out this operation automatically if it is needed.

3.3. Example Printouts. Figures 13 to 17 show the actual program printout for the example problem. In all cases as much information as possible has been printed, and the interested reader should have little difficulty in correlating the printed data with the previous discussions.

Figure 18 shows the FF2 printout for the same problem but with the FF2 tree size limited to 2 terms. This highly artificial situation illustrates the tree-splitting feature.

4. Conclusions. The complete generality of the procedure of inverting the GO process in order to identify the component fault combinations inducing a selected system event of interest is a powerful analytical capability. It is expected to see increasing use as the capability is more widely exploited.

The fact that the entire system joint distribution of events is created in a single GO run and that each of the n output events or any combination of them can be analyzed in detail to determine the fault combinations which induce them corresponds to generating and analyzing 2^n fault trees simultaneously without the necessity of creating them. The single GO chart provides all the necessary information.

One must recognize, however, that the fault sets retained are directly influenced by the threshold of retention value, PMIN, selected for the initial GO run since fault combinations whose likelihoods of occurrence are smaller than the PMIN value have not been explicitly retained. For this reason when the identification of minimal cut sets is the objective it may be desirable to select fault probabilities at some standard value like 1×10^{-2} or 1×10^{-3} then set the PMIN value in the GO3 run to pick up all the second or third or fourth order minimum cut sets. In so doing, one can eliminate the higher order sets, reduce the number of terms retained and increase the efficiency of identifying the desired fault combinations of specified order. This has been done on several real life problems with promising results.

```
FF THEORY EXAMPLE

     RUN ON 01/10/77 AT 13.43,50.

VALUES = 3, BIAS = 5000, CPS = 1, SIGNALS = 1, ERRORS = 25
OPERATOR DATA

OP_  _DATA_____

  1   5   1   1   $
  2   4   2   1   2   $
  3   3   3   2   3   $
//  0   3   $

SIGNAL DATA
            SOURCE OPER.
SIGNAL_   _NUM  TYPE  KIND  USING_OPERATORS_(-_IF_DELETED_AT)
    1      1     5     1     -2
    2      2     1     2     -3
    3      3     3     3

NUMBER OF OPERATORS      = 3
NUMBER OF SIGNALS        = 3
MAX NUMBER ACTIVE        = 1
MAX SIGNAL LIST SIZE     = 1
NUMBER OF SIGNAL VALUES  = 3
NUMBER OF WORDS/WORD     =30
NUMBER OF WORDS/TERM     = 1

FINAL SIGNALS  = 3
```

FIGURE 13. PROGRAM GO1 PRINTOUT.

```
FF THEORY EXAMPLE

     RUN ON 01/10/77 AT 13.43.52.

OPERATOR FILE --- (01/10/77 13.43.50.) FF THEORY EXAMPLE

RECORD_KIND_DATA_____

  1   1   5  2  0  .5  1  .5  $
  2   2   1  .9  .1  $
  3   3   3  .9  .05  .05  $
------------------------------------------------------------------
```

USE SUMMARY TABLE. ENTRY = KIND/TYPE (FREQUENCY)
    (FREQUENCY IS NEGATIVE FOR PERFECT KINDS.)

    1/5(1)   2/ 1(1)   3/ 3(1)

    NUMBER OF KINDS INPUT ----- 3
    NUMBER USED - NONPERFECT -- 3
    NUMBER USED - PERFECT ----- 0

FIGURE 14. PROGRAM GO2 PRINTOUT.

FF THEORY EXAMPLE
        RUN ON 01/10/77 AT 13.43.55.

OPERATOR FILE --- (01/10/77 13.43.50) FF THEORY EXAMPLE
KIND FILE ------- (01/10/77 13.43.52.) FF THEORY EXAMPLE

MAXIMUM SIGNAL VALUE (INFINITY) IS 2
MAXIMUM DISTRIBUTION SIZE IS 3000

RUN NUMBER 1
      PMIN = 1.0000E-08
      NEW = 0, INTER = 0, SAVE = 1
      FIRST = 0, LAST = 10000, TRACE = 0.

ANALYSIS DETAILS

OP	TYPE	KIND	SIZE
1	5	1	2
.5000000000		1(	1)
.5000000000		1(	1)
2	1	2	3
.1000000000		2(	2)
.4500000000		2(	0)
.4500000000		2(	1)
3	3	3	3
.1400000000		3(	2)
.4550000000		3(	0)
.4050000000		3(	1)

FINAL EVENT TABLE (INFINITY =   2)

                    SIGNALS AND THEIR VALUES

PROBABILITY	3
.14000000000	2
.40500000000	1
.45500000000	0

TOTAL PROBABILITY = 1.0000000000
TOTAL ERROR =   .0000000000

INDIVIDUAL SIGNAL PROBABILITY DISTRIBUTIONS

VAL.	3
0	.4550000000
1	.4050000000
2	.1400000000

FIGURE 15.   PROGRAM GO3 PRINTOUT.

FF THEORY EXAMPLE
        RUN ON 01/10/77 13.44.00.

OP FILE   - 01/10/77 13.43.50. FF THEORY EXAMPLE

KIND FILE - 01/10/77 13.43.52. FF THEORY EXAMPLE

INTER = 1, PRUNE = 0.

SELECTED EVENT DATA CARDS

3 2 $ ---------------------------------------

SELECTED EVENTS - SIGNAL NUMBER (VALUE)
   1.   3(2)

OP	TYPE	KIND	MIN	MOUT	PMIN
3	3	3	3	3	1.000000E-08
2	1	2	2	4	1.000000E-08
1	5	1	1	2	1.000000E-08

FIGURE 16.   PROGRAM FF1 PRINTOUT.

FAULT SET FINDER
FF THEORY EXAMPLE
RUN ON 01/10/77 AT 13.44.03.
INTER = 1, FIRST = 0, LAST = 4
USER-DEFINED MODE CODES

1 5 2 0 3 $_____

TIME = 4.117

```
0   1   1   2   1   1
0   2   2   4   1   1
0   3   3   3   1   0
```

TREE VANISHED AT OP   3.   NO 0-ELEMENT SETS

TIME = 4.136

NO 0-ELEMENT FAULT SETS EXIST

```
1   1   1   2   1   2
1   2   2   4   2   3
1   3   3   3   3   2
```

1-ELEMENT SETS.   NO.   OP(MODE)

```
              1       2(1)
              2       3(1)
```

TIME = 4.165

```
2   1   1   2   1   2
2   2   2   4   2   2
2   3   3   3   2   0
```

TREE VANISHED AT OP   3.   NO 2-ELEMENT SETS

TIME = 4.182

NO HIGHER-ORDER FAULT SETS EXIST.

FIGURE 17.   PROGRAM FF2 PRINTOUT.

FAULT SET FINDER
FF THEORY EXAMPLE
     RUN ON 01/10/77 AT 09.25.13.
INTER = 1, FIRST = 0, LAST = 4
USER-DEFINED MODE CODES

1 5 2 0 3 $_____

TIME = 4.008

```
0   1   1   2   1   1
0   2   2   4   1   1
0   3   3   3   1   0
```

TREE VANISHED AT OP.   3.   NO 0-ELEMENT SETS

TIME = 4.027

NO 0-ELEMENT FAULT SETS EXIST.

```
1   1   1   2   1   2
```

T OVERFLOW AT ICOM = 1, OP = 2, LEVEL = 1

```
1   2   2   4   1   2
1   3   3   3   2   2
```

1-ELEMENT SETS.   NO.   OP(MODE)

```
              1       2(1)
              2       3(1)
```

TIME = 4.06

PICK UP TREE AT OP = 2, LEVEL = 1

```
1   2   2   4   1   1
1   3   3   3   1   0
```

TREE VANISHED AT OP   3.   NO 1-ELEMENT SETS

TIME = 4.078

```
2   1   1   2   1   2
2   2   2   4   2   2
2   3   3   3   2   0
```

TREE VANISHED AT OP   3.   NO 2-ELEMENT SETS

TIME = 4.095

NO HIGHER-ORDER FAULT SETS EXIST.

FIGURE 18.   PROGRAM FF2 PRINTOUT WITH TREE SPLITTING.

REFERENCES

1. GO - A Computer Program for the Reliability Analysis of Complex Systems, KN-67-704, Wilson Y. Gateley, Dan W. Stoddard, R. Larry Williams, Kaman Sciences Corp., April 1968.

2. Joint Army/ERDA Sprint Warhead Section Safety and Reliability Analysis (U), RCS: SARQA 116, 15 October 1975, Picatinny Arsenal, Dover, New Jersey, SRD.

3. Joint Army/ERDA SPARTAN Warhead Section Safety and Reliability Analysis (U), RCS: SARQA 116, 30 September 1975, Picatinny Arsenal, Dover, New Jersey.

4. Reliability Analysis of an HTGR SCRAM System Including Human Interfaces, KSC-1037-1, March 1975, Kaman Sciences Corporation, Colorado Springs, Colorado, Noel J. Becar.

5. Results of Demonstration of a Reliability Assessment Methodology Using Two Coal Conversion Plant Models, Contract No. 14-32-001-1788, Office of Assistant Administrator - Fossil Energy, ERDA, 6 June 1975.

6. Kaman Sciences Corporation Input to NASA/ECON Low Cost Payloads Study, Kaman Sciences Corporation, Colorado Springs, Colorado, 23 August 1976, Noel J. Becar.

7. GO Manual (Draft), K77-27U(R), 14 April 1977, Kaman Sciences Corporation, Colorado Springs, Colorado under Project RP 818-1 with Electric Power Research Institute, Palo Alto, California.

8. Fault Finder Manual (Draft), K77-32U(R), 5 May 1977, Kaman Sciences Corporation, Colorado Springs, Colorado under Project RP 818-1 with Electric Power Research Institute, Palo Alto, California.

9. Vesely, W.E., and Narum, R.E., PREP and KITT: Computer Codes for the Automatic Evaluation of a Fault Tree, IN-1349, Idaho Nuclear Corporation, Idaho Falls, Idaho, August 1970.

10. Warrell, R.B., Set Equation Transformation System (SETS), SLA-73-0028A, Sandia Laboratories, Albuquerque, New Mexico, July 1973.

11. Fussell, J.B., Henry, E.B. and Marshall, N.H., MOCUS - A Computer Program to Obtain Minimal Sets from Fault Trees, Aerojet Nuclear Company, Idaho Falls, Idaho, March 1974.